Water Resources Development and Man

Impacts of Large Dams: A Global Asses
Series editors: Asit K. Biswas and Cecilia Tortajada

Editorial Board

Dogan Altinbilek (Ankara, Turkey)
Chennat Gopalakrishnan (Honolulu, USA)
Jun Xia (Beijing, China)
Olli Varis (Helsinki, Finland)

Cecilia Tortajada • Dogan Altinbilek
Asit K. Biswas
Editors

Impacts of Large Dams: A Global Assessment

With 51 Figures and 89 Tables

🕮 Springer

Editors
Cecilia Tortajada
Third World Centre
for Water Management
Atizapan, Mexico

Lee Kuan Yew School of Public Policy
Singapore

International Centre for Water
and Environment
Zaragoza, Spain
ctortajada@thirdworldcentre.org

Dogan Altinbilek
Civil Engineering Department
Middle East Technical University
Ankara, Turkey
hda@metu.edu.tr

Asit K. Biswas
Third World Centre
for Water Management
Atizapan, Mexico

Lee Kuan Yew School of Public Policy
Singapore
akbiswas@thirdworldcentre.org

ISBN 978-3-642-43316-0   ISBN 978-3-642-23571-9 (eBook)
DOI 10.1007/978-3-642-23571-9

Water Resources Development and Management ISSN: 1614-810X

© Springer-Verlag Berlin Heidelberg 2012
Softcover reprint of the hardcover 1st edition 2012
All rights reserved. This work may not be translated or copied in whole or in part without the written permission of the publisher (Springer Science+Business Media, LLC, 233 Spring Street, New York, NY 10013, USA), except for brief excerpts in connection with reviews or scholarly analysis. Use in connection with any form of information storage and retrieval, electronic adaptation, computer software, or by similar or dissimilar methodology now known or hereafter developed is forbidden.

The use in this publication of trade names, trademarks, service marks, and similar terms, even if they are not identified as such, is not to be taken as an expression of opinion as to whether or not they are subject to proprietary rights.

Cover illustration: deblik

Printed on acid-free paper

springer.com

Preface

Up to until the 1970s, it was generally assumed that large dams overwhelmingly contributed more benefits to the society compared to their costs. This perception started to change in the late 1970s. During the 1980s, the global debate on the benefits and costs of large dams became increasingly emotional, dogmatic and confrontational. While the initial debate started primarily in the United States, it subsequently engulfed many other countries. It became especially heated during the 1990s, when the pressure from primarily single-cause activist NGOs, mostly again from the United States, contributed significantly to the reduction of funding support for the construction of large water infrastructure projects in developing countries, especially from the World Bank and the Regional Development Banks. In fact, during the 1990s, to paraphrase Margaret Thatcher, former Prime Minister of the United Kingdom, all these Banks considered somewhat erroneously construction of large water infrastructure projects to be a 'sunset industry.' Not surprisingly, the World Bank lending for hydropower projects during the decade of the 1990s fell by an incredible 90%.

Concurrently, environmental and social concerns started to become increasingly important issues starting from about 1970. The National Environmental Policy Act (NEPA) was enacted by the United States on January 1970. It is the first such comprehensive environmental policy act in any country of the world. Its preamble states:

> To declare national policy which will encourage productive enjoyable harmony between man and his environment; to promote efforts which will prevent or eliminate damage to the environment and biosphere and stimulate the health and welfare of the man; to enrich the understanding of the ecological systems and natural resources important to the Nation … …

NEPA required that construction of any large project, including water infrastructure, could only proceed after an environmental assessment had been prepared and approved. It is in fact the first act in any country of the world which made it mandatory that all projects must prepare detailed environmental impact assessments before approval and funding can be authorized.

Environmental issues received a further international boost when the United Nations convened its first-ever megaconference, on the Human Environment, in June 1972, in Stockholm. It is worth noting that when the Stockholm Conference was convened, there were very few countries which had even an Environmental

Ministry, and even fewer which required mandatory environmental impact assessments before any large project could be approved. However, within a short period of two decades, environment became a mainstream subject and nearly all countries had a full-fledged Ministry, or at least institutional arrangements, to ensure that environmental issues received appropriate consideration during the project preparation and implementation phases.

As the environmental awareness of various countries during the 1970s evolved, a common perception started to develop, that is, small is beautiful and big is ugly. An important victim of this philosophy was large dams, irrespective of the fact that no country or region in tropical or sub-tropical climate has ever managed to make significant economic progress without harnessing adequately its water resources. This fact can be exemplified by the fact that countries like the United States and Australia have over 5,000 m^3 of storage per person, but countries like India and Pakistan have around 150 m^3 per person, and Ethiopia and Kenya only about 50 m^3 per person. Viewed in another way, dams on major rivers like the Colorado in the United States and Murray-Darling in Australia can hold some 900 days of river runoff, and the Orange River in South Africa for about 500 days. In contrast, the major peninsular rivers of India can store flows for 120 to 220 days, and countries like Pakistan can barely store enough water for about 30 days. Such skewed construction of water infrastructure has seriously hampered, and continue to hamper, the economic and social development of many developing countries. This is a fact which has still not received appropriate recognition.

Absence and delays in construction of properly planned dams have also contributed to serious energy shortages and balance of payment problems in many developing countries, especially those that have to import fossil fuels. Whereas OECD countries as a whole have developed nearly 70% of their economically viable hydropower potential, corresponding figures for countries like India and China are 30%, Pakistan about 10%, the African continent as a whole less than 5% and Nepal only about 1%.

Political leaders of all major arid and semi-arid countries when they were in their early development phase, ranging from President Roosevelt of the United States to Jawaharlal Nehru of India, Gamal Abdel Nasser of Egypt, and Kwame Nkrumah of Ghana, gave construction of large dams priority for the social and economic development of their countries. They realised that dams provide reliable sources of water for domestic, industrial and agricultural uses, contribute to hydropower generation, protect countries from the twin ravages of floods and droughts, and provide navigation. Prime Minister Nehru of India, while inaugurating the Bhakra-Nangal Project, expressed the general view of the leaders in such countries as 'dams are the temples of modern India.'

The public perception of the importance of large dams started to change in the 1980s. For example, a major Japanese newspaper like the Asahi Shimbun used to regularly castigate the Japanese Government for not building enough dams during the 1960s and 1970s. However, during the 1980s and later, its philosophy turned around 180°. It became virulently anti-dam, and focused on, and even exaggerated, only the negative impacts of such structures. This change in mindset happened for many reasons, only one of which will be noted here. Large dams, like any infrastructure development, have both societal benefits and costs. Many groups of people benefit from these structures but some others pay the costs. Unfortunately, those who receive these benefits are often diffused and may not be even aware of the fact that the benefits are accruing because of a specific dam. For example, hydroelectric power generated by a dam could simply be another source of power in an electricity grid, whose contributions to energy security many users may not know. Similarly, increased food availability at a reasonable price in the market could be due to reliable irrigation provided by a dam. But, an average person often may find it difficult to relate these benefits directly to the presence of a large dam.

In contrast, people who are adversely affected by dams are fully aware of the reasons of their problems, like people who have to be resettled. They are much smaller in number compared to the number and the range of the beneficiaries, but are very visible, easily identifiable and vocal. During the 1980s and later, they were aided by single issue anti-dam NGOs who were often articulate and mediagenic. These NGOs were significantly media savvy and were aware of the power of the media and thus managed to get widespread and consistent media coverage for their views compared to their pro-dam counterparts. These activists focused on these single issues and costs alone. For practical purposes they ignored the benefits that such hydraulic infrastructures could bring to the society. For overall social and economic improvement, a logical and balanced approach would have been to argue that those who pay the costs should be made the direct beneficiaries of the dams and that other negative impacts should be minimised and positive impacts should be enhanced. Such an approach would have maximised the net benefits to the society. However, for many different reasons, this was not the case.

The success of these anti-dam NGOs were such that by the 1990s, the World Bank and the Regional Development Banks became somewhat afraid of them and their media power and were reluctant to fund any project that had anything to do with the construction of large dams. In fact, one can argue that the debacle with the Sardar Sarovar Project in India became the World Bank's Vietnam in the area of dam. True to form, the Regional Development Banks simply followed the World Bank's footsteps.

In 1993, the World Bank established an Inspection Panel to investigate complaints from project-affected communities 'to investigate IBRD/IDA financed projects' 'to determine whether the Bank has complied with its operational policies and procedures (including social and environmental safeguards) and to address related issues of harm.'

The very first case that the Inspection Panel considered was a dam (Arun III Hydroelectric Project in Nepal) which the Bank declined to fund. It has been estimated that the probability of the Inspection Panel reviewing a project with a dam was 64 times higher than one without a dam. As John Briscoe, currently a Professor at the Harvard University and formerly a senior World Bank staff, has perceptibly noted the ambitious Bank managers realised very soon that the Bank was ruthless in punishing 'sins of commission' but basically ignored 'sins of omission'. Thus, if the managers could help it, they gave projects with dams a wide berth.

Tom Kenworthy, a Washington Post reporter, admirably summed up the then prevalent situation in 1997 on the special animus the environmental activists held for dams as follows: 'To them, there is something disproportionately and metaphysically sinister about dams. Conservationists who can hold themselves in reasonable check before new oil spills and fresh megalopolises mysteriously go insane at even the thought of a dam.'

During the decade of the 1990s, with the discussions on the benefits and costs of large dams becoming more and more acrimonious, the World Bank and the IUCN sponsored the World Commission on Dams (WCD) which was given the mandate to:

- Review the development effectiveness of large dams, and assess alternatives for water resources and energy development
- Develop internationally acceptable criteria, guidelines and standards for the planning, design, appraisal, construction, operation, monitoring and decommissioning of dams

The Commission was established in May 1998, and delivered its final report entitled 'Dams and development: a new framework for decision-making' in November 2000. The Commission, right from the very beginning, was hijacked by the anti-dam lobby, and was highly skewed against dams by the majority of its commissioners, Secretariat staff and consultants.

The publication of the WCD report coincided almost with the peak of the anti-dam movement. A major unexpected development never foreseen by WCD or its anti-dam, single purpose NGO allies, was that it united all the major developing countries, like Brazil, China, Ethiopia, India, Lao PDR, Nepal Pakistan, the Philippines, Sri Lanka, Turkey, and Vietnam. They all unanimously agreed that the WCD report was biased, and could not be accepted. The Water Resources Sector Strategy of the World Bank correctly noted that the 'multi-stage, negotiated approach to project preparation recommended by the World Commission on Dams is not practical and virtually preclude the construction of any dam'.

The Chinese Government probably summed up the views of most developing countries on the report as follows: 'very much biased to the developed countries and anti-dam activists and extreme environmentalists. We therefore retreated from the WCD in 1998. We think it would be more appropriate to change the title of the report into "Anti-dams and anti-development."'

Patrick McCully, a leader of the single issue of anti-dam NGO, International Rivers Network, has candidly discussed how the anti-dam NGO hijacked very successfully the WCD process, including successfully sidelining of the Governments, in his authoritative account of what actually happened in his paper 'The use of a trilateral network: an Activist's perspective on the formation of the World Commission on Dams' (American University International Law Review, Vol. 16, 2001, pp. 1453–1454).

The initial approach of the Asian Development Bank (ADB), in contrast to that of the World Bank, to the WCD report was positive. Shortly after the report was released, it convened a meeting in Manila which was opened with the statement that ADB intended to comply with the recommendations of the WCD. India refused to participate in the Manila meeting. Ramaswamy Iyer, an Indian anti-dam former bureaucrat whose country report on India was flatly rejected by the Indian Government, wrote to the chair of the WCD that 'The WCD process, far from narrowing differences, seems to have led to a greater divisiveness. Developing countries see it as yet another instance of the imposition on them by the developed counties of an agenda designed in the latter's interests'.

My own view is that the WCD process did contribute to at least three unexpected benefits for the developing world, though none of these were intended by the Commission itself or by its two godfathers. First, it contributed to a concerted action by the developing countries which were forced to unite by the biased report which otherwise may not have happened. With a combined voice, they could tell developed countries who had already constructed most of their large dams, that infrastructure construction is important for their socio-economic development, and that they need such structures to produce food, generate energy, employment and income, provide basic services and improve the overall quality of life of their citizens. This aspect, in their view, is not negotiable. The report turned out to be the catalyst that made this possible.

Second, the WCD report reinforced the essential requirement, if any was still needed, that it is imperative that people who have to be resettled because of the dams must be made their direct beneficiaries, and all environmental and social costs must be properly considered.

Third, many developing countries were tired by the 'paralysis by analysis' approach of the World Bank and the Regional Development Banks, especially during the 1990s. They wanted well-planned and well-designed dams to be built without unnecessary increase in costs and inordinate delays so that their people could enjoy the fruits of the infrastructure as soon as possible.

During the first decade of the twenty-first century, the debate on dams, though still polarised, is gradually becoming more balanced. It is slowly being realised that infrastructure is essential for the future accelerated development of developing countries. This 'new' perspective is reflected also by major donor institutions, like the World Bank, whose support to infrastructure has doubled, from around 20% in 2000 to some 40% in 2008. Its support to water projects increased almost 3.5 times in six years, from $1.8 billion in 2003 to $6.2 billion in 2009. The world as a whole is generally coming to appreciate the fact that large water infrastructure is essential for the economic development of the developing countries as long as social and environmental issues (both positive and adverse) are given appropriate considerations.

Within this slowly changing mindset on this issue, the Third World Centre for Water Management decided to undertake a series of objective case studies on the overall impacts of large dams. Leading objective and knowledgeable specialists were selected very carefully, and were requested to prepare case studies of positive and negative impacts of large dams, and their net impacts on the society. These case studies were discussed and analysed in two international workshops, first one in Istanbul and the second one in Cairo. Thereafter, all the authors modified their analysis in the light of the discussions. The analyses in the book are based on date and information available until the end of 2009.

The Istanbul Workshop was especially noteworthy. Through the direct personal intervention of my co-editor, Professor Dogan Altinbilek, the participants had the pleasure and privilege of listening to President Süleyman Demirel of Turkey, who is an eminent water resources expert and under whose leadership Turkey underwent a most remarkable water resources transformation. President Demirel outlined the history of water development in Turkey, and the roles large dams have played to foster the country's social and economic development.

A project of this breadth and magnitude could not have been completed without the strong support of the authors who prepared the various case studies. On behalf of the Third World Centre for Water Management and the Middle East Technical University, I would like to express our most sincere appreciation for their work and their continuous support until this book was completed. I am especially grateful to my co-editors, Professor Cecilia Tortajada and Professor Dogan Altinbilek, for all the hard work they did for the completion of the book. The work of Thania Gomez in formatting the manuscript in Springer's style is most appreciated.

Atizapán, Mexico and Singapore   Asit K. Biswas

Contents

1 **Impacts of Large Dams: Issues, Opportunities and Constraints** 1
Asit K. Biswas

2 **Indirect Economic Impacts of Dams** 19
Rita Cestti and R.P.S. Malik

3 **Resettlement Outcomes of Large Dams** 37
Thayer Scudder

4 **Greenhouse Gas Emissions from Reservoirs** 69
Olli Varis, Matti Kummu, Saku Härkönen, and Jari T. Huttunen

5 **Impacts of Dams in Switzerland** 95
Walter Hauenstein and Raymond Lafitte

6 **Hydrodevelopment and Population Displacement in Argentina** 123
Leopoldo J. Bartolome and Christine M. Danklmaier

7 **Impacts of Sobradinho Dam, Brazil** 153
Benedito P.F. Braga, Joaquim Guedes Correa Gondim Filho, Martha Regina von Borstel Sugai, Sandra Vaz da Costa, and Virginia Rodrigues

8 **The Atatürk Dam in the Context of the Southeastern Anatolia (GAP) Project** 171
Dogan Altinbilek and Cecilia Tortajada

9 **Impacts of King River Power Development, Australia** 201
Roger Gill and Morag Anderson

10 **Resettlement in China** 219
Guoqing Shi, Jian Zhou, and Qingnian Yu

11 *Officials' Office and Dense Clouds*: **The Large Dams that Command Beijing's Heights** 243
James E. Nickum

12	**Resettlement due to Sardar Sarovar Dam, India**	259
	C.D. Thatte	
13	**Impacts of Kangsabati Project, India**	277
	R.P. Saxena	
14	**Regional and National Impacts of the Bhakra-Nangal Project, India**	299
	R. Rangachari	
15	**Impacts of Koyna Dam, India**	329
	C.D. Thatte	
16	**Resettlement and Rehabilitation: Lessons from India**	357
	Mukuteswara Gopalakrishnan	
17	**Impacts of the High Aswan Dam**	379
	Asit K. Biswas and Cecilia Tortajada	

Index 397

Contributors

Dogan Altinbilek Civil Engineering Department, Middle East Technical University, Ankara, Turkey

Morag Anderson RDS Partners, Hobart, TAS, Australia

Leopoldo J. Bartolome Graduate Programme on Social Anthropology, National University of Misiones, UNAM, Posadas, Argentina

Asit K. Biswas Third World Centre for Water Management, Atizapan, Mexico
Lee Kuan Yew School of Public Policy, Singapore

Benedito P.F. Braga Department of Hydraulic and Sanitary Engineering, EPUSP, São Paulo, Brazil

Rita Cestti Sustainable Development Department for the Latin America and the Caribbean Region, The World Bank, Washington, DC, USA

Sandra Vaz da Costa Kaerl Lake Project, Fluor Canada, Calgary, Canada

Christine M. Danklmaier National University of Misiones, El Bolson, Rio Negro, Argentina

Joaquim Guedes Correa Gondim Filho National Water Agency of Brazil, Brasilia, Brazil

Roger Gill Hydro Focus Pty Ltd., Taroona, TAS, Australia

Mukuteswara Gopalakrishnan Indian Water Resources Society, New Delhi, India
New Delhi Chapter World Water Council, New Delhi, India
International Commission on Irrigation and Drainage, New Delhi, India

Saku Härkönen Uusimaa Centre for Economic Development, Transport and the Environment, Helsinki, Finland

Walter Hauenstein Swiss Association for Water Resources Management, Baden, Germany

Jari T. Huttunen Department of Environmental Sciences, University of Kuopio, Kuopio, Finland

Matti Kummu Water and Development Research Group, Aalto University, Espoo, Finland

Raymond Lafitte Ecole Polytechnique Fédérale de Lausanne, EPFL - ENAC - ICARE- LCH, Laboratoire de Constructions Hydrauliques, Lausanne, Switzerland

R.P.S. Malik International Water Management Institute, New Delhi Office, New Delhi, India

Agricultural Economics Research Centre, University of Delhi, Delhi, India

James E. Nickum Asian Water and Resources Institute at Promar Consulting, Tokyo, Japan

R. Rangachari Centre for Policy Research, Delhi, India

Virginia Rodrigues National Water Agency of Brazil, Brasilia, Brazil

R.P. Saxena Central Water Commission, Ministry of Water Resources, Government of India, New Delhi, India

Thayer Scudder California Institute of Technology, Pasadena, CA, USA

Guoqing Shi National Research Center for Resettlement (NRCR), Hohai University, Nanjing, China

Martha Regina von Borstel Sugai Water Resources Consultant of COPEL, Parana State Power Utility, Curitiba, Brazil

C.D. Thatte International Commission on Irrigation and Drainage (ICID), Pune, India

Cecilia Tortajada International Centre for Water and Environment, Zaragoza, Spain

Lee Kuan Yew School of Public Policy, Singapore

Third World Centre for Water Management, Atizapan, Mexico

Olli Varis Water and Development Research Group, Aalto University, Espoo, Finland

Qingnian Yu Land Management Institute, Hohai University, Nanjing, China

Jian Zhou National Research Center for Resettlement (NRCR), Hohai University, Nanjing, China

Chapter 1
Impacts of Large Dams: Issues, Opportunities and Constraints

Asit K. Biswas

1.1 Introduction

For nearly 5,000 years, water-retaining structures have been built in different parts of the world to ensure water is available for domestic and agricultural purposes throughout the year. From time immemorial, human beings have settled in the fertile plains of major rivers like the Nile in Africa, Euphrates-Tigris in Mesopotamia, and the Indo-Gangetic plain in the Indian subcontinent. In these areas, floods and droughts had to be managed to reduce losses to human and cattle populations and also to limit economic damage. During the past two centuries, hundreds of millions of people lived around rivers, which necessitated control of these rivers to provide assured water supply for domestic, agricultural and industrial purposes and to reduce flood and drought damages. Thus, the building of dams has gained steady momentum. More recently, after the 1930s, water requirements increased exponentially in countries where there was significant immigration, such as Argentina, Australia, Brazil, Canada and the United States, to satisfy the needs of their expanding populations. Globally, with the passage of time, water control and assured availability of water of appropriate quality became essential requirements for continuing economic and social development.

As human knowledge and experience advanced, it was possible to construct progressively larger and more complex water storage and distribution structures than ever before, especially towards the second half of the last century. Fortunately, these advances coincided with the rapid growth in global population during the post-1950 period, when more and more water was necessary to support ever-increasing human activities in the domestic, agricultural and industrial sectors. With very significant

---

A.K. Biswas
Third World Centre for Water Management, Atizapan, Mexico

Lee Kuan Yew School of Public Policy, Singapore
e-mail: akbiswas@thirdworldcentre.org

advances in technology, the human knowledge base and the global economy, and plentiful availability of water, it was possible to match the accelerating water demand by increasing its available supply in most countries till about 1990.

In addition, electricity requirements to support economic expansion and an ever-increasing global population have been rising in recent years, at a much higher rate than the water available. Since no large-scale generation of electricity is possible without water, water requirements for power generation have increased concomitantly. Furthermore, water-based transportation became over time an important means to move goods produced in one country to another where they were needed. Populations became progressively more dispersed over larger areas, and the rates of urbanisation increased. As a consequence, social settlements had to be protected from the ravages of regular droughts and floods through the construction of water control structures and better management practices. Water therefore became a critical component of development across the world for the nineteenth century as well as the first half of the twentieth century. This situation is still valid for most developing countries today.

A natural result of these and other related developments was that a large number of dams had to be built to satisfy the growing demands for, and control of, water for various purposes, including generating hydroelectricity to meet the burgeoning energy demands of the domestic, industrial and agricultural sectors. Hydropower became an important source of energy, so much so that in a country like Canada, the word 'hydro' became synonymous with electricity. From the 1930s to the 1980s, numerous dams were built all over the world for hydropower generation, flood control or multi-purpose water development.

1.2 Developments During the Post-1950 Period

Construction of large dams before the 1960s was very significant in the so-called developed world, which included western Europe, the United States, Australia, Canada, the former Soviet Union and Japan. Institutions such as the Bureau of Reclamation and the Corps of Engineers of the United States became famous all over the world because of their expertise in constructing and managing large dams to promote economic development and sustain human welfare. The Tennessee Valley Authority (TVA) of the United States was viewed with admiration worldwide for a considerable time after the 1940s because of its positive impacts on regional economic development. Also during this period, the TVA was generally seen through rose-coloured glasses and while its benefits were the subject of adulation, its weaknesses were not considered seriously, either within the United States or in the rest of the world. A few countries such as India tried to duplicate the TVA experience with its Damodar Valley Corporation (DVC). Not surprisingly, the DVC model did not work out too well for India, because of problems arising from technology transfer between two countries with significantly different physical, technical, social, cultural, economic and institutional conditions, and also because times and perceptions had changed in the intervening period between the establishments of the TVA and of the DVC. Similarly, the Murray-Darling Basin development in Australia was initially considered to be a successful project, only to be discarded as a development model in the post-2000 era when prolonged drought ravaged Australia. In other words, the models that were considered to be very good when they were first planned and constructed were later found to have many disadvantages. This is not only true in the area of dams development but also in other areas of development.

During the post-1950 period, many countries of Asia and Africa began to shed their colonial past. With their newly gained independence, there was an urgency to accelerate their national development processes, to which inadequate attention had been paid by colonisers during centuries of European rule. Accelerated social and economic development became an urgent necessity in all these countries to improve the standard of living of their people. Water was considered to be an important means to foster such development processes. Because of the major contributions dams could make to national development processes, construction of large dams often became a symbol of nation-building and national pride, and in many instances was considered to have contributed to national unity. Thus, the first Prime Minister of India, Jawaharlal Nehru, said that large dams were the new temples of modern India. Not surprisingly, the Bhakra and Hirakud dams in India, Volta Dam in Ghana, Kariba Dam in Zambia, and the High Aswan Dam in Egypt were all considered to be symbols of development and progress in these newly independent countries. It is also clear that these dams helped their national economies in a myriad of ways, many of which are still not fully known or understood. Eminent leaders of the time, such as President Gamal Abdel Nasser of Egypt and Prime Minister Kwame Nkrumah of Ghana, viewed these large structures as indicators of shedding the colonial past, and of postcolonial development. (Biswas and Tortajada 2002).

By 1975, the United States, Canada and most countries in Western Europe had essentially completed their programmes of constructing large dams. In addition, by this time the best and the more economic sites had been developed in these advanced countries. Of course, the situation was very different in the developing world, where much of the large water infrastructures could not be built for a variety of reasons. Thus, during the post-1975 period, the construction of large dams rarely occurred in the developed countries mentioned above; the focus shifted completely to developing countries such as Brazil, China, India, Indonesia, Malaysia, Thailand, Turkey, etc., where earlier progress had been insufficient. Japan is one of the very few developed countries where large dams continued to be built during the post-1975 period.

1.3 Developments During the Post-1975 Period

A major development of the period after 1975 was the gradual emergence of environmental and social movements, initially in a few select developed countries. Their environmental advocacy steadily contributed to major changes in people's mindsets.

In June 1972, in Stockholm, the United Nations convened the first of its large conferences of the decade on the human environment. This conference was a landmark for the environmental movement, even though it was boycotted by the then Soviet Union and the countries of Eastern Europe over political issues related to the status of East Germany. This conference was followed in rapid succession by similar UN mega-conferences on population (Bucharest, 1974), food (Rome, 1974), women (Mexico City, 1975), human settlements (Vancouver, 1976), water (Mar del Plata, 1977), desertification (Nairobi, 1977), science and technology for development (Vienna, 1979) and new and renewable sources of energy (Nairobi, 1981). All these conferences did include some discussions on water, and all of them also considered the environment in one form or another (Biswas and Tortajada 2009). The Stockholm Conference also resulted in the establishment of the United Nations Environment Programme. It was the first UN agency that was established in a developing country, and was expected to represent the environmental consciousness of the UN system. All these events cumulatively, and in their own ways, have had a considerable impact on the way social perceptions of large dams have evolved in one form or another.

When the Stockholm Conference was held in 1972, few countries had ministries concerned with the environment. Now, some 38 years later, it is difficult to find a single important nation that does not have an environment ministry or department. The environment has now rightly become a mainstream subject, and environmental impact assessments of large development projects have become mandatory in nearly all countries of the world.

While ensuring that environmental issues are properly considered in all development projects has been a most welcome improvement, it must also be admitted that projects to improve or maintain environmental conditions and people's quality of life in the developing world are now often prevented or delayed due to some vocal activists working on a single cause whose main objective is to prevent construction of all types of infrastructures irrespective of their overall benefits to society. These so-called environmentalists, who are primarily from the developed world, have exercised considerable power directly and through financial and intellectual support to their counterparts in the developing world. Many of these activists from the developed world, who already have a good standard of life, access to clean water, adequate food and energy, and a very good lifestyle, have often eschewed scientific and technical facts, manipulated available information, quoted data and information that are patently erroneous or out of context and always had their own hidden agendas.

For reasons that remain difficult to fathom, the construction of large dams became a lightning rod for many of these so-called environmental activist groups in recent years. In an era of 'small is beautiful', large was automatically deemed bad and ugly on ideological grounds, irrespective of desirability or benefits. Accordingly, and not surprisingly, a myth began to emerge that all large dams are bad, and also that the water problems of the developing world could be successfully and cost-effectively resolved in a timely manner only through small dams or rainwater harvesting structures.

There is no doubt that small dams and water harvesting techniques will help in some rural and smaller urban areas. Thus, given their technological, economic and socio-environmental desirability, their use in appropriate areas needs to be encouraged.

At the same time, however, it is important to recognise that small water structures alone will not be able to solve the complex water problems of large urban areas and major industries where demand for water is extremely high, and increasing, and where rainfall is scanty and erratic. Large and medium dams will be essential to continue to provide water for meeting the escalating needs of a steadily urbanising world for decades to come. Further, if future climatic regimes change, and/or climatic fluctuations intensify as compared to the present, countries will have to store more and more water to assure water security. This will require the construction of additional reservoirs, as well as more efficient use of stored water.

People in the western world have to realise that small dams alone cannot solve the water problems of the developing world. The same situation prevailed earlier in their own countries, where large dams had to be built to satisfy increasing water requirements. Having completed the construction of necessary large dams in the west, they are often opposed to the construction of large dams in the developing world, where social and economic needs for water-related activities are growing exponentially. It has to be realised that small can be beautiful, but under many conditions it could be inadequate or even ugly. Equally, big could be magnificent, but in a different context, it could also be bad and undesirable.

Whether a small or large dam is the most appropriate solution depends on many specific local conditions. Thus, what is desirable at this point is not a dogmatic and irrational debate between small and large dams, since both are necessary, but rather a judicious mixture of small and large dams which could solve the water problems of the developing world, and could simultaneously contribute to improvements in the quality of life of citizens. The decision regarding the most appropriate balance has to be made by the people of each developing country themselves, based on their own requirements and aspirations, and physical, social, economic, cultural and environmental conditions.

More specifically, these decisions must be made by those living in the river basins where development projects are being considered, and which may affect their future lifestyles positively and/or negatively. In the present era, where democracy is considered essential, these complex decisions must be made after a serious and informed debate, primarily among the people in areas where development is being contemplated. Thus, the decision whether or not to construct a large dam in a specific state in Brazil, India or China should be made by the people of that area, and must not be imposed on them by activists from the United States or Western Europe, irrespective of their motives, dogmatic beliefs and hidden agendas, as a form of neocolonialism. Nor should decisions be dictated by urban elites who are not from the region where such projects are being considered. People from outside the region can make a contribution to the debate, but they should not be allowed to manipulate the process so that their dogmatic views, whatever they may be, prevail in the end, unless that is also what the local people want. Sadly, this does not happen in many cases, and people from outside the development areas, and also from outside the countries, are influencing decisions because of their economic power and media influence. Such decisions are often detrimental to the people of the developing countries on a long-term basis.

1.4 Controversy During the Post-1985 Period

The construction of large dams became an even more controversial issue than ever before, especially after 1985. Proponents of large dams claim that they deliver many benefits, among which are increased and assured water availability for domestic and industrial purposes, increased agricultural production because of the availability of reliable irrigation water, protection from floods and droughts, generation of hydroelectric power, navigation and overall regional development which improves the quality of life of the people, particularly women. They argue that like any other large infrastructure development or national policy, dams have both benefits and costs. However, properly planned and constructed, the overall benefits of dams far outweigh their total costs, and thus society as a whole is better off with such well-planned dams. It should be noted that at present there is enough knowledge, experience and technology to plan, construct and manage new large dams properly.

In contrast, opponents argue that dams bring catastrophic losses to society, and the social and environmental costs far outweigh any benefits that dams may contribute. They claim that dams accentuate unequal income distribution since benefits accrue almost exclusively to the rich, while the poor slide further down the economic ladder. They also claim that the main beneficiaries of dams are construction companies, consulting engineers and corrupt politicians and government officials, who work in tandem to promote such dams. They argue that the poor do not benefit, instead they mostly suffer because of these structures.

1.4.1 Why This Controversy?

In recent years, the views of proponents and opponents of large dams have become polarised. In scientific and logical terms, both views cannot be correct.

There has never been a real dialogue between the two camps, especially on a continuing basis. For example, during the Second World Water Forum, held in The Hague in 2000, the pro-dam sessions discussed the benefits of dams, and the anti-dam sessions blamed all the ills of society on them. For the most part, proponents and opponents did not attend each other's sessions. Both sides went home thinking that the Forum basically endorsed their views as correct!

The situation was a little better at the Third World Water Forum in Kyoto, where the International Hydropower Association (IHA, a professional association promoting well-planned hydro development) and the then International Rivers Network (IRN, a non-governmental organisation whose sole *raison d'être* is to oppose the construction of all dams) arranged a debate on the benefits and the costs of large dams. By all accounts, the pro-dam group won this debate with arguments based on observed facts and scientific analyses, and not on polemics or hypotheses. IHA's presentation focused exclusively on the benefits and costs of the Atatürk Dam, and drew on observed facts and figures which were meticulously collected. In contrast, the IRN's generalised innuendos were extensively attacked by the audience for being one-sided, erroneous and highly economical with the truth. However, interactions of this kind between the two opposing camps have been very rare. Such discussions and debates should be encouraged, since only through such debates can a societal consensus on this complex issue be reached.

An important question that needs to be asked is why, in the twenty-first century, with major advances in science and technology, it has not been possible to answer the relatively simple question of the real costs and benefits of large dams, so that their net impacts and benefits can be determined authoritatively and comprehensively? The sterile debate on dams needs to be resolved conclusively once and for all, so that appropriate water development policies can be formulated and implemented, especially in developing countries, which will maximise their overall social and economic welfare. Prima facie, it should not be a difficult question to answer, scientifically and objectively. However, vested interests have stood against its resolution.

It should be noted that the world of development is complex, with scientific uncertainties, regional variations, vested interests, dogmatic views and hidden agendas. The issue of dams is no exception, and not surprisingly, it has fallen victim to this complex interaction of forces.

Many factors have fuelled the current controversy, some of which are real, but some are artificial and manufactured. The main reasons for this controversy will be briefly discussed in the next section.

1.5 Vested Interests

There is no doubt that many people have a personal interest in this debate, irrespective of which side they are on. Much has been written and said about the construction and consulting companies that are associated with the planning, design and construction of large dams, and accusations have been made about their financial contributions to political parties, who are often the final arbiters and decision-makers in democratic societies. There is also no doubt that construction and development of large dams is a capital-intensive activity, and many people benefit economically from this process. The anti-dam lobby often portrays the pro-dam lobby as being interested in the construction of dams only because of the financial benefits obtained through the planning and construction processes. Unfortunately, the voices of people from varied sectors of society who benefit from dams, like farmers and others who use the hydropower generated by dams, are seldom heard in this debate.

In contrast, those non-governmental organisations (NGOs) which are against dams (there are numerous pro-dam NGOs as well, but they generally are not as media-savvy as the anti-dam NGOs, and are thus not as visible) mostly portray themselves as little 'Davids' who are pitted against the well-heeled 'Goliaths' of the pro-dam lobby, who have direct connections to the corridors of power. It is true that many grassroots NGOs have made useful contributions in highlighting the plight of people who have to be resettled as a result of large development projects (dams, new towns, airports, highways and so on). However, many of the main activist-NGOs in the anti-dam lobby have now become financially powerful, mainly with support from several international funding agencies, primarily from the United States. Their self-portrayal as small or weak organisations is primarily aimed at the media and for publicity purposes, and is far from the truth.

The current power of NGOs can be surmised from some research by the Johns Hopkins University Centre for Civil Society (1999) which indicated that, globally, the non-profit sector (excluding religious organisations) has become a $1.1 trillion industry, employing some 19 million fully paid employees. This represents the world's eighth largest economy. A global assessment of NGOs carried out by a reputable NGO, SustainAbility (2003: 2), has pointed out that NGOs 'that once largely opposed—and operated outside—the system' are becoming integral to the system. They are no longer small or even outsiders, as many would like to portray themselves. Nor are their motives and activities transparent, which is exactly what they accuse their opponents of.

International activist anti-dam NGOs are at present no exception to the findings mentioned above. They have become adept at playing the system to promote their own agendas, at least in terms of obtaining funds from various institutions, and generating extensive media publicity for their unitary causes. The anti-dam lobby has also become financially powerful. Anecdotal evidence can confirm this. For example, the only institution concerned with the debate on dams that participated in the Third World Water Forum in Kyoto and brought its own recording team was an NGO belonging to the anti-dam lobby.

There is no doubt that there are extremists in both the pro-dam and the anti-dam lobbies, who have their own vested interests and hidden agendas, and thus their views and statements need to be carefully analysed for accuracy, generalisations based on limited or no facts, and innuendos. Truth has often become a casualty in this bitter fight between the two camps.

Sebastian Mallaby (2004a), columnist and editorial writer for *The Washington Post*, assessed two dam projects that have been consistently opposed by activists: a dam on the Nile at Bujagali, Uganda, and the Qinghai Project in Tibet, China. In a very detailed analysis of local opposition to the proposed large dam at Bujagali, Mallaby found that contrary to the extensive claims of a Californian anti-dam NGO, the people who are to be resettled are quite happy to accept the terms. The only people who objected to the dam were those living outside its perimeter since they would not benefit from its generous relocation payout.[1]

Mallaby went on to argue that:

> The story is a tragedy for Uganda. Clinics and factories are being deprived of electricity by Californians whose idea of an electricity crisis is a handful of summer blackouts. But it is also a tragedy for the fight against poverty worldwide, because projects in dozens of countries are similarly held up for fear of activist resistance. Time after time, feisty internet-enabled groups make scary claims about the inequities of development projects. Time after time, Western public raised on the stories of World Bank white elephants believe them.
>
> (Mallaby 2004a: 52)

---

[1] On this issue, see also Mallaby (2004b).

1 Impacts of Large Dams: Issues, Opportunities and Constraints

He concluded that 'NGOs claim to campaign on behalf of the poor, yet many of their campaigns harm the poor' (Mallaby 2004a: 52). Based on our own experience from different parts of the world, we have to agree with Mallaby.

1.6 Complex Issues with No Single Answer

The sweeping generalisations of the two groups, made primarily to justify their positions, for the most part do not survive scrutiny. In the cacophony of arguments, what is often forgotten is that the issues involved are complex, and that there is no single answer that could apply to all the dams of the world, constructed or proposed, irrespective of their locations and sizes. Nor can one view be everlasting in any country: it will invariably change with time.

What has been forgotten in the current debate on dams is that neither the statement 'all dams are good' nor 'all dams are bad and thus no new ones should be constructed' are correct and applicable to all dams. Depending on the criteria of 'good' selected, it has to be admitted that there are both good and bad dams. Furthermore, the needs vary from one country to another, from one period to another and often from one region to another, especially within large countries such as Brazil, China or India, depending on climatic, economic, social and environmental conditions.

It is equally important to recognise that countries are at different stages of economic development, and thus their needs for dams also vary. Also, an industrialised country like the United States has developed nearly all of its best and most economic dam sites. In contrast, most potential sites in sub-Saharan Africa (with the exception of South Africa) have yet to be developed. A country like Nepal has a similar amount of hydropower potential as that already developed in the United States. However, Nepal has developed only about 4% of its hydro potential. Thus, what may appear to be a logical and efficient solution for the United States at present is unlikely to be the best and most suitable solution for conditions in Nepal.

There is a time dimension to these arguments as well. For example, during the 1960s, 1970s and 1980s, Asahi Shimbun, one of the newspapers with the largest circulation in Japan, routinely took the Japanese government to task for not building enough dams. This has changed dramatically in the past two decades—the same newspaper now routinely takes the government to task for building dams. From being a pro-dam newspaper, it has now become an anti-dam newspaper. Such changes in perception and attitude often happen over time.

In the area of dams, as in most other complex development-related issues, there simply is not 'one size that fits all'. Each large dam is unique: it has its own sets of benefits and costs. In the final analysis the decision to construct a new dam should be based on its overall benefits to society. Both the proponents and opponents in the dam debate have ignored this simple fact.

1.7 Climatic Differences

A major scientific issue that has been totally ignored in the current debate are the very significant climatic differences between developed and developing countries, especially in terms of rainfall distribution across the year, and from one year to another. This is an important issue, because storage is more important for developing countries because of the rainfall patterns which are more erratic, compared to the developed countries of the temperate zones.

Very few development experts, including water experts, have appreciated the importance and relevance of climate patterns for economic development. This lack of understanding is especially difficult to understand in the case of water experts, since one of their main concerns is precipitation. As early as 1951, Galbraith, an eminent economist, noted that if 'one marks off a belt a couple of thousand miles in width encircling the earth at the equator, one finds within it no developed countries' (Galbraith 1951: 693). The same year, a United Nations report (1951) noted that if the industrialised countries are marked on a map, it will be seen that they are located in the temperate zone. In other words, developed countries are located in temperate zones, but developing countries are found in tropical and subtropical climate areas (Biswas 1984).

Another important issue that has received scant attention is the distribution of rainfall in the tropics and subtropics compared with the temperate zones. The annual rainfall averages mask the very significant differences in the patterns of rainfall distribution between developed and developing countries. For example, if the annual average rainfalls of three cities are compared, two in developing countries (Sokoto on the southern border of the Sahel, in Nigeria, and New Delhi in India) and London, United Kingdom, they are somewhat similar: 57, 71 and 67 cm, respectively. However, if their distributions over the year are considered, the patterns are totally different. For example, London, a temperate zone city, can be characterised by a low but reasonably uniform monthly rate of rainfall over the year, varying from a high of 61 mm in October to a minimum of 35 mm in April. Similarly, rainfall retained in the soil across the year is reasonably uniform.

However, the rainfall pattern is very different for Sokoto. Nearly 36% of annual average rainfall occurs during the month of August alone. More than 92% of average rainfall occurs within the 4-month period of June to September. There is no rainfall during the 5 months of November to March, and very little in April and October (10 and 13 mm, respectively). Not surprisingly, Sokoto has a significantly lower rate of rainfall retention in the soil throughout the year, compared with London. In fact, the highest rainfall retention rate in the soil in Sokoto is 42% (September), which is lower than the lowest retention rate in London that occurs in August. Thus, water management strategies for London and Sokoto have to be very different, even though their annual average rainfalls are somewhat similar. Irrigation, the largest user of water in the world as a whole, has to be very different as well. Sokoto cannot manage year-long agricultural production without storing water during the rainy months, which can then be progressively released as required over the year during the dry months. In contrast, in the case of climatic regimes like London with its more uniform precipitation and high soil moisture retention rates, the need for irrigation water is significantly lower. In fact, no irrigation is needed for high soil moisture retention rates as noted in London.

Even monthly rainfall figures may lead to a misleading comparison. For example, the average number of rainy days in New Delhi is about 40%. However, during the rainy days, rainfall does not occur uniformly over a period of 24 h. It has been estimated that New Delhi receives nearly 90% of its annual rainfall in less than 80 h, though these hours are not necessarily consecutive. One of the rainiest towns of India, Cherrapunji, receives much of its annual rainfall of 10,820 mm during the southwest monsoon, between June and August. This immense rainfall occurs mostly in about 120 h. Because this vast quantity of water cannot be properly stored, Cherrapunji, in spite of its very substantial rainfall, currently faces serious water problems during the dry months of the year.

Overall, India receives nearly most of its annual rainfall in less than 100 h. Because of this very skewed pattern of rainfall distribution, water management strategies in India have to be different as compared with countries in temperate climates where rainfall is significantly more regular and predictable.

Because of the very high seasonality of rainfall in most developing countries, the critical issue is the storage of such immense quantities of rainfall over very short periods, so that they can be used over the entire year. In addition, the inter-annual fluctuations in rainfall are also high in such countries, which means that the incidences of floods and droughts are much more frequent than in the temperate zone. Thus, what is needed for countries of the developing world in the tropical and subtropical regions are cost-effective, socially acceptable and environmentally sound solutions for storing large volumes of precipitation over a comparatively short period, so that the stored water can be used during the dry periods each year, and also inter-annually during prolonged periods of droughts. The technical complexities of water management in the developing countries of the tropics and subtropics are therefore significantly more complex than those in developed countries located in temperate zones. This simple fact has been totally ignored in the debate on large dams.

Because of climatic differences, developing countries must consider all possible alternatives available for storing water during the periods of intense rainfall, so that these can be made available whenever necessary to satisfy human needs. The alternatives to smoothen out these wide inter- and intra-annual fluctuations in rainfall may include dams (small, medium and large), groundwater recharge and storage, as well as rainwater harvesting. Water problems can only be resolved by taking such a holistic approach. None should a priori be excluded.

It is sad that the current debate on dams has become increasingly dogmatic and emotional. Participants may often hear what their opponents are saying, but they do not listen. The alternatives available are often not 'either/or' solutions; what is needed is the consideration of which alternatives will work best, where and under what conditions. The focus of discussions should be on how best to meet the water needs for all segments of society cost-effectively, efficiently and reliably, on a long-term basis.

For the most part, the current debate on 'dams or no dams' is an irrelevant one. It is far more important to objectively assess the societal needs for water, and then take steps to meet them in the best and the most socially, economically and environmentally acceptable way. Depending on the prevailing conditions of the location under consideration, the most efficient alternative may be the construction of a large dam, rainwater harvesting, a mixture of these two and/or other solutions. There is simply no single solution that will be appropriate for all climatic, physical, social, economic and environmental conditions, for all countries of the world and also for all periods in history.

In the real world of water resources management, small may not always be beautiful. By the same token, large could sometimes be magnificent, but on other occasions it could be a disaster. Each alternative must be judged on its own merit and within the context in which it is to be applied. Once the right solution has been identified for the specific location in question, the scheme should be planned, designed, implemented and managed as efficiently, equitably and quickly as possible.

It should also be noted that the economies of developed countries are at present no longer dependent on water or climatic fluctuations. Accordingly, if there are droughts and floods, these often result only in temporary inconvenience for some of the people but not serious long-term damage to the country as a whole. In contrast, availability of drinking water in developing countries often depends on rainfall because of lack of infrastructures and their poor operation and maintenance. Furthermore, agriculture continues to be a very important factor for survival in the developing world, in terms of both food and employment generation. Prolonged droughts also mean low reservoir levels, and thus lower hydropower generation. This often results in regular power cuts and voltage reductions, which seriously disrupt industrial production and contribute to a poorer quality of life of the people. Industrial production and employment suffers. Reduced agricultural production and the disruption of industrial activities as a result of power shortages contribute directly to serious human hardships. The absence of proper water management thus produces a 'lose-lose' situation all around. On the whole, developed countries are largely immune to this process at present. They may have been vulnerable some 50–100 years ago, but their economies are now significantly more diversified and resilient to successfully meet the current climatic vagaries. The situation may become more complex and difficult in the future, once the impacts of climate change on water management can be properly assessed.

1.8 Absence of Post-Construction Assessment of Dams

One of the major reasons why the current non-productive debate on dams has thrived is the absence of objective and detailed ex-post analyses of the physical, economic, social and environmental impacts of large dams, 5, 10 or 15, years after their construction. At present, thousands of studies exist on environmental impact assessments (EIAs) of large dams, some of which are very good while others are not worth the paper on which they are written. It should be realised that all EIAs are invariably predictions, and until the dams become operational, their impacts (types, magnitudes, and spatial and temporal distributions) are not certain, and thus remain in the realm of hypotheses. Even the very best assessment can perhaps accurately forecast only about 70–75% of identified impacts in terms of time, space and magnitude. It is not possible to identify 25–30% of the impacts that will actually occur after the dams have become operational. For an average EIA of a large dam, some 30–50% of its impacts (positive or negative) are not accurately identified at present, in terms of types, magnitude, spatial distribution and timings.

In addition, all environmental assessments must include both positive and negative impacts. A two-pronged approach is needed which will include identification and assessment of positive benefits and recommendations of measures which will maximise them, as well as an assessment of the negative impacts and policy actions that should be taken to minimise them. Only such a comprehensive approach can ensure that the net benefits to society can be maximised. Exclusive consideration of negative impacts while conducting EIA, as is widely practised at present, is a fundamentally flawed procedure, which will seldom contribute to the maximisation of overall benefits of any project. Such methods will yield information only on part of the story, and will not provide a sound, logical and scientific basis for making rational and long-term decisions.

As mentioned above, while thousands of EIAs were conducted for large dams prior to their construction, assessment of actual impacts of large dams 5, 10 or 15 years after their construction are very few and far between. Some have claimed that the World Commission on Dams (WCD) prepared numerous such assessments of large dams in different parts of the world. Unfortunately, most of these analyses are superficial and often skewed to prove the dogmatic and one-sided views of the authors who undertook the studies, the majority of whom belonged to the anti-dam lobby. These assessments can therefore neither be considered objective nor accurate. It is possible that among these highly biased assessments, there are a few good case studies. Unfortunately, however, no rigorous peer reviews of these case studies were ever carried out and as a consequence, any valuable cases among these reports, if they exist, remain very well hidden. The authors of five of these case studies submitted their papers for possible publication in the *International Journal of Water Resources Development*. All five papers were rejected for their poor quality by peer reviewers. Thus, the WCD case studies of assessments of the real impacts of large dams from different parts of the world are of very limited use to the water and development professional, irrespective of the current rhetoric of their supporters whose numbers are shrinking as time passes.

Because of this unfortunate current situation, the Third World Centre for Water Management carried out a comprehensive impact assessment (positive and negative) of three large dams that have been operational for a minimum of 10 years: the High Aswan Dam in Egypt, Atatürk Dam in Turkey, and the Bhakra-Nangal Project in India. This analysis also included the perceptions of people in areas affected by the dams, both beneficiaries as well as those who had to pay some costs, for example, people who had to be resettled. Two of these in-depth case studies, on the Bhakra-Nangal Project

(Rangachari 2006), and the High Aswan Dam (Biswas and Tortajada 2011) are now available. A comparison of these two comprehensive analyses with similar studies carried out by the WCD would likely show the superficial quality of the WCD studies and their inherent biases.

1.9 World Commission on Dams

Much has been said and written on the WCD. Views on the process and the report have ranged from admiration to outright dismay. An objective assessment of the process and an assessment of the real impacts (positive and negative) of the WCD have yet to be made. However, some comments on the commission and its report would be appropriate here.

In April 1997, the World Bank and the International Union for Conservation of Nature (IUCN) convened a meeting at the IUCN headquarters in Gland, Switzerland, ostensibly to discuss an internal World Bank review on large dams and the need for a more detailed study. This review concluded that 'the finding that 37 of the large dams in this review (74%) are acceptable and potentially acceptable, suggests that, overall, most large dams were justified' (World Bank 1996).

The two sponsors arbitrarily chose participants for this meeting. The only consideration appears to have been that the participants represented diverse groups of interests. However, the reasons why a specific person or institution and not another from the same interest group was selected remain a mystery. According to the list of participants available, 38 people attended this workshop, of which 12 (nearly a third) represented the two sponsors alone.

This group unilaterally decided to establish an international commission to review the effectiveness of large dams and develop standards, criteria and guidelines. The group, which was originally selected without an adequate rationale, subsequently became a self-appointed 'reference group'. In addition, some members of the World Bank and IUCN unilaterally formed an Interim Working Group. The net result was the creation of a World Commission on Dams, the 'mandate' of which, according to its own pronouncement, was to develop a report that would be submitted to the two sponsors, the reference group, and the 'international community'. What constituted the 'international community' was not spelt out.

The chairman and commissioners were selected through a completely opaque process. The criteria and reasons for selection are still unknown. Since one of the authors of this chapter has been associated with several world commissions before, it is useful to compare the WCD process with two earlier commissions, the Independent Commission on International Development Issues (the Brandt Commission) and the World Commission on Environment and Development (the Brundtland Commission).

The Brandt Commission owed its formation to the personal interest of Robert McNamara, then President of the World Bank, and it had the moral backing of the United Nations. Unlike the WCD, the Brandt Commission did not pretend to be representative of all stakeholders. It consisted of a group of eminent persons

1 Impacts of Large Dams: Issues, Opportunities and Constraints

from both the North and the South, and was chaired by a very well-known and highly regarded international figure, the former German Chancellor Willy Brandt. The individual members of this commission were also respected international figures who, because of their own accomplishments in various areas, brought credibility and gravitas to the Brandt Commission, which most of the WCD members sorely lacked.

The Brundtland Commission had an even better mandate compared to the Brandt Commission, since this initiative came directly from the Secretary-General of the United Nations. The UN General Assembly unanimously adopted a resolution in 1983 to establish this Commission. It was chaired by the former Prime Minister of Norway, Gro Harlem Brundtland, also a well-known development personality. In retrospect, both these commissions had very modest long-term impacts in terms of the implementation of their recommendations on a global basis. This, unfortunately, has been the case for most of the world commissions.

It is important to view the WCD within an overall perspective of global development-related events of the past 25 years, especially because of the highly exaggerated claims made regarding its effectiveness and impacts by many supporters. Unfortunately, it was neither a unique exercise nor a totally new initiative, but in fact the continuation of a well-established trend. In addition, it had a somewhat dubious origin, which leads one to seriously question its legitimacy and objectivity.

There are some fundamental differences between the three commissions discussed above, among which are the following:

- The Brandt and the Brundtland commissions both had mandating authorities, the Brundtland more so than the Brandt. In contrast, while the WCD had a mandate there was no mandating authority. The WCD basically consisted of 26 individuals (excluding the World Bank and the IUCN staff members) who took it upon themselves to start a 'World Commission'. Some participants later decided not to be actively associated with the commission itself. Because the WCD had no mandating authority, its recommendations have not been binding on any party. Even the World Bank, one of the two godfathers, made some initial positive comments on the WCD process but then showed very little interest in changing its policies to reflect the recommendations of the WCD report.
- Irrespective of the claims by the WCD and by its supporters that the process was transparent, democratic and unique, the authors' interactions with the Secretariat suggested the contrary. It was somewhat opaque, secretive, autocratic and ad hoc.
- The legitimate question that has not been raised so far, and has certainly not been answered, is who or what gave the arbitrarily selected 26 persons, and the 12 staff members of the World Bank and the IUCN present at the Gland meeting, the right to set up an international commission, and give it an 'international mandate'. How, by what authority, and through what processes was the WCD made representative so that it earned the right to speak for all stakeholders? In fact, many of the major stakeholders played absolutely no role in its deliberations and were also not considered for membership.
- According to any logical criteria, the WCD was not a truly representative body of its stakeholders, irrespective of claims to the contrary. For example, the WCD

had commissioners from the NGO community who spoke in the name of indigenous and tribal people who would be displaced as a result of dam construction. However, the commissioners were neither indigenous nor tribal. It is not clear who or what gave them the right to represent indigenous people; they were the self-appointed representatives of a large group. However, the WCD did not consider including NGOs that represented farmers, whose agricultural production would increase because of irrigation provided by dams. As a general rule, the number of farmers who are affected by a dam is far higher than the number of people who are displaced. It is a strange understanding of democracy, transparency and representation of all stakeholders, when the largest stakeholder, deliberately or otherwise, is not invited to participate. Nor did the WCD invite those who would receive assured water supply, which they earlier lacked. Democracy involves the consideration of pluralism, and pluralism cannot be one-sided as was the selection of commissioners by the WCD and as was reflected in its work.

Since the process employed by the WCD was seriously flawed, its report has had unsurprisingly modest impacts so far, if any, on the countries that are building dams, or on the international funding institutions that finance such projects. The real question that has yet to be asked, let alone answered, is: would the world have been any different, now or 10 years hence, if the WCD had not been established? The authors' view is that it would not have mattered very much one way or another!

1.10 Conclusions

The continuing controversy on dams is a dogmatic and emotional debate. To the extent that it raises new issues which need to be carefully considered and addressed, this debate should be welcomed. To the extent, however, that it is a debate between vested interests, any progress resulting from it is likely to be somewhat limited, or even futile. The debate needs to be refocused. What is necessary is to consider the overall architecture of the water development system that will fulfil the objectives and meet the needs of societies in developing countries: poverty alleviation, regional income redistribution and environmental protection. Within this overall framework, it is imperative to determine how best to supply the water needs of society, in a cost-effective, equitable, timely and environmentally friendly manner. The world of development is complex, and there will always be instances of trade-offs resulting from a major policy, programme or project. These trade-offs should be considered objectively, accurately, honestly, sensitively and in a socially acceptable manner. Within such an overall architecture, the best solution for water development must be sought for each specific case. This may warrant construction of a large dam in a specific location, but it may just as well require a different solution, such as local rainwater harvesting, in another location. Until the site-specific needs, conditions and requirements have been carefully assessed and considered, a 'dam' or 'no dam' solution should not be imposed a priori, especially by people from outside the region.

In the final analysis, the alternatives selected may require the construction of properly planned and designed dams, which could be large, medium or small, or rainwater harvesting, or any number of other appropriate options such as water conservation and the improvement of water-use efficiencies. The solutions selected must not be dogmatic, and should always reflect the needs of the areas under consideration. In terms of water development, it is important to remember that small is not always beautiful and large is not always magnificent. Solutions must be specifically designed to solve the problems at hand. The current emotional debate on dams is somewhat akin to a solution-in-search-of-a-problem, where the a priori solution becomes 'dams' or 'no dams', depending on the lobbies concerned and their vested interests. Such a process, if it is allowed to continue, may prove to be scientifically unacceptable, socially disruptive and environmentally dangerous, especially on a long-term basis.

No single pattern of water development is the most appropriate for all countries at any specific time, or even in the same country because of differing climatic, technical, economic, social or institutional conditions. Water development patterns will vary with time, knowledge and experience, as will development paradigms. Nothing is permanent. Countries are often at different stages of development. No two countries are identical. Their economic and management capacities are not identical; climatic, physical and environmental conditions are often dissimilar; institutional and legal frameworks for water management differ; and, social and cultural conditions vary significantly. Under these heterogeneous conditions that change with time, no single solution can be appropriate for all countries for all times.

In addition, the world is changing at a rapid pace, and its water management concepts and processes must change as well. Past experiences can only be of limited help in terms of water management, especially since these changes, unlike in the past, will come from forces external to the water sector. Among these driving and overarching forces are concurrent rapid and extensive urbanisation and ruralisation in developing countries, accelerated globalisation and free trade, advances in technology such as biotechnology and desalination, and a continuing communication and information revolution. All these and other associated changes will affect water management in a myriad ways, some of which can be anticipated at present while others are still mostly unpredictable and thus likely to be somewhat unexpected. In this vastly changing and complex world of water management, there should be no room for sterile debates on 'dam or no dam'.

The main question facing the developing countries of Asia, Africa and Latin America is not whether large dams have an important role to play in the future, but rather how best they can be planned, designed and constructed where they are needed so that their performance, in economic, social and environmental terms, can be maximised and their adverse impacts can be minimised. At the same time, it is necessary to ensure that those who may have to pay the costs of implementing dams (e.g. the people who have to be resettled) are made the direct beneficiaries of these projects. It will not be an easy task to accomplish, but it is nevertheless essential to make progress in that direction.

References

Biswas AK (1984) Climate and development. Tycooly International, Dublin

Biswas AK, Tortajada C (2002) Development and large dams: a global perspective. Int J Water Resour Dev 17(1):9–21

Biswas AK, Tortajada C (2009) Impacts of megaconferences on the water sector. Springer, Berlin

Biswas AK, Tortajada C (2011) Hydropolitics and impacts of High Aswan Dam. Springer, forthcoming

Galbraith JK (1951) Conditions for economic change in underdeveloped countries. Am J Farm Econ 33(4 Part 2):689–696

Johns Hopkins University Center for Civil Society Studies (1999) Global civil society–dimensions of the non-profit sector. Johns Hopkins University, Baltimore

Mallaby S (2004a) NGOs: fighting poverty, hurting the poor. Foreign Policy 1(5):50–58

Mallaby S (2004b) The World's Bankers: a story of failed states, financial crises, and the wealth and poverty of nations. Penguin Press, New York

Rangachari R (2006) Bhakra-Nangal Project, socio-economic and environmental impacts. Oxford University Press, Delhi

SustainAbility (2003) The 21st century NGO in the market for change. Sustainability, London

United Nations (1951) Measures for the economic development of under-developed countries. United Nations, New York

World Bank (1996) The World Bank's experiences with large dams: an overview of impacts. World Bank, Washington DC

Chapter 2
Indirect Economic Impacts of Dams

Rita Cestti and R.P.S. Malik

2.1 Introduction

Dam projects generate a vast array of economic impacts—both in the region where they are located, and at inter-regional, national and even global levels. These impacts are generally evaluated in terms of additional output of agricultural commodities, hydropower, navigation, fishing, tourism, recreation, prevention of droughts and reduction in flood damages, and are referred to as direct impacts. The direct impacts, in turn, generate a number of indirect and induced impacts as a result of:

- Inter-industry linkage impacts, including both backward and forward linkages, which lead to increase in the demand for and outputs of other sectors.
- Consumption-induced impacts arising as a result of increase in incomes and wages generated by the direct outputs of the dam.

To illustrate, hydropower produced from a multipurpose dam provides electricity for households in urban and rural areas and for increased output of industrial products (e.g. fertilisers, chemicals, machinery). Changes in the output of these industrial

---

This chapter is based on a larger study on the subject sponsored by the World Bank and carried out by a number of researchers. For details see Ramesh Bhatia, Rita Cestti, Monica Scatasta and R.P.S. Malik (eds), Indirect Economic Impacts of Dams: Case Studies from India, Egypt and Brazil, Academic Foundation, New Delhi and World Bank, 2008.

R. Cestti (✉)
Sustainable Development Department for the Latin America and the Caribbean Region,
The World Bank, Washington, DC, USA

R.P.S. Malik
International Water Management Institute, New Delhi Office, New Delhi, India

Agricultural Economics Research Centre, University of Delhi, Delhi, India

Table 2.1 Values of variables required for the estimation of a project multiplier

Definition of project multiplier = $\dfrac{\text{Regional value added with project minus regional value added without project}}{\text{Value added of agriculture and electricity with project minus value added of agriculture and electricity without project}}$

commodities require increased inputs from other sectors such as steel, energy and chemicals, among others. Similarly, water released from a multipurpose dam provides irrigation that helps increase output of agricultural commodities. Increases in the output of these commodities require inputs from other sectors such as energy, seeds and fertilisers. Further, increased output of some agricultural commodities encourages the establishment of food processing and other industrial units. Thus, increased outputs of both electricity and irrigation from a dam result in significant backward linkages (i.e. demand for higher input supplies) as well as forward linkages (i.e. providing inputs for further processing).

Increased outputs of industrial and agricultural commodities generate additional employment, wages and incomes for households. Higher incomes result in higher consumption of goods and services which, in turn, encourage production of various agricultural and industrial commodities. Further, changes in output generated by the project may affect prices of direct project outputs, inputs, substitutes, complements and factors of production. Changes in wages and prices have both income and substitution effects on the expenditure and saving decisions of different owners of factors, which further impacts the demand for outputs within the region and throughout the economy. Induced impacts reflect the feedbacks associated with these incomes and expenditure effects, and also include any impacts of changes in government revenues and expenditures that result from the project.

The magnitude of indirect impacts of a dam via the inter-industry linkages and consumption-induced impacts depend on the strength of linkages amongst various sectors within the given economy. Multiplier analysis provides an approach for quantifying the magnitude of inter-industry linkages and consumption-induced effects in relation to purely direct impacts. As shown in Table 2.1, in order to estimate a project multiplier value for a dam (say the Bhakra Dam), for the numerator we need to estimate the regional value added (for the region where it is located, in this case Punjab, a state in India) under the 'with project' situation as well as under the 'without project' situation. For the denominator, we need to estimate the value added from the sectors that are directly affected by the major outputs of the dam (namely, agricultural output, hydroelectricity, water supply, etc.). A multiplier value of 1.90, for instance, indicates that for every dollar of value added generated directly by the project at maturity, another 90 cents are generated in the form of indirect or downstream effects.

Practitioners and policy analysts, though conscious of these indirect impacts of dams, have for a long time felt the need for their proper appraisal and quantification. The World Commission on Dams (WCD) Report and numerous other studies have discussed the importance and difficulties of evaluating indirect impacts of dams.

According to the WCD Final Report, 'a simple accounting for the direct benefits provided by large dams—the provision of irrigation water, electricity, municipal and industrial water supply, and flood control—often fails to capture the full set of social benefits associated with these services. It also misses a set of ancillary benefits and indirect economic (or multiplier) benefits of dam projects' (WCD 2000). As noted by the Operations Evaluation Department of the World Bank, 'dams providing water for irrigation also produce, in general, substantial benefits stemming from linkages between irrigation and other sectors of production. Unfortunately, there are no estimates available on the indirect benefits of the projects reviewed in the OED report' (World Bank 1996).

The present chapter attempts to move one step forward in bridging this information gap and addressing the felt need of improving our understanding of the impacts of dams. The chapter puts forward a methodological framework for estimation of multipliers and illustrates this framework by providing quantitative estimates of the indirect economic impacts and multipliers in respect of three large dams located in different parts of the World: Bhakra Dam (India), High Aswan Dam (Egypt), and Sobradinho Dam and the set of cascading reservoirs (Brazil). The salient features of the dams selected for these case studies are presented in Table 2.2.

2.2 Analytical Framework for the Estimation of Multipliers

A number of analytical tools for estimating multiplier effects have been suggested in the literature (Bell et al. 1982; Hazell and Ramasamy 1991; Haggblade et al. 1991; Hoffman et al. 1996; Aylward et al. 2000). These tools, which are essentially in the nature of multi-sector models, include: (i) input–output (I/O) models, (ii) social accounting matrices (SAM)-based models and (iii) computable general equilibrium (CGE) models. We briefly describe the three analytical tools below.

2.2.1 Input–Output (I/O) Models[1]

The core around which all economy-wide, multi-sector models are built is the input–output model pioneered by Leontief (1953, 1970). Input–output analysis is a way to trace the flow of production among the sectors in the economy, through the final

---

[1] Semi-input-output (S-I/O) models represent a variant of I/O models whereby a distinction is made between tradable and non-tradable goods. The former are assumed to have an exogenously set domestic level of output, so that any change in demand will be reflected in a change in exported quantities. In terms of domestic production, therefore, the whole brunt of demand shocks is borne by non-tradable goods. The implication of this distinction is that induced, consumption-based impacts will reverberate throughout the economy only via adjustments in non-tradable goods and their inter-industry linkages. This characterisation is important, in that it refines the representation of the specific regional structure of production, reducing the risk of overestimating induced impacts that are not felt by regional sectors.

Table 2.2 Salient features of selected dams

Dam and location		Height of dam/type of dam/size of reservoir	Area irrigated/ production of foodgrain per year	Installed capacity (MW)/annual generation (kWh)	Urban water use and development	Persons benefited	Persons resettled
Bhakra Dam (B), Pong Dam (P) and Nangal Dam (N)[a]	India	225 m/concrete and earth-filled/18 BCM	10.3 million ha/27 million tons grain	2,800 MW/14 billion kWh	Water for cities in two states and Delhi	34 million	32,000 (B); 100,000 (P)
Aswan High Dam	Egypt	111 m/rock-filled/164 BCM	2.65 million ha/ production of foodgrain per year is not available	2,100 MW/8–10 billion kWh	Aswan city: Value of production quadrupled	40 million	100,000
Sobradinho Dam (S) and Cascade of Reservoirs (C)[b]	Brazil	41 m × 8.5 km 4,214 km² (area)/34.1 BCM	São Francisco Basin: 400,000 ha (120,000 ha are public but private is growing faster); high-value crops/ production of foodgrain per year is not available	1,050 MW (S) 9,800 MW (C)/ annual generation (kWh) is not available	Water for cities and towns in numerous Brazilian states	40 million (with hydropower) 5.5 jobs/ha (incl. induced urban empl.)	120,000 (C)

BCM billion cubic metres

[a]Bhakra Dam system includes Bhakra Dam, Pong Dam and Nangal Dam

[b]The set of cascading reservoirs include Itaparcia (Luis Gonzaga) Dam, the Paulo Afonso I-IV Complex, Moxoto (Apollonio Sales) Dam and Xingo Dam. The latter is not part of this analysis since it became operational in 1995

domestic or export demand. The essence of I/O analysis is that it captures the interrelatedness of production arising through the flow of intermediate goods among sectors. The fundamental input–output problem is that of a planner who wants to determine the appropriate adjustments in economic quantities throughout the economy in order to achieve a specific final output (Hewings 1985; Dervis et al. 1982).

I/O models are based on an accounting framework that records all inter-industry flows at the chosen level of disaggregation, final demand (household, government, investment and export), factor remuneration and total imports. Columns in an I/O table record payments from the column sector to other industries, factors and imports. Their totals represent total gross inputs for a sector. Rows represent all receipts for a row sector from the columns (other industries and final demand categories), with their totals representing total gross output. Inter-industry linkages are based on a fixed coefficient Leontief matrix, which eliminates any substitution possibilities for producers. Similarly, factor wages are fixed value-added proportions of total gross output.

These models evaluate indirect and induced economic impacts by computing multipliers embodying the impact of a unitary change in one sector's output—due to an exogenous change in final demand—onto the output of other sectors, income and employment. The existence of a multiplier depends on drawing unused or underused resources into more productive economic activities (Haggblade et al. 1991), so that the presence of such underutilised resources in the region of interest is crucial for the existence of multiplier impacts as estimated by this class of models. Leontief or output multipliers only reflect the degree to which industrial sectors are linked with each other and the strength of such linkages, but tell us nothing about the larger or smaller impact on a regional or national economy of increased demand for the output of any of those sectors. The main limitation of this model is that it assumes linearity in production and cost-determined prices independent of demand.

2.2.2 Social Accounting Matrices (SAM) Models

More recently, I/O linear models have been applied to databases that extend the I/O table to include the distribution of income among 'institutions' (i.e. household categories, firms, government), to better represent their expenditure, and to distinguish between production activities and produced commodities. A SAM is an economy-wide data framework that represents the circular flow of income and expenditure in the economy of a nation or region. Again, each cell represents a payment from a column account to a recipient row account. A distinction between SAM-based multiplier analysis and I/O models is that SAM explicitly traces the distribution of factor incomes to institutions. Thus, such models are able to account for the way in which initial asset distribution and factor endowments interact with the structure of production in determining final outcomes, particularly for welfare analysis. This capacity is also enhanced by the fact that SAMs generally comprise numerous household groups.

Each cell of the SAM can be seen as a 'block' representing sets of transactions. As row and column totals must be equal, expressions linking row and column elements with their totals can be used to form a system of linear equations that embodies market, behavioural and system relationships. The first has to do with the accounts relative to goods and factor markets. Behavioural relationships regard the budget constraints of the economic agent or the institution represented in the SAM. Finally, system relationships regard the capital and rest of the world accounts, where macroeconomic (internal and external) balances are represented. This system can then be used to estimate project multipliers.

2.2.3 Computable General Equilibrium (CGE) Models

As policy makers' capacity to directly control quantity variables—as is implicitly assumed by linear I/O and SAM-based linear models—declines, and the use of market incentives to affect them becomes more important, it is crucial to understand how markets respond to different sets of policy interventions. CGE models are well suited to this purpose, providing a framework where endogenous prices and quantities interact to simulate the workings of decentralised markets and autonomous economic decision-makers. Following the pioneering efforts of researchers (Adelman and Robinson 1978; Taylor et al. 1980), standard CGE models have been extensively used to study policy impacts on income distribution, growth and structural change in developing economies. A CGE model is a system of simultaneous non-linear equations that provide a complete and consistent picture of the 'circular flow' in an economy, capturing all market-based interactions among economic agents.

Four features distinguish CGE models from Leontief's input–output modelling tradition: (i) price endogeneity, as opposed to quantity adjustments, to reach an equilibrium, (ii) price-responsive input and output substitutability—perfect or imperfect—through the use of non-linear supply and demand equations (Robinson 1989), (iii) the abandonment of the perfect dichotomy between traded and non-traded goods from traditional I/O models and (iv) factor supply constraints, which generate output supply constraints.

2.3 Choosing the Right Analytical Tool and the Analytical Models Used in the Case Studies

From the point of view of the analysis of indirect and induced impacts of dam projects, the choice of analytical tool should not always favour the most sophisticated tool, but rather be driven by the assumptions regarding the mechanisms through which impacts are transmitted in the specific region of interest—particularly factor mobility. When prices are assumed fixed, as in I/O or SAM-based multiplier analysis, all adjustments occur through quantity changes. In the absence of supply constraints, adjustments occur via impacts on labour or capital employment and

2 Indirect Economic Impacts of Dams

Table 2.3 Case studies and methodologies for the estimation of multiplier effects and income distribution impacts of dams

Case study	Methodology	Outputs
Bhakra Dam, India	Social accounting matrix (SAM)-based multiplier model and linear programming model	Value-added multipliers and income distribution and poverty reduction impacts
High Aswan Dam, Egypt	Computable general equilibrium (CGE) model coupled with mathematical programming model for the water sector	Value-added multipliers and income distribution
Sobradinho Dam and Reservoirs, Brazil	Semi-input/output model for multiplier analysis only	Value-added multipliers

inter-regional factor migration. The presence of idle labour or capacity in the system—either locally or in other regions, if the model is inter-regional—is thus crucial for the existence of quantity-driven multiplier impacts as estimated by these models.

On the other hand, a variable-price model, such as a standard CGE,[2] implies the presence of supply constraints, so that for at least one factor the aggregate levels of factor employment are fixed. In this case, a change in sectoral demand results in relative price changes, determining substitution effects among inputs and among outputs, with factor reallocation across sectors in the regional economy. If available, a CGE could also be used to compute SAM-based, fixed-price multipliers analysis, making it possible to highlight the differential impacts that can be seen when considering changes in relative prices, factor mobility and wage differentiation.

Often the selection of a suitable analytical tool for multiplier analysis of a dam critically depends on the availability of I/O tables or SAM databases and models for a region. For estimation of the indirect and induced impacts, for multipliers and income distributional analysis, we have used different analytical techniques in the three case studies (Table 2.3).

2.4 Summary Results of Case Studies

2.4.1 Multiplier Effects of the Bhakra Dam, India

The Bhakra Dam in the northern part of India has contributed significantly to the increases in the output of agricultural commodities and electricity over the last 45 years or so. Additional gross irrigated area has been of the order of about 7 million ha. Total foodgrain production in the Bhakra command area during the year 1996–1997 was of the order of 27 million tons—an additional output of 24.6 million tons compared to the food output in the early 1960s. The hydropower stations installed in

---

[2] A Social accounting matrix can feed into a standard CGE model.

the Bhakra system have a combined generating capacity of 2880 MW, which currently generate about 14,000 million units (kWh) of electricity in a year. These increases have inevitably generated downstream growth in many other sectors of the regional economy as well as in other parts of the country.

Substantial marketed surplus of foodgrains in the states of Punjab and Haryana have provided foodgrains for urban poor all over India at relatively low food prices through 'fair price shops'. Such surplus foodgrains have been used for 'food for work' programmes in many states thus helping the rural poor, especially during drought conditions, and generating multiplier effects. Hundreds of thousands of migratory workers from underdeveloped areas of Bihar and Uttar Pradesh have sent large amounts of remittances creating further multiplier effects in those regions.

The case study employs a SAM-based analytical framework to estimate regional value-added multipliers arising from inter-industry linkages of production and consumption-induced effects. The SAM-based model has been used to estimate the outcomes of 'with project' and 'without project' scenarios. Simulation results obtained for 'without project' situations are compared with the values of relevant variables as estimated in the SAM for 1979–1980 representing the 'with project' situation. The analysis has been done for a year (1979–1980) for which adequate data were available from a detailed study (Bhalla et al. 1990) of the input–output structure of the Punjab economy.

The aspects of the dam that have been analysed include: changes in area irrigated, changes in supplies of electric power and changes in yields and production technology (primarily in fertiliser use) that would have been likely in its absence. Relevant variables comprise all the elements of the regional SAM, assuming fixed prices. The effects are divided into direct and indirect. The indicators capturing the effects include production, trade, as well as disaggregated household incomes and their distribution.

The multiplier has been estimated under two alternative scenarios about the groundwater availability in the absence of dam. Table 2.4 presents a summary of the estimates of regional value added under 'with' and 'without project' scenarios and of how these have been used to estimate multiplier value under the two assumptions.

As shown in Table 2.4, the multiplier values under the two scenarios work out to 1.90 and 1.78. Thus, for every rupee (Rs) of additional value added directly by the project in agricultural and electricity sectors, another Rs 0.90 (in the first scenario) and Rs 0.78 (in the second scenario) were generated in the form of downstream or indirect effects.

The income distribution impacts of the Bhakra Dam have been analysed by comparing the differences in aggregate income levels of various household categories under 'with' and 'without project' scenarios and by assessing direct and indirect components of income differences under the two situations. Figures 2.1 and 2.2 summarise the results obtained.

The results obtained signify the following:

- The dam provides income gains to all categories of households, including urban households, and that percent increase in income of agricultural labour households is higher than that of landed households. This poorest group (agricultural labour) registered a 65% increase in income as compared to a rural average

2 Indirect Economic Impacts of Dams

Table 2.4 Estimated values for multiplier effects of the Bhakra Dam, India

Definition of multiplier	Regional value-added with project minus Regional-value added without project
	Value-added of agriculture and electricity with project minus Value-added of agriculture and electricity without project
Value under assumption I (million Indian Rupees)	$\dfrac{42{,}379 - 32{,}878}{15{,}343 - 10{,}343} = 1.90$
Value under assumption II (million Indian Rupees)	$\dfrac{42{,}379 - 30{,}729}{15{,}343 - 8{,}807} = 1.78$

Note: Assumption I: Assuming that under 'without project' situation, groundwater use will be at 50% of the use in 1979–1980 (under 'with project' situation) and additional thermal power equal to 50% of hydro output will be available

Assumption II: Assuming that under 'without project' situation, groundwater use will be at 50% of the use in 1979–1980 (under 'with project' situation) and no additional power will be available from thermal sources

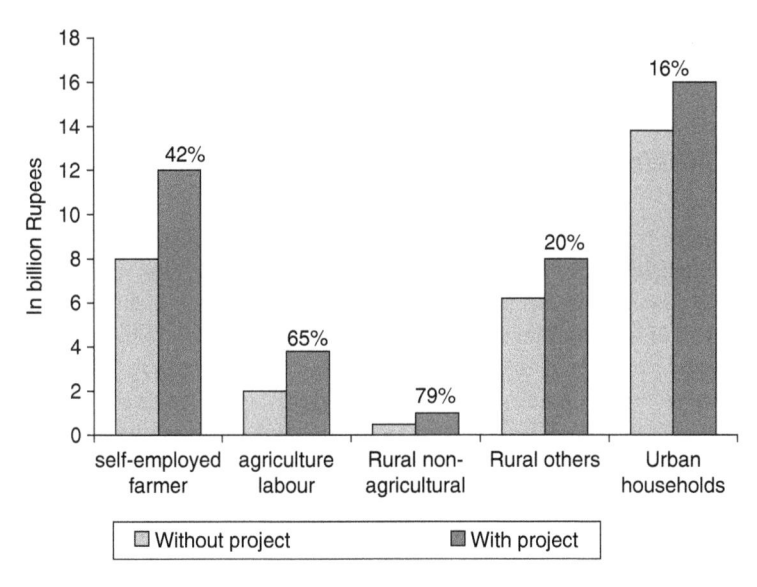

Fig. 2.1 Income of different types of households with and without Bhakra Dam, India. (The percentage figures indicate increase in income under the 'with project' scenario)

increase of 38% (under the 'with project' scenario as compared to the hypothetical 'without project' scenario).

- The gains from dams for different categories of households emanate from different sources. While landed households, agricultural labour and rural non-

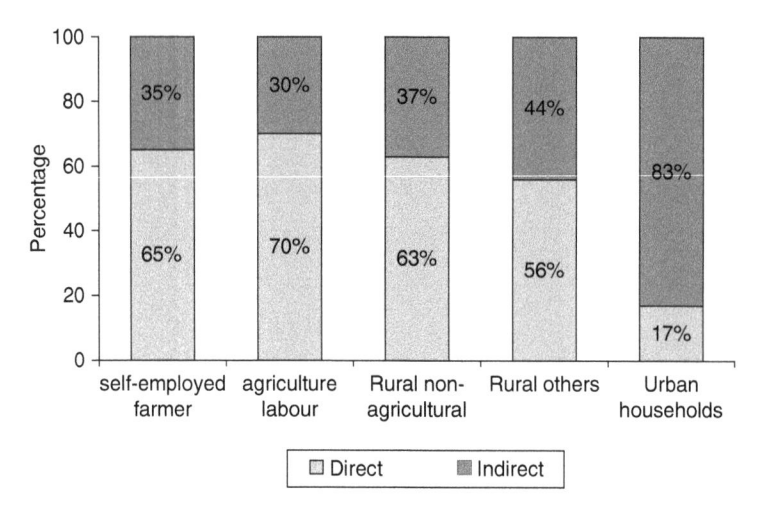

Fig. 2.2 Income gains to households by sectors affected directly and indirectly by the Bhakra Dam, India

agricultural labour households derive larger gains from the sectors directly affected by the project, rural other households and urban households derive larger gains from indirectly impacted sectors.

2.4.2 Multiplier Effects of the High Aswan Dam, Egypt

The economic benefits of providing highly reliable and non-flooding water supply to Egypt through the High Aswan Dam have been as follows: (i) It has saved Egypt from devastating floods resulting in lost summer harvests, damage to infrastructure and potential loss of life. (ii) Dam water has been used to reclaim 1.3 million new *feddans*.[3] (iii) Perennial irrigation of one million *feddans* has replaced basin irrigation. (iv) Rice and sugar cane production has increased considerably. (v) It has enabled the average generation of 8 billion kWh used in industry and the electrification of all towns and villages in Egypt. (vi) It has facilitated navigation up and down the Nile all the year round. Table 2.5 shows the actual crop areas in 1995, 25 years after the High Aswan Dam, compared with pre-dam areas. It shows major increases in areas of wheat, maize, rice and sugar cane cultivation, and a major decrease in the area for cotton.

This case study assesses the impact of the High Aswan Dam in Egypt from an economy-wide perspective. The primary analytical tool used is a CGE model. The study also shows that SAM multiplier analysis—characterised by the assumption that the prices of commodities, factors and foreign exchange are fixed—is a special case of CGE analysis. The model is built around a 1996–1997 SAM for Egypt

---

[3] A feddan is equal to 1.038 acres.

2 Indirect Economic Impacts of Dams   29

Table 2.5 Cropped area, before and after High Aswan Dam, Egypt

Crop	1970	1995
Wheat	1387	1829
Maize	1727	1906
Millet	469	346
Rice	799	1276
Cotton	1751	884
Sugar cane	122	274

complemented by a wide range of additional information. The analysis consists of a set of comparative-static simulations under alternative assumptions about the workings of the economy. The simulations are used to assess how Egypt's economy would have performed in 1996–1997 without the dam and how the economy, with and without the dam, would have been affected by year-to-year variations in Nile flows.

The scenarios without the dam consider the impact of changes in the supplies of irrigated land and water, changes in the supplies of electric power, changes in yields and production technology (primarily changes in fertiliser use) and real costs associated with the investment relative to other investments (in flood control and hydropower) that would have been likely in its absence. The 'without dam' scenarios also consider the implications of the fact that the performance of Egypt's economy in each year would have depended on Nile flow levels. Monte Carlo simulations have been used to assess the impacts of their stochastic nature in the absence of the dam. The CGE model is calibrated to account for water requirements for specific crops. In addition, the model explicitly models water balances across seasons.

Three simulations have been considered. In simulation 1 (SIM 1), all factor prices adjust to clear factor markets, assuming exogenously specified aggregate employment of labour and capital. In SIM 2, the labour wage is fixed and labour supply adjusts freely to clear the labour market—there is no aggregate employment constraint. In SIM 3, both the wage and the return to capital are fixed, assuming unlimited supplies of labour and capital are available at the fixed wages.

A CGE model with all factor prices fixed, and hence with no supply constraints, operates like a fixed-price, multiplier model. Because factor prices are fixed, output prices must also be fixed, given standard cost functions. Output is completely demand determined. In fact, this model is not completely demand driven because land and water are assumed to be in fixed supply. The result is a kind of 'constrained multiplier model', which will behave like a SAM multiplier model. In SIM 2, with only labour unconstrained, one would expect the multipliers to be smaller than in SIM 3, where both labour and capital are unconstrained. SIM 2 seems like a reasonable specification for a country in which there is excess labour. In SIM 1, all aggregate factor supplies are fixed, and the model will operate like a standard neoclassical CGE model.

The results of value-added multipliers under the three scenarios are presented in Table 2.6. The value-added multipliers range between 1.22 and 1.4 in the three simulations. The multiplier value of 1.4 implies that for every Egyptian pound (LE) of value added in directly impacted sectors, another 0.4 LE of value added is generated through inter-industry linkages and consumption-induced effects.

Table 2.6 Multiplier results for the High Aswan Dam (HAD)

	'With HAD' minus 'without HAD'		
	SIM 1	SIM 2	SIM 3
Value added of directly impacted sectors[a] (LE billion)	3.1	5.3	22
Total value added	3.8	6.5	30.9
Multiplier	3.8/3.1 = 1.23	6.5/5.3 = 1.22	30.9/22 = 1.40

[a]Directly impacted sectors—agriculture, hydropower, navigation and tourism

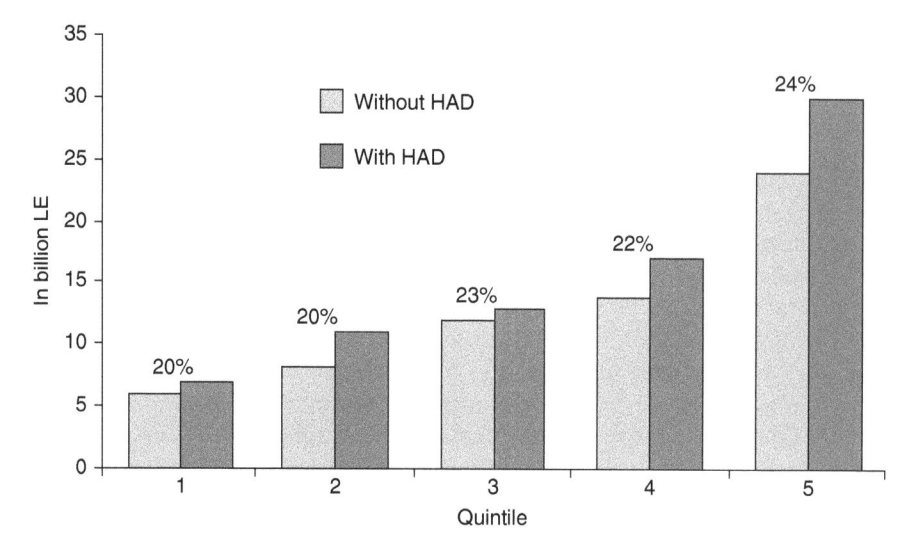

Fig. 2.3 Rural consumption by quintile, High Aswan Dam (HAD). (The percentage figures represent the percent difference between with and without HAD)

In order to trace the income distribution impacts of High Aswan Dam, we look at the consumption levels of different quintiles with and without the High Aswan Dam. Figure 2.3 shows consumption by quintile of the rural population with and without the dam, and Fig. 2.4 shows the corresponding quintile-wise consumption of the urban population. In the 'with project' situation, the lowest 20% (first quintile) of the rural population in Egypt accounted for 10.5% of the total rural income. The gain for this group is a 20% increase compared to a rural average increase of 22% (under the 'with project' scenario as compared to 'without project' scenario). Thus the gain for the poorest group is less than the rural average.

The above analysis thus suggests the following:

- The High Aswan Dam was, and is, a good investment, yielding significant annual net returns. The model analysis of net benefits is conservative, ignoring some benefits, which are significant but difficult to model (e.g. the elimination of major damages from periodic serious flooding).
- For agriculture, the existence of the dam allowed Egypt to pursue policies that distorted agricultural production, yielding a cropping pattern that favoured

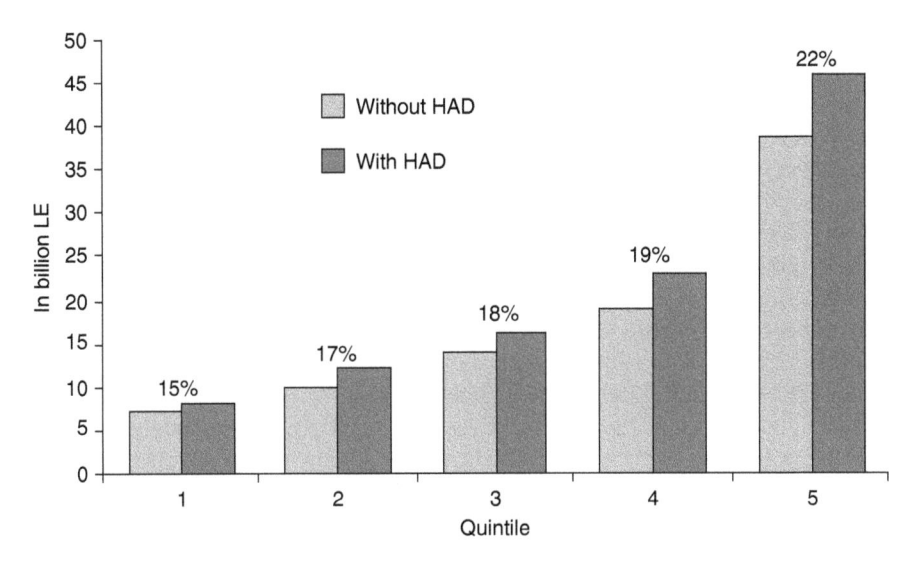

Fig. 2.4 Urban consumption by quintile, High Aswan Dam (HAD)

low-value crops that made inefficient use of both water and land. Reducing summer water and using the remaining water efficiently would yield increases in the value of agricultural production. With or without the dam, eliminating these distortions would greatly increase efficiency in the Egyptian agriculture.

Given the distortions in agricultural production, the largest gains from the dam arise from non-agricultural sources: hydropower, transportation and tourism.

2.4.3 Multiplier Effects of Sobradinho Dam and Cascade of Reservoirs in Brazil

This case study focuses on the complex of large dams and reservoirs built along the Sub-Médio São Francisco River, in one of the driest regions of Brazil's semi-arid north-east. The construction of the Sobradinho Dam in 1973–1979, was one of the most important factors in transforming the region's economy, society and landscape. The Sobradinho Lake has a surface area of over 4,000 km^2, stores 34,000 million cubic metres (MCM) and aliments a hydropower plant with a 1 GW capacity. The flow stabilisation that it provided enabled the construction of downstream reservoirs that form a hydropower complex with an installed capacity of almost 10 GW. These plants are operated by the São Francisco River Hydropower Company (CHESF), which serves over 40 million people in an area of about 1.2 million km^2—14.3% of Brazil's territory. The study characterises more precisely the benefits from hydropower generation by the selected dams, estimating the induced growth impacts generated by the creation of the north-north-east electric *pólo*.[4]

---

[4] It refers to regional development districts.

The dams also provided reliable water supply for investment in large irrigation projects and reduced risks for private irrigators. This has transformed agriculture production in the region (Nishizawa and Uitto 1995), with significant increases in planted area and output for high-value crops—especially permanent crops, such as mango and grapes—between 1975 and 1995. The Petrolina/Juazeiro district, located along the Sobradinho Reservoir, is now the major producer of table grapes in the country, accounting for 80% of Brazil's grape exports and 70% of mango exports, with its products headed primarily for Europe and the USA. In addition, integration between small, medium and large farmers has increased, generating both on-farm and off-farm employment (de Janvry and Sadoulet 2001) for both land-owning and landless rural populations. Recent analyses have shown that significant positive spillovers for smaller producers are taking place (Rodrigues 2001; Vergolino and Vergolino 1997). Agricultural growth has also spurred indirect and induced benefits through its linkages with the rest of the regional economy, benefiting populations in what is a largely urbanised area (around 60%).

The study uses an I/O model to estimate multipliers associated with the various products of the dams. The model is based on a 1992 I/O table for north-east Brazil, disaggregated at 31 productive sectors and 39 final products. It is important to note that prior to the construction of the Sobradinho Dam, the north-east economy of Brazil was characterised by the presence of considerable unused or underused resources. This is a crucial condition for the existence of quantity-driven multiplier impacts as estimated by fixed-price models (Haggblade et al. 1991).

The model analyses the impacts of eliminating one or more of the studied dams on the 1992 economy—a year of normal operation for them. The analysis does not attempt to produce a full counterfactual describing alternative development paths the region might have followed in the absence of these dams. The study simulates nine alternative 'without project' scenarios by combining three alternative scenarios each about hydropower availability and irrigation availability. The three alternative scenarios about power generation are as follows:

HLO: Sobradinho is not built, with no impacts on the output of the downstream stations existing in 1992 (i.e. they are not negatively affected by less reliable flows, nor do they increase their share in generation); no additional thermo or hydropower generation is installed; power generation declines by 14%.

HME: Sobradinho and Itaparica are not built, and we assume that this would force Paulo Afonso IV to operate at 50% of its 1992 capacity; we also assume that both small hydropower developments and thermal capacity would increase by a factor of 2.5; power generation declines by 43%.

HHI: The extreme scenario assumes that Sobradinho and Itaparica dams are not built, that this prevents the operation of Paulo Afonso IV and that no additional small hydro or thermal capacity is added to the system; the worst-case scenario sees a 66% decline in power generation.

Regarding availability of irrigation in 'without project' cases, the following alternative scenarios were formulated:

2 Indirect Economic Impacts of Dams

Table 2.7 Value-added multipliers under supply-constrained combined scenarios

Scenario	HLO+ ALO	HLO+ AME	HLO+ AHI	HME+ ALO	HME+ AME	HME+ AHI	HHI+ ALO	HHI+ AME	HHI+ AHI
Multiplier	2.078	2.029	2.007	2.101	2.075	2.059	2.105	2.087	2.074

ALO: Only public irrigation projects in the Sub-Médio São Francisco—downstream of Sobradinho and upstream of Xingó—are not undertaken if the dams are not built, resulting in a 41,000 ha reduction in irrigated area.

AME: All irrigated land in the Sub-Médio São Francisco area (141,000 ha) becomes rain fed.

AHI: The extreme value adopted for a 'without project' case is a 75% decline in the basin's irrigated area (225,000 ha).

Combining these sets of assumptions, we obtain a grid of nine alternative combined 'without project' scenarios. The estimated multiplier values in the nine cases are given in Table 2.7.

The simulation results of a supply-constrained semi-I/O model for Brazil's northeast macro-region show that the large dams located in the Sub-Médio São Francisco have generated significant indirect and induced effects in the region. Value-added multiplier values are close to 2.0 in most scenarios. For every unit of value generated by the sectors directly affected by the dams, another unit could be generated as an indirect impact in the region. It may, however, be noted that under the unconstrained supply assumption, value-added multipliers are 10–18% larger. Under this assumption, the multipliers range between 2.28 and 2.4, which are comparable with other case studies in the present chapter.

Beyond the magnitude of these impacts, what matters is the information they provide regarding the structure of benefits that can be attributed to the dams. The results suggest that larger overall impacts might have been achieved if more attention had been given to multipurpose use of their water.

2.4.4 Value-Added Multipliers: A Comparison of the Case Studies

Although in all three case studies there is a common objective of estimating indirect economic impacts of dams, the results obtained are not strictly comparable due to differences in methodology, data sets used and underlying conditions prevailing in the three sites. We nevertheless present in Table 2.8 a summary comparison of the estimated values of value-added multipliers for the three case studies.

For the Bhakra Dam in India, estimates of multipliers range between 1.78 and 1.9 depending on assumptions about the impact of seepage from canals on the availability

Table 2.8 Value-added multipliers: comparative results of three case studies

Case study	Country	Methodology	Regional value added or income multiplier values under alternative assumptions
Bhakra Dam	Northern India	Social accounting matrix (SAM)-based model	1.78–1.9
High Aswan Dam	Egypt	Computable general equilibrium (CGE) model	1.22–1.40
Sobradinho Dam and the set of cascading reservoirs	Brazil	Semi-input/output model	2.28–2.40

of groundwater for irrigation and the availability of additional thermal power in the absence of the Bhakra Dam. In the case of the High Aswan Dam, the value-added multiplier values range between 1.22 and 1.4 in the three simulations. The multiplier values in the case study of Sobradinho Dam (and the set of cascading reservoirs) in Brazil range from 2.28 to 2.4 under the assumptions of unconstrained supply of labour and capital. The multiplier values for Brazil under supply-constrained scenarios for selected sectors, the value-added Type II multiplier values are close to 2.0 in most scenarios (Type II multipliers, in addition to the direct and indirect impacts, also include inducedimpacts).

2.5 Conclusions

The results on value-added project multipliers suggest that in addition to having direct impacts, large dams have significant indirect and induced impacts, which must be accounted for and taken into consideration in project evaluation. The income distribution impacts, in respect of two of the three dams which permitted such an analysis, also suggest that gains from the dams are shared by all sections of society, including people living in urban areas.

In both the High Aswan Dam and the Bhakra Dam, rural households gained more (in percentage terms) than urban households when income levels under 'without' and 'with project' scenarios are compared. In the case of High Aswan Dam, the income gains for the lowest income groups (poorest) in the 'with project' situation as compared to a 'without project' situation was 20%, in contrast to a rural average increase of 22%. Thus, the gains for the poorest group were slightly less than the rural average. In the Bhakra case study, however, rural agricultural labour households that account for 23% of the total rural population, gained a 65% increase in income as compared to a rural average increase of 38% (under the 'with project' scenario compared with a hypothetical situation where the project had not been undertaken). Thus, the estimated gain for the poorest group was much higher than that for the average, and was also higher than that for landed farmers. Such a result signifies an important implicit message: the dams act as a powerful vehicle for poverty alleviation.

References

Adelman I, Robinson S (1978) Income distribution policy in developing countries. Stanford University Press, Los Angeles

Aylward B, Berkhoff J, Green C, Gutman P, Lagman A, Manion M, Markandya A, McKenney B, Naudascher-Jankowski K, Oud B, Penman A, Porter S, Rajapakse C, Southgate D, Unsworth R (2000) Financial, economic and distributional analysis. Thematic review III.1. Prepared as an input to the World Commission on Dams, Cape Town. Available at www.dams.org

Bell C, Hazell P, Slade R (1982) The project evaluation in regional perspective: a study of an irrigation project in Northern Malaysia. Econ J 9(372):959–961

Bhalla GS, Chadha GK, Kashyap SP, Sharma RK (1990) Agricultural growth and structural changes in the Punjab economy: an input-output analysis. International Food Policy Research Institute (IFPRI), Washington, DC

Bhatia R, Cestti R, Scatasta M, Malik RPS (eds) (2008) Indirect economic impacts of dams: case studies from India, Egypt and Brazil. Academic Foundation for World Bank, New Delhi

de Janvry A, Sadoulet E (2001) World poverty and the role of agricultural technology: direct and indirect effects. J Dev Studies 38(4):1–26

Dervis K, Demelo J, Robinson S (1982) General equilibrium models for development policy. Cambridge University Press, Cambridge

Haggblade S, Hammer J, Hazell P (1991) Modeling agricultural growth multipliers. Am J Agric Econ 73(2):361–374

Hazell PBR, Ramasamy C (1991) The Green Revolution reconsidered: the impact of high-yielding rice varieties in South India. The Johns Hopkins University Press, Baltimore

Hewings GJD (1985) Regional input-output analysis. Sage Publications, Beverly Hills

Hoffman S, Robinson S, Subramanian S (1996) The role of defense cuts in the California Recession: computable general equilibrium models and interstate factor mobility. J Reg Sci 36(4):571–595

Leontief W (1953) Studies in the structure of the American economy. Oxford University Press, New York

Leontief W (1970) The dynamic inverse. In: Carter AP, Brody A (eds) Contribution of input-output analysis. North Holland, Amsterdam, pp 17–46

Nishizawa T, Uitto JI (1995) The fragile tropics of Latin America: sustainable management of changing environments. United Nations University Press, Tokyo

Robinson S (1989) Multisectoral models. In: Chenery H, Srinivasan TN (eds) Handbook of development economics. Elsevier Science, Amsterdam, pp 886–935

Rodrigues L (2001) Potencial da Agricultura Irrigada como Indutora do Desenvolvimento Regional: o Caso do Projeto Jaíba no Norte de Minas Gerais. Rev Econ Nordeste 32(2):206–232

Taylor L, Bacha EL, Cardoso EA, Lysy FJ (1980) Models of growth and distribution for Brazil. Oxford University Press, New York

Vergolino TB, Vergolino JR (1997) Reforma do Estado: Efeitos sobre a Renda e o Emprego na região do Sub-Médio São Francisco (Petrolina/Juazeiro). Rev Econ Nordeste 28:447–459, No. Especial: Economia Agricola, Recursos Naturais e Meio Ambiente

WCD (2000) Dams and development: a new framework for decision-making. The report of the World Commission on Dams, issued on November 16, 2000. Available at www.dams.org/report

World Bank (1996) The World Bank's experience with large dams: a preliminary review of impacts. Operations Evaluation Department Report No. 15815. The World Bank, Washington, DC

Chapter 3
Resettlement Outcomes of Large Dams

Thayer Scudder

3.1 Introduction

The adverse social impacts of most large dams continue to be unacceptable. They also reduce a project's potential benefits. This is especially the case with resettlement which some experts (including Asit Biswas and Robert Goodland, former Chief Environmental Adviser of the World Bank Group) consider to be the most contentious issue associated with large dams. Fortunately, there is potential for helping resettling communities to become project beneficiaries.

This chapter is organised in four sections. The first discusses a policy-relevant theoretical framework for understanding what happens during the resettlement process. The second section deals with the magnitude of resettlement and what we know about the record to date with resettlement outcomes. It emphasises research by the World Bank as well as John Gay's and my statistical analysis of resettlement outcomes associated with 50 large dams distributed around the world. The third section explains why outcomes over the past 50 years continue to impoverish the majority. The fourth and final section deals with the many opportunities that are available for helping the majority of resettlers to become project beneficiaries which, as ICOLD's 1995 *Position Paper on Dams and Environment* emphasises, should be a requirement associated with the construction of all large dams.

Most of the issues discussed in this chapter, including the statistical analysis of resettlement outcomes, are dealt with in more detail in my 2005 *The Future of Large Dams: Dealing with Social, Environmental, Institutional and Political Costs.*

T. Scudder
California Institute of Technology, Pasadena, CA, USA

3.2 Theoretical Frameworks of the Resettlement Process

3.2.1 Introduction

Research on development-induced displacement and resettlement in recent years has led to the formation of its own subfield, with contributions from many countries, within the social and policy sciences. Guggenheim's partially annotated bibliography on 'involuntary resettlement caused by development projects' contained approximately 800 sources when published in 1994. Following an international meeting in Brazil in 2000, 60 resettlement specialists from over 20 countries formed the International Network on Displacement and Resettlement (www.displacement.net) whose website has been receiving increasing use.

Research institutions addressing development-induced resettlement include China's National Research Centre for Resettlement at Hohai University and Oxford University's Refugees Study Programme. Currently, the World Bank and the Asian Development Bank house the largest number of resettlement specialists, with a series of reports from the World Bank being especially important.

World Bank reports include Cernea's edited *The Economics of Involuntary Resettlement: Questions and Challenges* (1999b) and the Environment Department's 1994 *Resettlement and Development: The Bankwide Review of Projects Involving Involuntary Resettlement 1986–1993*; also the Bank's Operations Evaluation Department's *Early Experience with Involuntary Resettlement* (1993), *Recent Experiences with Involuntary Resettlement* (World Bank 1998) and Picciotto et al. (2001) *Involuntary Resettlement: Comparative Perspectives*.

Other World Bank sources include Cernea and McDowell's edited *Risks and Reconstruction: Experiences of Resettlers and Refugees* (2000), the Africa Technical Department's *Involuntary Resettlement in Africa* edited by Cynthia Cook (1994) and the Economic Development Institute's *Managing Projects that Involve Resettlement: Case Studies from Rajasthan, India* by Hari Mohan Mathur (1997). In addition to the above reports that incorporate information from a number of projects, the large majority of which involve dam-induced resettlement, there are an increasing number of studies of specific cases including PhD dissertations and published accounts. A representative sample dealing with projects in Asia, the Middle East, Africa and Latin America is included in the references.

3.2.2 A Policy-Relevant Theoretical Framework of the Resettlement Process

3.2.2.1 Introduction

A substantial body of policy-relevant theory exists on the global experience with development-induced and, more specifically, dam-induced, involuntary community resettlement. Recently I have combined the two major theoretical frameworks into

3 Resettlement Outcomes of Large Dams

Table 3.1 The four-stage process for achieving successful resettlement

Stage 1: Planning for resettlement prior to physical removal
Stage 2: Coping with the initial drop in living standards that tends to follow removal
Stage 3: Initiation of economic development and community formation activities that are necessary to improve living standards of first-generation resettlers
Stage 4: Handing over a sustainable resettlement process to the second generation of resettlers and to non-project authority institutions

a single theory that also incorporates other important contributions (Scudder 2005). Dating back to the late 1970s, the first is my four-stage framework which focuses on how a majority of resettlers can be expected to behave over two generations during a successful resettlement process that enables them to become project beneficiaries (Scudder 1981, 1985, 1993, 1997, 2005; Scudder and Colson 1982). The second framework, developed during the 1990s, is Cernea's 'impoverishment risks and reconstruction model' (Cernea 1996, 1997, 1999a; Cernea and McDowell 2000).

3.2.2.2 The Four-Stage Framework

The first stage (planning and recruitment) of the four-stage framework identifies who must resettle because of a dam project and includes planning for their future (Table 3.1). Physical removal and the years immediately after removal are the focus of the second stage (adjustment and coping). That stage requires special attention because during those initial years the global experience is for living standards of the majority to worsen. At this point, it is important to emphasise why the phrase 'the majority' is mentioned throughout this chapter. That is because one can expect a small minority of individuals and households to be able to take advantage from the start of the new opportunities available for raising their living standards. Their initiative is to be encouraged for if they succeed they could play an important role in demonstrating to the majority future courses of action for livelihood improvement.

There are several reasons why living standards for the majority can be expected to drop during Stage 2. One relates to the long planning horizon for a major dam project. During that time span, governments and other agencies are unlikely to improve infrastructure within project areas with the major exception of road construction accessing the dam site. They are even less likely to build new schools and other social services or upgrade development staffing. As for those who must resettle, often, as in the case of the binational Swaziland and South African Maguga Dam, affected people are told not to improve their housing or to initiate new ventures for raising their living standards. For such reasons, living standards can be expected to drop leaving people worse off than would have been the case without the project. At the same time, their living standards can be expected to worsen in comparison to their neighbours whose resettlement is not required.

A second reason for the drop in living standards for a majority of resettlers is related to two impacts of removal on the majority. The first is that resettlement is an overwhelming experience. During preparation for a move, labour migrants—as in the case of Nubian resettlers in both Egypt and the Sudan in connection with the

High Aswan Dam, tend to return home to assist their relatives who forego whatever remittances or savings those wage earners would have been able to supply. Following physical removal, resettlers must build or accommodate themselves to new housing and also to new, usually larger, communities. They must also prepare and plant new farms or commence other occupations, adjust to new neighbours from the host population and, almost inevitably, adjust to an increased presence of government administrators. At the same time, temporary jobs become less available as the dam construction phase draws to a close, while all household members usually find that they have reduced access to common property resources.

The second impact of removal adversely affecting living standards is caused by the multidimensional stress and by the reduction of cultural inventory that accompanies involuntary resettlement. Stress has physiological, psychological and socio-cultural components which are synergistically interrelated. Physiological stress is associated with increased illness and death rates. Children are especially at risk where formerly dispersed communities are concentrated in larger communities with inadequate water supplies (an all too frequent characteristic of resettlement) and inadequate protection against water-borne diseases and diseases such as measles.

Psychological stress has been associated with higher death rates among the elderly. It has two distinct components. One is Fried's 'grieving for a lost home syndrome' (1963). The other is anxiety about the future. Both can be expected to occur in rural societies of peasants and ethnic minorities with strong cultural ties to their homeland and to their homes. Especially affected are the elderly and women.

Socio-cultural stress can have many manifestations due to the departure from a preferred homeland and problems associated with resettlement areas. Leaving behind cemeteries, religious sites, especially those tied to the land, and structures, and familiar economic and social routines are examples. In attempting to cope with the unfamiliar, resettlers try to cling to the familiar by moving with kin (including those from whom they had previously separated due to disputes), remodelling or building new homes that replicate old ones and following the same occupations in the same way.

When such attempts to transfer old practices are maladaptive, as some will be in a different physical environment or where the project authority is emphasising a different production system, socio-cultural stress will be involved. The same applies to a reduction in cultural inventory. Because resettlement is associated with an increased government presence, resettlers find it more difficult to re-establish illegal activities, such as cultivation of marijuana or poaching. Further stress occurs where the host population and government officials ridicule customs and rituals associated with such important events as births, marriages and funerals. The host population, as was initially the case with Kariba resettlement, may warn resettlers not to practise funeral drumming or to re-establish shrines which, resettlers are told, will alienate the spirits of the new lands. Host populations may also expect resettlers to obey their own officials. Even where such host objections are not present, the uncertainty associated with how to transfer a socio-cultural system from a familiar to an unfamiliar site is stressful.

Stage 2 comes to an end only if the majority is able to re-establish their former living standards. Unfortunately, that has not occurred with the large majority of large dams analysed to date; rather resettlement is characterised by an increase in impoverishment. Several indicators are useful in ascertaining when the second stage, seldom less than 2 years in duration following removal, has ended. Not only have the majority regained whatever former self-sufficiency they had, but their behaviour shifts from being risk adverse to a willingness to take risks. They have also come to terms with their new environment and with the host population. Indicators, other than economic self-sufficiency, include cooperating to build community institutions and infrastructure including funeral societies and other community organisations as well as churches, mosques, temples and other religious structures. Local features of the environment may be given names and the resettlement experience immortalised in poetry, song and dance.

During Stage 3 community formation and economic development living standards improve. Now the majority is contributing to the stream of project benefits rather than to project costs which can result when impoverished resettlers have little option but to degrade their natural resource base by clearing forested land for hillside cultivation or for making charcoal. Since the majority of resettlers prefer to remain within the reservoir basin (i.e. to move the least distance, geographically and sociologically) this can increase rates of reservoir sedimentation hence reducing project life.

I am fascinated by the fact that throughout the world, Stage 3 households follow the same investment strategies as they develop their economies. Once food self-sufficiency is achieved, cereal crops are treated as cash crops with other cash crops such as cotton and coffee added. Livestock management is also developed, including dairy and sale of meat. Education of children receives more emphasis on the assumption that future jobs will further diversify and improve the household economy and/or provide a securer retirement for aged parents. The household production system is further diversified by adding a range of non-farm sources of income and employment. Some add additional rooms to their housing for rental or for opening small shops. Others practise a range of skills including small-scale production of clothing with one or more sewing machines and workers, carpentry, masonry and traditional healing. Success enables a small minority of entrepreneurs to expand their businesses into nearby villages, towns and, occasionally, urban centres including the capitol. Note that throughout this process all family members, including husbands, wives and children become involved in more productive activities that also raise their social status within the household and the community.

It is also interesting that as incomes rise, a majority of households can be expected to spend that income in remarkably similar ways. The quality of housing is improved with additional rooms added, wells are sunk where possible and sanitary facilities improved. Productive equipment such as ploughs, tractors with trailers, equipment for grinding cereal staples and sewing machines are purchased. Lighting within houses is improved and is characterised by a shift from candles to paraffin lanterns, petromax lamps and, where available, electricity. A similar trend characterises cooking fuels. As for household furnishings, they too have a remarkable similarity in different parts of the world. They include a set of stuffed chairs with an accompanying couch, table and chairs, a glassed-in cabinet within which a tea or coffee set is displayed along with family trophies purchased during pilgrimages or other family jaunts, and a radio followed by a radio cassette player and battery/electricity-operated TV as income rises. Walls within the room in which guests are entertained are also adorned in similar fashion, with, predictably, a wall clock, calendars and framed pictures of family members which have been taken on such prominent occasions as school graduations or weddings.

A resettlement process cannot be considered successful unless it is sustainable at least into the second generation. Stage four (handing over and incorporation) deals with two types of transitions. The first involves handing over the management of settler affairs at household and community levels to a second generation of leaders with the capacity to compete for their share of development resources at district, provincial and national levels. The second type involves the responsible project authority or authorities handing over their assets to the relevant resettler, departmental and other agencies. Where irrigation is provided, such resettler institutions as water user associations should take over increased responsibilities for water management within each irrigation system. Local institutions should also be able to maintain water supplies within the community, as well as feeder roads. At the departmental level, such ministries as public works, education and health should take over responsibility for the maintenance of major roads and the construction of new ones and for staffing schools and clinics. Non-governmental organisations (NGOs) should also be encouraged to play a role commensurate to that played in other areas.

3.2.2.3 Cernea's Impoverishment Risks and Reconstruction Model

While only a small minority of dam-induced resettlement schemes have successfully moved through all four stages, an analysis of those that have, show the potential for increasing the proportion of successful cases in the future. Cernea's impoverishment risks and reconstruction model, though applicable to that large majority of failed resettlement projects, also illustrates the potential for improving living standards by first identifying the major impoverishment risks and then detailing the reconstruction approach necessary to avoid them.

Michael Cernea retired from the World Bank at the end of 1996 as Senior Adviser for Social Policy and Sociology. During his 22-year Bank career, he pioneered the drafting of the Bank's first operational guidelines on development-induced resettlement (World Bank 1980) which became a model for those of other regional banks, OECD and a number of countries including China. In 1994, he was the instigator and senior author of the Bank's *Resettlement and Development: The Bankwide Review of Projects Involving Involuntary Resettlement 1986–1993*. The resulting review was unique in that it drew on knowledge supplied by special working groups in each of the Bank's regions along with a number of special studies completed by the Legal Department and various central sectoral departments. This review along with research by colleagues outside the Bank provided the database out of which Cernea's Impoverishment Risks and Reconstruction Model grew.

The Cernea model advanced thinking about development-induced displacement and resettlement in three ways. First, by mining a large number of case studies, it drew out and emphasised eight characteristic impoverishment risks that time and again have been associated with a failed resettlement process. As presented in the model, those risks are landlessness, joblessness, homelessness, marginalisation, increased morbidity and mortality, food insecurity, loss of access to common property and social disarticulation.

Among the less obvious risks, marginalisation 'occurs when families lose economic power and spiral downward; it sets in long before displacement, when new investments in the condemned areas are prohibited' (Cernea 1999a:17). Food insecurity is due to the inability of rural resettlers, who continue to constitute the large majority of reservoir resettlers, to clear, plant and harvest enough land to feed their families during the year or years immediately following removal. Loss of access to common property refers to such essential community resources as grazing and use of forest resources for fuel, building materials and a wide variety of non-timber forest products. Social disarticulation is a result of a tendency of communities, kinship groupings and households to be broken up at the time of physical removal. It 'means dismantling of structures of social organisation and loss of mutual help networks' (Cernea 1999a:18).

Cernea's framework makes a second major contribution by elaborating the necessary policies not just for avoiding impoverishment but for improving living standards. The third, and most original, contribution is to apply the type of risk assessment to resettling communities that donors, governments and project authorities use for specific projects. In Cernea's words:

> Conventional risk analysis focuses only on the risks to capital investments, but not on the various kinds of 'post-normal risks' (Rosa 1998) that displacement imposes on affected people… Moreover it is also common practice for governments to provide guarantees against various risks incurred by investors in infrastructure projects… Yet when the same private investments *create* risks to such primary stakeholders as the residents of the project area…the state does not provide comparable protection against risks to these affected people… The current methodology of risk analysis at project level must be broadened to recognise risk distribution among all project actors and address equitably the direct risks to area people as well. (1999a:15–16)

3.2.3 Combining the Two Analytical Frameworks with Contributions from Other Researchers

Combining and broadening the two analytical frameworks makes sense for a number of reasons. Both are based on the premise that dam-induced resettlement has the potential to improve the living standards of the majority to a much greater extent than in the past. The two frameworks also complement each other. Cernea's concentrates on how planners and project authorities can achieve a successful resettlement outcome by avoiding specific impoverishment risks. Mine predicts how a majority of resettlers can be expected to behave over two generations when appropriate policies and plans are adequately implemented. Other researchers have made valuable contributions to broadening the usefulness of the two frameworks by placing more emphasis on specific issues that need be addressed by policy makers and planners. These include the wider political economy at the national and international level that influences policies and plan implementation for such important development projects as large dams and the institutional context for plan implementation. They also include important issues that need be dealt with if the resettlement process is to produce a viable society for resettler households and communities. Especially important are issues of human rights and cultural reconstruction and policies and plans that are gender sensitive, and that facilitate the re-emergence of strong political leadership within resettled communities.

Wali (1989) and Ribeiro (1994) emphasise the wider political economy in their respective analyses of Panama's Bayano Dam and Argentina and Paraguay's Yacyreta Dam. Wali's work illustrates how government policies for the regional development of the project area can adversely influence the resettlement process in a number of ways, including giving immigrants a comparative advantage over resettlers in responding to project-related opportunities. Ribeiro discusses how political considerations dominated the decision-making process that led to the sitting and construction of Yacyreta.

Once a decision is made to proceed with a dam, the nature of the institutional structure for proceeding with planning and implementation has major implications for how resettlement and broader issues of regional development, of importance to resettlers, are dealt with. More often than not, a single ministry (irrigation, energy, or water affairs for example) is in a dominant position that is apt to exclude a broader range of development goals unless an overseeing structure with political clout and finance is in place (Scudder 2005:244–254). Kariba, for example, was constructed on the Zambezi as a uni-purpose scheme to produce hydropower for the Federal Power Board of the Central African Federation. The significant potential of the project for the development of irrigation below the dam and within the reservoir basin was ignored. More specifically, Rew addresses organisational structures for dealing with resettlement and with the implications of different management styles for resettlement outcomes (Rew 1996; Rew et al. 2000).

The final report (2000) of the World Commission on Dams (WCD) deals with dam impacts on project-affected people, including resettlers, as a human rights issue in Chap. 7. Reference to the 1947 UN Declaration of Human Rights and other UN covenants was intended to emphasise a global framework for sustainable development and to remind nations as signatories of their responsibilities toward affected people in project planning and implementation.

Downing (1996) draws attention in his writing on violations of human rights as an impoverishment risk for resettlers. He also stresses important cultural issues that are ignored in donor guidelines and by project authorities which, if properly analysed, would reveal the existing unsatisfactory global record with resettlement to be even more unacceptable. He discusses questions that relate to a people's 'social geometry'; for example, how 'involuntary displacement forces people to re-examine primary cultural questions which, under routine circumstances, need not be considered. Key among these is... Who are we? Where are we?' (Downing 1996:33, 36). Such questions require resettlers to re-examine and where necessary to reconstruct their identities. That takes time and contributes to the multidimensional stress that characterises the initial years following removal.

Gender issues are dealt with in detail in Colson's *The Social Consequences of Resettlement* (1971) which deals with the Kariba experience and remains, in my opinion, the best case study of resettlement impacts. More recent sources on gender issues include Koenig (1995), Bilharz (1998), Parasuraman (1999) and Mehta and Srinivasan (2000). Bilharz also deals in detail with how the unsuccessful fight to stop construction of the Kinzua Dam in eastern United States, and the resettlement that followed, sped up the emergence of a younger and better educated group of leaders. The successful fight to stop construction of Arizona's Orme Dam strengthened the Yavapai Nation and its leadership (Khera and Mariella 1982) while the successful fight of the James Bay Cree Nation against Hydro-Quebec's Great Whale Project not only unified nine previously separate communities but also led to new Hydro-Quebec policies requiring the informed consent of affected communities where new projects are planned, and incorporation of affected people as share holders in joint ventures and as co-project managers (Egré et al. 2002).

3.3 The Magnitude of Resettlement and Resettlement Outcomes

3.3.1 The Magnitude of Resettlement

Controversy over the estimates of how many people have been relocated worldwide because of large dams continues, and this is in itself an indictment of the lack of attention paid to the resettlement process. That is even the case with World Bank-financed projects. A 1996 Bank post-project evaluation found that no resettlement figures at all were available in connection with seven of 50 large dams. What figures are available tend to be underestimates. Hence the *Bankwide Review* of 192 Bank-financed projects for the period 1986–1993 (of which dam projects accounted for 63% of those resettled) found that, 'the total number of people to be resettled is 47% higher...than the estimate made at the time of appraisal' (World Bank 1994, Section 2:2), which is when budgets are formulated. During Gay's and my statistical analysis of 50 projects, initial and final resettlement figures were only available in 20 cases. In those cases 'earlier estimates of numbers of resettlers on which planning and budgeting was based were approximately 50% of the final tally' (Scudder 2005:22).

The most reliable, as well as objective, figures gathered at the time of physical removal were collected by the World Bank in the mid-1990s in connection with its *Bankwide Review*. For the 1987–1996 period, Cernea (1996:6) presents figures that indicate that 3,000 of over 40,000 large dams caused the involuntary resettlement of 32 million people. Figures in the *Bankwide Review* for 4,000–5,000 large dams built during 1980–1993 raise that figure to 56 million resettlers. This figure falls within the 40–80 million estimate of the WCD for the total number of people moved in connection with 45,000 large dams, an estimate that might well be an underestimate. Those who continue to disagree are being irresponsible, unless they can produce case material which suggests that World Bank and WCD estimates are overestimates.

3.3.2 Resettlement Outcomes

Within the World Bank, awareness has increased over the years concerning the impoverishing impact on the majority of large dams that have been evaluated. In 1994, the *Bankwide Review* stated 'that although the data are weak, projects appear often to not have succeeded in re-establishing resettlers at a better or equal living standard and that unsatisfactory performance still persists on a wide scale' (World Bank 1994:x). In 2001, following the analysis of seven Bank-financed projects, the authors of the Bank's *Involuntary Resettlement: Comparative Perspectives* stated in a section heading in bold print that 'The Income Restoration Record is Unsatisfactory' (Piccioto et al. 2001:9).

During the period 2001–2004, John Gay and I undertook what is the largest statistical analysis of resettlement outcomes in connection with large dams (see next section). John Gay and I found no statistical evidence that outcomes for the majority have improved over time. In 36 (82%) of the 44 cases where documentation was adequate, the impact of the project was to worsen the living standards of the majority not just economically but socially and culturally as well. In fact, social and psychological marginalisation had the highest association with an adverse outcome of the eight impoverishment risks that Michael Cernea associates with development-induced resettlement. What about the other 18%? Living standard improvement was documented for only three (7%) of 44 cases. Those were the High Aswan Dam, Sri Lanka's Pimburetewa Dam and Costa Rica's Arenal Dam. Restoration characterised another five cases (11%). They were China's Shuikou and Yantan Dams, Nigeria's Kainji, the Ivory Coast's Kossou and Sri Lanka's Victoria. Fortunately, those eight cases cover a wide variety of opportunities including irrigation, annual and perennial rain-fed crops, and non-farm employment. As will be discussed in the last section, they, along with components of other projects, show the potential that exists for a majority of resettlers to raise their living standards and in the process contribute to the stream of project benefits.

3.4 Why Impoverishing Outcomes Continue

3.4.1 Introduction

Chapter 3 of my book on the future of large dams (Scudder 2005) presents an exhaustive analysis of resettlement outcomes based on what remains the most detailed statistical analysis of dam-induced resettlement. The sample size was 50 large dams. That is relatively small because of the lack of adequate data on resettlement as a topic and, more specifically, on resettlement outcomes—the absence of data is just one of many examples of the persisting lack of attention paid to the issue. While WCD covered resettlement in its Cross-Check Survey of 150 dams (Clarke 2000), in only 68 (54%) of the 123 replies received from project authorities was resettlement even mentioned and in only 12 of those 68 cases (18%) were 'valid resettlement data' received. As for World Bank surveys, they too were inadequate for purposes of analysis since they were restricted to a still smaller number of Bank-financed projects.

The purpose of the 50 dam survey was to analyse outcomes and to test the utility of the Scudder and Cernea theoretical frameworks to explain those outcomes. Table 3.2 lists the dams that were involved, host country or countries, completion date, number of resettlers, date when last data was collected and outcome. The large majority was in Asia, Africa and the Middle East, and Latin America since that is where most future large dams will be built. Starting with the construction of the Grand Coulee Dam in 1939 in the United States, 22 were completed before 1980, 19 during the 1980–1990 period and the remaining nine either completed more recently or still under construction.

3.4.2 The Design of the 50 Dam Survey and the Nature of the Database

The 50 dam survey dealt only with households that were physically displaced by dam construction and reservoir formation since data were virtually non-existent for those displaced by such associated project works as roads, transmission lines and irrigation canals, or those who lost their land and other natural resources to the reservoir and the dam site, but not their homes. For coding purposes, 185 data items were listed. The first 42 dealt with such general issues as location, dam and reservoir size, date of completion and dam purpose, number of resettlers, nature of host population, government, the project authority and supporting institutions (including donors), financial costs and source of funds, and NGOs involvement. The next 24 items dealt primarily with resettlement policy issues including the extent to which attempts were made to reduce the numbers of resettlers, synchronisation of resettlement with the construction planning and implementation timetable, political will

Table 3.2 The 50 dams

Name of dam	Country/countries	Date of completion	Number of reservoir resettlers	Date when last data collected	Outcome
Kariba	Zambia[a]	1958	34,000	2002	Four
Aswan	Egypt[a]	1967	50,000	1999	One
Kainji	Nigeria	1968	44,000	1991	Two
Narayanpur	India	1982	30,600	1997	Four
Shuikou	China	1993	67,000	1997	Two
Yantan	China	1992	43,176	1997	Two
Kinzua	USA	1964	550	1997	Three
Katse	Lesotho	1995	1,470	2002	Four
La Grande	Canada	1995	One village	N/A	N/A
Grand Coulee	USA	1939	2,000	1999	Three
Nam Theun 2	Laos	Planned only	6,000	N/A	N/A
Pak Mun	Thailand	1994	1,205	2002	Four
Zimapán	Mexico	1993	2,452	1997	Four
Nangbeto	Togo	1987	10,600	1997	Four
Itaparica	Brazil	1988	26,000	1997	Four
Garrison	USA	1953	1,625	2001	Three
Fort Randall	USA	1952	95	2002	Four
Oahe	USA	1962	2,100	1994	Three
Kedungombo	Indonesia	1988	24,000	1997	Four
Khao Laem	Thailand	1985	11,694	1999	Four
Kpong	Ghana	1982	5,697	1992	Four
Pantabangan	Philippines	1973	13,000	1988	Four
Bayano	Panama	1976	4,123	1994	Four
Tucurui	Brazil	1984	23,924	1999	Four
Manantali	Mali	1988	9,535	1992	Four
Mohale	Lesotho	2002	2,000 est.	N/A	N/A
Pong	India	1974	150,000	1994	Four
Tarbela	Pakistan	1976	96,000	1999	Four
Morazan	Honduras	1985	3,618	1995	Four
Hirakud	India	1958	>110,000	1988	Four
Ukai	India	1972	52,000	1982	Four
Arenal	Costa Rica	1980	2,500	1983	One
Ramial	India	1988	5,000	1995	Two
Yacyreta	Argentina/Paraguay	Ongoing construction	>68,000	N/A	N/A
Nan Ngum	Laos	1972	3,500	2001	Four
Cahora Bassa	Mozambique	1975	>42,000	2002	Four
Aleman	Mexico	1952	19,000	1999	Four
Ceyhan	Turkey	1984	5,000	2000	Four
Kossou	Ivory Coast	1972	75,000	1995	Two
Chixoy	Guatemala	1985	1,500	2002	Four
Pimburetewa	Sri Lanka	1971	120	2001	One
Victoria	Sri Lanka	1984	29,500	2001	Two
Alta	Norway	1987	None	N/A	N/A
Sardar Sarovar	India	Ongoing construction	>200,000	N/A	N/A

3 Resettlement Outcomes of Large Dams

Table 3.2 (continued)

Name of dam	Country/ countries	Date of completion	Number of reservoir resettlers	Date when last data collected	Outcome
Kiambere	Kenya	1988	7,500	1995	Four
Saguling	Indonesia	1986	13,737	1999	Four
Cirata	Indonesia	1988	27,978	1996	Four
Norris	USA	1936	14,249	2001	Three
Cerro Oro	Mexico	1989	26,000	1999	Four
Akosombo	Ghana	1964	78,000	2002	Four

One improved living standards for the majority, *Two* restored living standards for the majority, *Three* restored or improved living standards for the majority but not project related, *Four* living standards for the majority worsened, *N/A* not applicable or relevant
[a]Though Kariba involved both Zambia and Zimbabwe, only Zambian resettlement was analysed. In the High Aswan Dam case, analysis was restricted to Egyptian Nubians

and institutional and staff capacity to implement resettlement policy, and nature and adequacy of funding and monitoring.

Items 67 through 112 dealt with the planning process associated with Stage 1, while items 113–130 dealt with physical displacement and efforts to restore living standards (Stage 2) with emphasis on Cernea's eight impoverishment risks. Items 131–147 dealt with community formation and economic development (Stage 3), while items 148–156 dealt with handing over and incorporation (Stage 4). Items 157–175 involved an overall assessment of outcomes in each case until the time of last data collection, regardless of the number of stages completed. Eight of the last 10 items dealt with downstream impacts, followed by length of river and date of last data.

The small sample size required limiting efforts at quantification to the use of frequencies, means and correlational analysis with significance suggested by a value of $p<0.01$ or below, and a possibly significant relationship by values between $p<0.05$ and $p<0.01$. Since confounding factors can influence what appears to be a positive relationship between two variables, emphasis is placed on trends where a number of possibly related results all point in the same direction.

Once coding began, it was obvious that sufficient data for analysis of many variables of assumed importance were not available in the sources used. That included data on the productivity of the reservoir fishery, and the nature and use of the reservoir drawdown area—two resources that could provide important opportunities to resettlers. Because of data inadequacies and relatively recent NGO activism in regard to the 44 dams analysed, it was also not possible to assess NGO impacts on outcomes. On the other hand, data were lacking for only 11 (6%) of the 171 items of relevance to the analysis of reservoir displacement. Six of those items dealt with resettlement costs as a percentage of total costs, and with estimated number of resettlers at different stages in the planning process. Though such data were available for less than one-third of the 50 cases, they were still sufficient in some cases to compare with results from World Bank and World Commission on Dams' surveys.

3.4.3 Data Sources and Biases

3.4.3.1 Data Sources

Case material analysed came from a number of sources. They included the publications and reports of historians, and environmental and social scientists. Twelve PhD dissertations were especially useful. Also important were reports by the World Bank's Operations Evaluation Department on nine Bank-financed dams, other World Bank reports, six dam-specific WCD case studies and the WCD database which I mined during a March 2002 visit to the Commission's Cape Town office. NGO reports and the Internet provided more recent material on specific projects. Especially useful were NGO submissions at the various WCD consultations and the International River Network's *World Rivers Review* from 1986 to 2004. As for the Internet, rarely was I disappointed when searching for more information on a specific dam.

3.4.3.2 Data Biases and Significance

Because selection of cases is based primarily on data availability, the resulting sample is unique, as are the results based on its analysis. There is also the issue of coder bias. Because of the need for accuracy due to the small size of the sample, I did all the necessary background reading and coding. Where I had personal familiarity with a project (14 of the 50 cases), I believe the data and the coding are accurate. In the other cases, I tried to reduce personal bias by coding 'no data' where I was uncertain of the correct interpretation. I use the word 'interpretation' intentionally, since some interpretation, and hence possible bias, was also involved in deciding how to code a particular item. Two examples are how levels of education and living standards compared with those in other rural areas within the same country.

There are two important questions concerning the usefulness of results—as they relate to the 50 dam sample, on the one hand, and as they relate to data on the existing 50,000 large dams that have caused displacement on the other hand. In regard to the first question, data analysis indicates trend consistency that supports the conclusions that follow about the nature of resettlement outcomes and the reasons for those outcomes. In regard to the second question, what evidence was available suggests that there was no significant bias in the 50 dam survey toward worse or better outcomes. On the contrary, the global lack of attention paid to resettlement issues by governments and project authorities that is reported in broader World Bank (1994 and regional reports) and WCD documents suggests that outcomes in other projects would not be that different (WCD 2000; see also Adams 1999; Bartolomé et al. 2000; Clarke 2000).

Even though the 50 dam survey contained a disproportionate number of the very large dams that represent only a small minority of the large dams covered in ICOLD's 1998 global survey, John Gay and I found no statistically significant evidence that the size of dams and number of resettlers affected outcomes. Nor did we find

3   Resettlement Outcomes of Large Dams   51

a significant difference in outcomes between World Bank-financed and other dams. Indeed, perhaps the disproportionate number of very large dams such as the High Aswan Dam, Volta, and Kainji in the 50 dam sample would be more apt to attract international attention, including funding and supervision from multilateral and bilateral donors, which could lead to improved outcomes. In that case, the 50 dam survey would contain a larger proportion of favourable outcomes. We found, however, no significant difference in outcomes between World Bank-financed and other dams, for while evidence in World Bank reports indicates that the Bank's guidelines have reduced the amount of impoverishment that would otherwise have occurred, the Bank's record in the 50 dam survey was no better in regard to restoring or improving living standards.

3.4.4 Resettlement Outcomes

Four types of outcomes are analysed. The most favourable outcome is one where a majority of resettlers raise their living standards as a result of project planning and implementation. A second outcome notes cases where project initiatives enable a majority to restore their living standards. Concerning the second outcome, it is important to distinguish between a policy, such as the World Bank's, that allows restoration as a satisfactory outcome and the actual achievement of living standard restoration. I do not consider a restoration policy an acceptable one since an increasing amount of research documents that such a policy can be expected to leave the majority worse off. There are several reasons. Though a balance between compensation and development is essential to restore living standards, project authorities implementing a restoration policy tend to overemphasise compensation activities, as do World Bank policies, at the expense of the necessary development ones. A restoration policy also ignores the worsening of living standards that occurs immediately following large dam resettlement. While I include restoration as an outcome definition primarily to allow a more detailed analysis of differing resettlement experiences, the only acceptable policy—for reasons explained later—is to improve living standards as advocated by ICOLD and the WCD.

The third and fourth outcomes include cases where project impacts worsen the living standards of the majority. The difference between the two outcomes is that in one, a majority was still able to restore or improve their livelihoods by taking advantage of non-project-related opportunities, while in the other lower living standards have continued.

3.4.5 Outcome Changes over Time

One response to critiques of unsatisfactory resettlement outcomes is to counter that at least outcomes have improved in more recent projects. Guidelines certainly have improved over time, but have those improved guidelines been reflected in improved policies and planning? And if they have, have those improved policies and plans been reflected in improved outcomes?

To test such questions, John Gay and I analysed how policy and planning varied over time in relationship to outcomes. Three policy-relevant analyses were carried out. In the first two, policy categories were 'no policy', 'cash compensation only', 'restoration' and 'restoration with development'. Cross tabulations, first with the three time periods (before 1980, 1980–1990 and 1991–2005) and then with results before 1991 and 1991 and after, were not significant.

To analyse possible improvements in planning in relationship to outcomes over time, we created an index based on capacity, funding, political will and opportunities. It showed no significant improvement in outcomes over the three time periods. Also analysed was the nature of the resettlement outcome during the three different time periods. While the results seem to suggest that outcomes during the 1980–1990 and 1991–2005 periods were actually worse than those before 1980, we found no significant evidence of changes in implementation outcomes over time.

3.4.6 Why Success, Why Failure

3.4.6.1 Introduction

The five most important reasons for unsatisfactory outcomes were lack of staff numbers and expertise, lack of finance, and lack of political will on the part of implementing agencies, and for those being resettled, lack of compensation and development opportunities and lack of participation (Tables 3.3–3.8). Though each may be a necessary condition for planning and implementing an adequate resettlement process, each alone is insufficient. The nature of the project authority, presence or absence of unexpected events, conflicts between resettlers, immigrants and hosts, and impoverishment risks also play an important role in influencing project outcomes.

3.4.6.2 Capacity

Staff expertise and number were emphasised in assessing capacity for planning and implementing resettlement. Financial resources were dealt with separately. A major lack of planning capacity occurred in 27 (66%) of 41 cases with adequate data (Table 3.3). Resettlement outcomes were adverse in all 27 of these cases. By way of contrast, positive outcomes occurred in three of the four cases where adequate capacity was coded. The negative case was Ghana's Kpong Dam where funds and opportunities were lacking. Positive outcomes also occurred in seven of ten cases with a minor lack of capacity. Outcomes in the 36 cases where information was coded on staff numbers and expertise are shown in Table 3.4.

3 Resettlement Outcomes of Large Dams

Table 3.3 Adequacy of planning capacity, cross tabulation ($p<0.000$)

	Outcome Adverse	Positive	Total
Major lack of capacity	27	0	27
	100%	0%	100%
Minor lack of capacity	3	7	10
	30%	70%	100%
Capacity adequate	1	3	4
	25%	75%	100%
Total	31	10	41
	75.6%	24.4%	100%

Table 3.4 Numbers and expertise of staff, cross tabulation ($p<0.001$)

Staff numbers and expertise	Outcome Adverse	Positive	Total
Numbers and expertise inadequate	23	1	24
	95.8%	4.2%	100.0%
Numbers adequate but expertise lacking	3	3	6
	50.0%	50.0%	100.0%
Expertise adequate but numbers inadequate	1	0	1
	100.0%	0.0%	100.0%
Numbers and expertise adequate	1	4	5
	20.0%	80.0%	100.0%
Total	28	8	36
	77.8%	22.2%	100.0%

It is hard to underemphasise the importance of expertise. A case in point is Mexico's Zimapán Dam where the project authority and the World Bank 'tried to make Zimapán a model project for resettlement… The CFE [the project authority] met the new demands for the Zimapán project by creating a group of 84 professionals… The majority of them came directly from university and lacked work experience… It seems obvious that this group of young professionals had not been sufficiently prepared for the field. During informal talks, they claimed that they had not been trained in participatory methods, poverty analysis nor social situation analysis' (Aronsson 2002:114–116).

3.4.6.3 Adequacy of Funding

'Timely availability of adequate funds is a severe constraint in a large number of projects; it may be the single most powerful explanatory operational variable behind the failure to implement resettlement operations well', according to the authors of the World Bank's *Bankwide Review* of projects involving involuntary resettlement (1994: Section 6:11). Table 3.5 deals with adequacy of funding in the 43 cases with adequate information. Inadequate funds were coded in 25 (58%) of those cases in only one of which a positive outcome occurred. In that case (the Ivory Coast's Kossou Dam) inadequate funding and a minor lack of capacity were partially offset

Table 3.5 Adequacy of funding, cross tabulation ($p<0.001$)

	Outcome Adverse	Positive	Total
Funds inadequate	24	1	25
	96.0%	4.0%	100%
Physical removal funds adequate	2	0	2
	100%	0.0%	100%
Funds increased for resettlement and rehabilitation purposes	4	5	9
	44.4%	55.6%	100%
Funds adequate for resettlement and rehabilitation purposes	3	4	7
	42.9%	57.1%	100%
Total	33	10	43
	76.7%	23.3%	100%

Table 3.6 Adequacy of political will, cross tabulation ($p<0.001$)

	Outcome Adverse	Positive	Total
Inadequate political will	19	1	20
	95.0%	5.0%	100.0%
Political will a response to donor requirements	2	0	2
	100.0%	0%	100.0%
Political will to at least restore living standards	7	2	9
	77.8%	22.2%	100.0%
Political will to implement resettler development	3	7	10
	30.0%	70.0%	100.0%
Total	31	10	41
	75.6%	24.4%	100.0%

by strong backing from the President of the Republic who shared ethnicity with the majority of those resettled.

Planners need to realise that adequately implemented resettlement is expensive. Estimated costs per capita for Laos' Nam Theun 2 project, which currently involves the World Bank Group, are US$ 3,819 (NTEC 2002: Section 6:11). As numbers of resettlers increase, funding the resettlement process can become a major project cost; indeed, in the case of China's Three Gorges Project the major project cost. A major reason for underestimating financial costs is due to underestimating the number of affected people during the initial feasibility studies.

3.4.6.4 Political Will

The political will to implement an adequate resettlement process continues to characterise only a minority of dam projects, with lack of government commitment to resettlement identified by the World Bank's *Bankwide Review* as one of a small number of recurring deficiencies. In the 50 dam survey, inadequate political will occurred in 54% of the 41 cases for which data was adequate (Table 3.6).

3   Resettlement Outcomes of Large Dams

Table 3.7 Opportunities in relationship to outcome, cross tabulation ($p<0.003$)

	Outcome Adverse	Positive	Total
Inadequate opportunities implemented	31	6	37
	83.8%	16.2%	100.0%
Adequate opportunities available	0	3	3
	0.0%	100.0%	100.0%
Adequate opportunities, not project related	1	1	2
	50.0%	50.0%	100.0%
Total	32	10	42
	76.2%	23.8%	100.0%

3.4.6.5   Opportunities

The construction of large dams has the potential to provide a wide range of opportunities, the nature of which can be expected to vary from project to project, for enabling resettlers to raise their living standards and contribute to the stream of project benefits. Dealt with in more detail in the last section, they include involving affected people as shareholders and co-project managers. Specific development opportunities include irrigation and such associated multiplier effects as employment in agro-industries, rain-fed agriculture, reservoir drawdown cultivation and reservoir capture fisheries and aquaculture, natural resource management and tourism, job training for employment in rural and urban industries, and a wide range of informal sector economic activities. On the other hand, it is important for project authorities and politicians not to overemphasise employment opportunities during dam construction and subsequent operation. Case histories show that only a relatively small number of affected people are employed. Even if they have the necessary skills, their labour is needed 'at home' to prepare for and carry out household resettlement.

Among 42 cases coded, adequate project-related opportunities were implemented in only 3 (7%) cases (Table 3.7). Those involved the Arenal, High Aswan and Shuikou Dams.

3.4.6.6   Resettler Participation in Project Planning

John Gay and I developed an index of resettler participation that combined the scores for the extent to which resettlers participated in selection of resettlement sites, size of social units to be relocated together, choice of development options and nature of social services. Each variable was coded on an ascending scale from no to full participation with the resulting index from a lowest 'no participation' in any of the four variables to 'full participation' in all four. Mean values were then calculated for adverse and for positive outcomes. Significantly different (Table 3.8), the means suggested that outcomes were influenced by resettler participation.

Table 3.8 Resettler participation in relationship to outcome ($p < 0.005$)

Outcome	Mean	N	SD
Adverse	7.0	23	3.4
Positive	11.0	9	3.1
Total	8.6	32	3.7

Table 3.9 Cross tabulation on institutional responsibility in relationship to outcome ($p < 0.001$)

	Outcome		Total
	Adverse	Positive	
No specific unit	5	0	5
	100.0%	0.0%	100.0%
Project authority	3	5	8
	37.5%	62.5%	100.0%
Project authority and other government units	18	0	18
	100.0%	0.0%	100.0%
Other government agency	7	3	10
	70.0%	30.0%	100.0%
Total	33	8	41
	80.5%	19.5%	100.0%

3.4.6.7 Institutional Responsibility for Resettlement

Situations analysed for dealing with institutional responsibility for planning and implementing resettlement were 'no specific unit', 'the project authority working alone', 'the project authority cooperating with other institutions' and 'other government institutions'. Only project authorities working alone were associated with more positive than negative outcomes (Table 3.9). Keeping in mind that the number of cases is small, nonetheless, 'making a single agency responsible for all aspects of project planning, financing, and implementation certainly simplifies what would otherwise be complicated and potentially non-cooperative inter-institutional relationships. Furthermore, where other agencies are given unasked for resettlement responsibilities, case studies indicate inadequate commitment, funding and staffing capacity are to be expected' (Scudder 2005:63).

3.4.6.8 Unexpected Events and Resettler Ability to Compete with Host Populations and Immigrants

The 50 dam survey identified two other problems that planners and researchers alike have overlooked. One was the importance of unexpected environmental, policy-related and political events in influencing the nature of outcomes. Coded as major in 26 (59%) of 44 cases, they had a significant impact on outcomes (Table 3.10). An unexpected increase in the frequency of drought, for example, had an adverse effect on the Kariba resettlement. In the case of Sri Lanka's Mahaweli Project, decreased rainfall has reduced water supplies to irrigated holdings especially in the

3   Resettlement Outcomes of Large Dams

Table 3.10 Unexpected events in relationship to outcome, cross tabulation ($p<0.008$)

	Outcome		
	Adverse	Positive	Total
Unexpected events of major importance	24	2	26
	92.3%	7.7%	100.0%
Unexpected events of minor importance	10	8	18
	55.6%	44.4%	100.0%
Total	34	10	44
	77.3	22.7%	100.0%

project's westernmost irrigation system. Even more important in the Mahaweli case have been the unexpected adverse impacts of the fundamentalist religious beliefs of the first Mahaweli Chairman and of changes in the presidency on project returns and resettler livelihoods (Scudder 2005:157–159).

The resettlement literature is full of examples of what is predictable conflict between resettlers and host populations who end up competing for the same land and natural resources, jobs and other development opportunities, social services and positions of political and religious influence. Where arable land is limited, which is usually the case, conflicts are apt to intensify over the years, especially when the second generation of hosts find that land that had been intended for their use is now occupied by resettlers.

What had not been as well documented was that immigrants as individuals and representatives of companies who come to exploit dam-related opportunities present an even greater threat to resettlers (as well as, it should be added, to the host population). Inability to compete with immigrants was mentioned in 20 (43%) of 47 cases. Inability to compete and integrate with hosts was reported in 14 (32%) of 44 cases. Immigrants can out-compete resettlers for opportunities because they tend to be better educated and have better connections to government officials and politicians. They also tend to have greater access to financial resources with the result that they have a comparative advantage, for example, in catching and marketing reservoir fish, undertaking commercial agriculture and establishing tourist facilities around the reservoir. To avoid conflict and further impoverishment of resettlers, planners need to ensure that affected people have enforceable legal rights over the opportunities necessary for improving their living standards. An example from Kariba in what is now Zambia was a government decision to restrict reservoir fishing to the resettler and host population for a number of years, and to provide the necessary credit and training for them to exploit the reservoir's potential.

3.4.7   Impoverishment Risks

The frequency with which Cernea's impoverishment risks were associated with specific cases in the 50 dam survey illustrates how often unfavourable and unacceptable outcomes occur. John Gay and I created a well-being index based on

Table 3.11 Resettler 'well-being' in relationship to outcome ($p<0.000$)

Nature of outcome	Mean	N	SD
Adverse well-being	6.6	31	2.4
Positive well-being	10.9	9	2.2
Total	7.6	40	2.9

Table 3.12 Landlessness in relationship to outcome ($p<0.006$)

	Outcome		
	Adverse	Positive	Total
Landlessness a problem	32	6	38
	84.2%	15.8%	100.0%
Landlessness not a problem	2	4	6
	33.3%	66.7%	100.0%
Total	34	10	44
	77.3%	22.7%	100.0%

avoidance of five of Cernea's risks. It had a significant relationship to outcome (Table 3.11), as did each of the individual risks (Tables 3.12–3.16).

3.5 Improving Resettlement Outcomes

3.5.1 Introduction

So what can be done to improve outcomes so that resettlement also contributes to the stream of project benefits? State-of-the-art policies are essential. A crucial aim is that the desired outcome of the resettlement process must be to improve rather than merely restore resettler living standards. Another is that an options assessment process dealing with water resource and energy development should involve potential resettlers. In regard to specific opportunities, improvement can be fostered in two very different ways. One is to involve affected people in project management and ownership. The other is to provide specific income generating and livelihood improvement activities.

3.5.2 Improving Living Standards

Improvement is emphasised in ICOLD's November 1995 *Position Paper on Dams and Environment*. While the World Bank also states that improvement is the most desirable outcome, the Bank's April 2004 Revision of its safeguard policy on involuntary resettlement, which has influenced policies of the regional banks and the OECD countries as well as countries like China, allows governments and project authorities the fall-back position of restoring incomes and living standards. That is

Table 3.13 Joblessness in relationship to outcome, cross tabulation ($p<0.005$)

	Outcome Adverse	Positive	Total
Joblessness a problem	28	5	33
	84.8%	15.2%	100.0%
Joblessness not a problem	3	5	8
	37.5%	62.5%	100.0%
Total	31	10	41
	75.6%	24.4%	100.0%

Table 3.14 Ability to restore food self-sufficiency in relationship to outcome, cross tabulation ($p<0.01$)

	Outcome Adverse	Positive	Total
Household must buy food	29	4	33
	87.9%	12.1%	100.0%
No major change	2	1	3
	66.7%	33.3%	100.0%
Food self-sufficiency achieved	2	4	6
	33.3%	66.7%	100.0%
Total	33	9	42
	78.6%	21.4%	100.0%

Table 3.15 Marginalisation in relationship to outcome, cross tabulation ($p<0.000$)

	Outcome Adverse	Positive	Total
Marginalisation a problem for the majority	25	0	25
	100.0%	0.0%	100.0%
Marginalisation a problem for a minority	2	2	4
	50.0%	50.0%	100.0%
Marginalisation not a problem	6	8	14
	42.9%	57.1%	100.0%
Total	33	10	43
	76.7%	23.3%	100.0%

Table 3.16 Common property resources and outcome, cross tabulation ($p<0.007$)

	Outcome Adverse	Positive	Total
Planners unaware of common property importance	24	3	27
	88.9%	11.1%	100.0%
Planners with some awareness of common property importance	3	5	8
	37.5%	62.5%	100.0%
Total	27	8	35
	77.1%	22.9%	100.0%

the position that governments and project authorities are apt to take, as in the case of the Lesotho Highlands Water Project. Yet it is a position that has been shown by research not only to not restore incomes but to leave the majority impoverished, which is why I consider that Bank policy to be partly responsible for failed resettlement.

There are a number of explanations for such a result, of which five are mentioned here. First, planners assume wrongly that a compensation policy alone can restore living standards rather than a balance between compensation and development initiatives. The Bank's safeguard policy is at fault here since 'compensation' is mentioned 19 times while 'development' is mentioned only four times. Second, a restoration approach fails to take into consideration the fact that living standards for a majority of resettlers tend to drop during the long planning process (often over 10 years) that precedes construction of a large hydro project and during the initial years immediately after physical removal. Third, pre-project surveys carried out to establish a benchmark against which restoration can be measured are known to underestimate income and living standards which have already been lowered due to project-related cessation of investments in the area. Fourth, the Bank's safeguard policies deal only with direct economic and social impacts. Ignored are a wide range of socio-cultural effects associated with forced removal from a preferred homeland, the psychological stress affecting the elderly and women and increased rates of illness and death that have been reported in resettlement areas. Fifth, resettlement tends to be associated with increased cash expenditures because rural resettlers, who remain the large majority in connection with large hydro projects, are moved to less fertile soils which require costly inputs to provide equivalent yields, have less access to common property resources for grazing, fuel, building materials and for foraging, and become more dependent on credit and the risk of indebtedness.

Ted Downing (2002:14), research professor at the University of Arizona, also explains why the World Bank policy 'institutionalises a negotiating system that potentially violates human rights'. Moreover, after mentioning that involuntary resettlement is accompanied by impoverishments risks, the Bank's policy, unlike the policies of the WCD, fails 'to propose measures to address them. Instead, it falls back on the same flawed economic analysis and methodologies that have been responsible for decades of unacceptable performance' (Downing 2002:13). Failed performance has been acknowledged even by such senior World Bank officials as Robert Goodland and Michael Cernea and within the Bank's Operation's Evaluation Department (OED). OED's 2001 *Involuntary Resettlement: Comparative Perspectives* (Picciotto et al. 2001) that dealt only with large dams, concluded that 'the record on restoring—let alone improving—incomes has been unsatisfactory' (Picciotto et al. 2001:9). More to the point, the authors, one of whom was OED's Director, concluded for dam projects that 'Above all, displacees must be beneficiaries of the project. Merely aiming to restore standards of living and lifestyles common to isolated river valleys can be a dead-end development strategy' (Picciotto et al. 2001:140).

3.5.3 Participatory Options Assessment

The resettlement process needs to start with a participatory and open process of options assessment that includes the major stake holders and risk takers, including those who may be required to resettle should a dam be selected as the preferred alternative. A very interesting 2003 World Bank report deals with both the costs and the benefits of such a process. The principal cost is that options assessment takes up valuable time and increases initial financial costs. The World Bank report estimates that an adequate strategic options assessment with full stakeholder involvement would take about 2 years, with a cost of US $2 million and up to US $9 million if construction was delayed during that time period.

But what about the benefits if incomplete options assessment results in disputes that delay construction at a later date? As of 2003, the report's authors estimate that delays with the completion of India's Sardar Sarovar project have cost, and I quote, 'more than $1 billion in aggregate' and in the case of Argentina's and Paraguay's Yacyreta an annual cost in power benefits that 'exceeds US $200 million' (World Bank 2003:30–31). I was the World Bank's principal resettlement consultant for the Sardar Sarovar Project throughout the 1980s. I believe that if the three states involved and the central government had been willing to implement the World Bank's resettlement guidelines and the provisions of the 1979 Narmada Water Disputes Tribunal, that project which is still incomplete would have been completed before the turn of the century.

3.5.4 Resettlers as Project Managers and Owners

Two different management and ownership approaches have been implemented in Uruguay and Canada. The Uruguay example involves Ita Dam resettlers as managers; they were allocated funds with which they planned and implemented better community infrastructure at a lower price than that completed by the project authority. The Canadian examples are more ambitious and pioneer what I hope will become an increasingly assessed approach. They involved joint ventures with Native American communities that include all project-affected people as opposed to just resettlers, and that also involve a willingness to acknowledge that such communities have rights to their land and water resources.

In one case, Hydro-Quebec, with 49.9% of shares and a Native American group with 50.1%, has formed a limited partnership. Operational since May 2000, it involves a run-of-the-river facility with an installed capacity of about 10 MW. It is owned by a company under the people's local government which sells all electricity generated to Hydro-Quebec during a 20-year period that is extendable for another 20 years. Seventy-five percent of project funding came from a long-term bank loan, with the people's company 'mandated to conduct the feasibility studies, obtain all the governmental authorisations, have the project built under a turnkey contract and operate the facility' (Egré et al. 2002:57). In addition to profit sharing, the joint venture allows the Native Americans 'to design a project according to their priorities and in the long term reinvest the profits in a manner that supports the economic development of their community' (WCD 2000:257, footnote 11).

3.5.5 Specific Development Opportunities

3.5.5.1 Introduction

Improving livelihoods require the availability of a wide range of specific opportunities. Breadth is important. Whether coping with poverty or trying to further develop an upward trend in lifestyle, the global experience is that rural communities favour diversified household production systems that include both farm and non-farm occupations. The benefits of such diversified production systems extend beyond risk-aversion and/or income generation because they also give spouses as well as children and other dependents the opportunity to contribute in various ways to the household economy and to their own status within the household. I have described available opportunities in detail in chapter four of the *Future of Large Dams*. Here I describe briefly a number of them.

3.5.5.2 Irrigation and Associated Multiplier Effects

Irrigation with associated multiplier effects is one of the most important opportunities that can be provided by a large dam for raising living standards. Half of the dams in the 50 dam survey had a significant irrigation component. Yet in only three of the 23 cases was a serious attempt made to integrate a majority of resettlers within the project's irrigated command area. This is yet another example of the lack of attention paid to resettlement in spite of the fact that those three cases (the High Aswan Dam and two projects in Sri Lanka) were among that minority of cases where living standards were either improved or restored.

Irrigation can enable resettlers to raise their living standards in two quite different ways. One is through the cultivation of high-value crops; the other comes from the non-farm enterprise development and employment generation associated with the multiplier effects of well-planned and well-implemented irrigation. Such effects should be an important objective of water resource development projects. Yet, in spite of their potential (Bhatia et al. 2008), they are seldom planned for.

As for the cultivation of high-value crops, they are necessary simply because the double cropping of cereal grains is unable to move most farmers beyond subsistence due to such factors as a halving of food grain prices since the 1960s and unfavourable rural–urban terms of trade. A policy emphasis on the double cropping of paddy, for example, is a major reason why Sri Lanka's multi-billion dollar Accelerated

Mahaweli project, one of the world's largest water resource development projects, has been unable to move a majority of hundreds of thousands of settlers and resettlers beyond subsistence.

The tragedy is that the potential was there, as illustrated by Sri Lanka's Minneriya project. There, small-scale farmers pioneered the cultivation of such higher value crops as chillies, onions and tobacco. Increased income enabled them to diversify their household economies into a range of non-farm activities, while their increased consumption of pumps, two wheel tractors and a wide range of other goods and services helped the growth of what may well be the most dynamic rural town in the Sri Lankan dry zone.

3.5.5.3 Rain-Fed Agriculture

Throughout the tropics, rain-fed tree crops have been able to improve resettler livelihoods. In Thailand, what are perhaps the two most successful resettler outcomes are associated with a farming system based on rubber (Suwanmontri 1996, 1999a, b). One case, involving the Rajjaprobha Dam, is especially instructive for illustrating how tree crops raised living standards in the 1990s. Within 8 years of dam commissioning, net incomes from rubber sales that were controlled for inflation had increased almost fivefold and were much higher than those of a household sample surveyed in a neighbouring non-affected village.

This case also illustrates a number of other best practices. The project authority was also responsible for planning and implementing resettlement. Careful soil surveys were completed. The resettlement site was located within 4 km of the dam site, with electricity and waterworks provided. A living allowance was provided while the rubber trees were maturing. Plans complemented rubber with other agricultural and non-farm opportunities. Farm opportunities included poultry farming and cultivation of fruit trees and vegetables, while an agricultural cooperative and credit facilities were provided. Non-farm opportunities included such programmes as 'training in reservoir fishery work' (Suwanmontri 1999b:13).

3.5.5.4 The Reservoir Drawdown Area

The benefits associated with utilisation of the drawdown area remain one of the most overlooked opportunities. Except where too steep or rocky, drawdown zones have value for agriculture and grazing by domestic stock and game. That is especially the case throughout much of Africa and in the semi-arid zones of South Asia where the drawdown area reaches its greatest extent toward the end of the dry season when other grazing is scarce.

The best examples of planning for drawdown utilisation have been in Africa and Indonesia. In Zambia, the drawdown area behind the Kariba Dam has become the major producer of maize in the reservoir basin during recent drought years. It also supports one of the best livestock grazing areas in the country. In Ghana's Volta

Reservoir Basin, the maximum drawdown area during a 'normal' year has been estimated at over 100,000 ha. Residual moisture can support cropping for 40–60 days. In Nigeria's Kainji Lake Basin, the resettlers have adapted their pre-dam system of small pump irrigation of onions to the reservoir margin (Roder 1994). Resettlers there have also pioneered a new system of land use in which they harvest a colonising species of grass as fodder during the dry season and for sale to pastoralists. As an indication of its importance is an estimate that the total crop available is 'in excess of 110,000 t of utilisable standing crop' (Morton and Obot 1984:694).

In Indonesia, resettlers have pioneered drawdown cultivation at least since the 1970s. In the 1980s, local farmers pioneered drawdown cultivation behind the Wonogiri and Jatiluhur dams, while planning for resettlement for the Saguling and Cirata projects combined drawdown cultivation with fish cage culture. In those cases two crops a year can be grown, with local farmers experimenting with the interplanting and sole cultivation of different cash and food crops under varying soil and moisture conditions. Yields are high and 'benefits in terms of production, social equity and employment can be considerable' according to Winarto (1992:461). Winarto also notes that the farmers believe that the way in which they use the land does not have an adverse effect on water quality and siltation. The fertility of the land reduces the need for fertilisers while intercropping reduces insect damage and hence the need for pesticides. They also believe that the minimum tillage and plant cover associated with intercropping, and the bunds associated with rice cultivation, reduce soil erosion and hence siltation.

3.5.5.5 Reservoir Capture Fisheries and Aquaculture

Reservoir fisheries provide another major opportunity for improving the living standards of project-affected people. The same applies to natural resource management and tourism. Yet in both cases, unless special attention is paid to resettler participation, immigrants are apt to monopolise those benefits. There are many cases, with better trained immigrant fishers dominating the Lake Volta fishery in Ghana, a single buyer exploiting fishers on Laos' Nan Ngum Reservoir, and wealthy and influential outsiders dominating the cage fishery in Indonesia's Saguling and Cirata reservoirs. Elsewhere, including Costa Rica, Zimbabwe and Thailand, governments establish national parks on land formerly used by resettlers with their minimal involvement, while tourism is dominated by outsiders who also acquire, legally and illegally, prime reservoir frontage for building vacation homes.

3.5.5.6 Rural Employment Generation and Enterprise Development

My last example involves ownership of, and employment in, rural industries. As part of a national policy to slow down rural to urban migration, the Chinese have been the most successful in shifting farm households into non-farm village industries. Their approach requires careful examination in regard to its sustainability, adaptability to dam resettlement and applicability to other countries. The results to date in villages not required to resettle have been impressive. While advising on Three Gorges resettlement in the 1980s, I visited several villages in the middle and lower Yangtze Basin in which a purely agricultural economy had been converted within a single decade to one in which a wide range of village industries had become the major source of employment. Employed youth of both sexes, now better educated than their parents, were said to prefer such non-farm work within their communities. Out-migration was minimal to non-existent at that time.

Inadequate time has gone by, however, to know how competitive such industries will be over the longer term. Even less time has gone by to assess the ability of the Chinese to apply the same policy to dam-induced resettlement. Early evaluations of their record with village industries in connection with the Shuikou project are encouraging but too early to be convincing. The intentions of resettling large numbers of Three Gorges' farm families in non-farm enterprises are just that—intentions only.

More common is the approach of the Lesotho Highlands Water Project where job training is intended to help resettlers learn such crafts as carpentry, masonry and tailoring as well as to acquire training for poultry rearing and baking. To date, results after 8 years in connection with the Katse Dam project are not encouraging. A major problem is that most of those trained have not been able to establish commercially viable businesses for marketing their new skills. Another problem, common to all sorts of training institutions, has been failure of training staff to follow up trainees in the field to help them deal with unanticipated and other problems. A further constraint has been the persistent failure of project authorities or other agencies to institutionalise the necessary credit facilities. Such problems continue to have global applicability.

Acknowledgements I would like to dedicate this chapter to the memory of fellow WCD commissioner Jan Veltrop who, as an engineer, pointed out that this chapter would be more influential if it included the first detailed statistical analysis of resettlement outcomes. The result of that suggestion was the analysis that John Gay and I subsequently completed on 50 large dams.

References

Adams A (1999) Social impacts of an African dam: equity and distributional issues in the Senegal River Valley. Contributing paper to WCD Thematic Review 1.1, Cape Town

Aronsson IL (2002) Negotiating involuntary resettlement: a study of local bargaining during the construction of the Zimapán Dam. Occasional papers 17, Uppsala University, Uppsala

Bartolomé LJ, de Wet C, Mander H, Nagaraj VK (2000) Displacement, resettlement, rehabilitation, reparation and development. WCD Thematic Review 1.3 (prepared as an input to the World Commission on Dams), Cape Town

Bhatia R, Cestti R, Scatasta M, Malik RPS (eds) (2008) Indirect economic impacts of dams: case studies from India, Egypt and Brazil. Academic Foundation for World Bank, New Delhi

Bilharz JA (1998) The Allegany Senecas and Kinzua Dam: forced relocation through two generations. University of Nebraska Press, Lincoln

Cernea MM (1996) Understanding and preventing impoverishment from displacement. In: McDowell C (ed) Understanding impoverishment: the consequences of development-induced displacement. Berghahn Books, Oxford, pp 13–32

Cernea MM (1997) Hydropower dams and social impacts: a sociological perspective, paper. Social Assessment Series No. 44. World Bank Environment Department, Washington, DC

Cernea MM (1999a) Why economic analysis is essential to resettlement: a sociologist's view. In: Cernea MM (ed) The economics of involuntary resettlement: questions and challenges. World Bank, Washington, DC, pp 5–30

Cernea MM (ed) (1999b) The economics of involuntary resettlement: questions and challenges. World Bank, Washington, DC

Cernea MM, McDowell C (eds) (2000) Risks and reconstruction: experiences of resettlers and refugees. World Bank, Washington, DC

Clarke C (2000) Cross-check survey: final report. A WCD Survey prepared as an input to the World Commission on Dams, Cape Town. www.dams.org

Colson EF (1971) The social consequences of resettlement. Manchester University Press, Manchester

Cook CC (ed) (1994) Involuntary resettlement in Africa. World Bank Technical Paper Number 227, Africa Technical Department Series. World Bank, Washington, DC

Downing TE (1996) Mitigating social impoverishment when people are involuntarily displaced. In: McDowell C (ed) Resisting impoverishment – tackling the consequences of development-induced impoverishment. Berghahn Books, Oxford, pp 33–48

Downing T (2002) Creating poverty: the flawed economic logic of the World Bank's Revised Involuntary Resettlement Policy. Forced Migr Rev 12:13–14

Egré D, Roquet V, Durocher C (2002) Benefit sharing from dam projects – phase 1: desk study. Vincent Roquet and Associates for the World Bank, Montreal

Fried M (1963) Grieving for a lost home. In: Duhl L (ed) The urban condition. Basic Books, New York, pp 151–171, Ft. McDowell Yavapai Nation

Guggenheim S (1994) Involuntary resettlement: an annotated reference bibliography for development research. Environment working paper no. 64, World Bank, Washington, DC

ICOLD (International Commission on Large Dams) (1995) Position paper on dams and the environment. ICOLD, Paris

ICOLD (International Commission on Large Dams) (1998) Register of large dams. ICOLD, Paris

International River Networks (1986–2004) World Rivers Review (Formerly International Dams Newsletter), Berkeley

Khera S, Mariella PS (1982) The Fort McDowell Yavapai: a case of long-term resistance to relocation. In: Hansen A, Oliver-Smith A (eds) Involuntary migration and resettlement: the problems and responses of dislocated people. Westview Press, Boulder, pp 159–177

Koenig D (1995) Women and resettlement. In: Gallin R, Ferguson A (eds) The women and international development annual. Westview Press, Boulder, pp 21–49

Mathur HM (1997) Managing projects that involve resettlement: case studies from Rajasthan, India. Economic Development Institute (EDI working papers). World Bank, Washington, DC

Mehta L, Srinivasan B (2000) Balancing pains and gains: a perspective paper on gender and large dams. Contributing paper for WCD Thematic Review 1.1. World Commission on Dams, Cape Town

Morton AJ, Obot EA (1984) The control of *Echinochloa stagnina* (Retz,) P, Beauv by harvesting for dry season livestock fodder in Lake Kainji, Nigeria – a modelling approach. J Appl Ecol 21:687–694

Nam Theun 2 Electricity Consortium (NTEC) (2002) Nam Theun 2 hydroelectric project resettlement action plan. NTEC, Vientiane, Lao PDR

Parasuraman S (1999) The development dilemma: displacement in India. Institute of Social Studies, The Hague

Picciotto R, Van Wicklin W, Rice E (eds) (2001) involuntary resettlement: comparative perspectives. World Bank Series on Evaluation and Development, vol 2. Transaction Publishers, New Brunswick

Rew AW (1996) Policy implications of the involuntary ownership of resettlement negotiations: examples from Asia of resettlement practice. In: McDowell C (ed) Resisting impoverishment – tackling the consequences of development-induced impoverishment. Berghahn Books, Oxford, pp 201–222

Rew AW, Fisher E, Pandey B (2000) Addressing policy constraints and improving outcomes in development-induced displacement and resettlement projects. Refugee Studies Centre, University of Oxford, Oxford

Ribeiro GL (1994) Transnational capitalism and hydropolitics in Argentina: the Yacyretá High Dam. University Press of Florida, Gainesville

Roder W (1994) Human adjustment to Kainji Reservoir in Nigeria: an assessment of the economic and environmental consequences of a major man-made lake in Africa. University Press of America, Lanham

Rosa EA (1998) Metatheoretical foundations for post-normal risk. J Risk Res 1(1):15–44

Scudder T (1981) The development potential of new lands settlement in the tropics and subtropics: a global state-of-the-art evaluation with specific emphasis on policy implications. Institute for Development Anthropology, Binghamton

Scudder T (1985) A sociological framework for the analysis of new lands settlements. In: Cernea MM (ed) Putting people first: sociological variables in rural development. Oxford University Press for the World Bank, New York, pp 145–185

Scudder T (1993) Development-induced relocation and refugee studies: 37 years of change and continuity among Zambia's Gwembe Tonga. J Refugee Studies 6(2):123–152

Scudder T (1997) Chapters on social impacts and resettlement. In: Biswas AK (ed) Water resources: environmental planning, management and development. McGraw Hill, New York, pp 623–665, 667–710

Scudder T (2005) The future of large dams: dealing with social, environmental, institutional and political costs. Earthscan, London

Scudder T, Colson E (1982) From welfare to development: a conceptual framework for the analysis of dislocated people. In: Hansen A, Oliver-Smith A (eds) Involuntary migration and resettlement: the problems and responses of dislocated people. Westview Press, Boulder, pp 267–287

Suwanmontri M (1996) An analysis on involuntary resettlers. R & R Research Series, No. 3

Suwanmontri M (October 1999a) Agriculture-based resettlement design and management for hydrodam: the Thai experiences. Paper presented at the 2nd International Conference, World Council of Power Utility, China

Suwanmontri M (October 1999b) Establishment of a resettlement concept. Paper presented at the 2nd International Conference, World Council of Power Utility, China

Wali A (1989) Kilowatts and crisis: hydroelectric power and social dislocation in Eastern Panama. Westview Press, Boulder

WCD (World Commission on Dams) (2000) Dams and development: a new framework for decision-making. The report of the World Commission on dams. Earthscan, London

Winarto YT (1992) The management of secondary consequences in dam projects: the case of drawdown agriculture in Indonesia. World Dev 20(3):457–465

World Bank (1980) Social issues associated with involuntary resettlement in bank-financed projects. Operational Manuel Statement 2.33. World Bank, Washington, DC

World Bank (1993) Early experience with involuntary resettlement. Operations Evaluation Department. World Bank, Washington, DC

World Bank (1994) Resettlement and development: the bankwide review of projects involving involuntary resettlement 1986–1993. Environment Department. World Bank, Washington, DC

World Bank (1998) Recent experiences with involuntary resettlement. Operations Evaluation Department. World Bank, Washington, DC

World Bank (2003) Stakeholder involvement in options assessment: promoting dialogue in meeting water and energy needs – a sourcebook. World Bank, Washington, DC

World Bank (2004) Involuntary resettlement. OP 4.12. World Bank, Washington, DC

Chapter 4
Greenhouse Gas Emissions from Reservoirs

Olli Varis, Matti Kummu, Saku Härkönen, and Jari T. Huttunen

4.1 Introduction

It is difficult to conceive of another aspect of water resources development that evokes as much emotion, public concern and challenge for policy makers as dam and reservoir construction. Some want to store water, generate electricity, irrigate or ameliorate living in other ways by constructing dams. Others oppose this because dams and reservoirs destruct valuable ecosystems, displace people from river valleys where they have dwelled for millennia and so on.

The starting point is that water is distributed unevenly—natural supply does not coincide with human demand. This problem is particularly pressing in climatic zones which have a strong seasonal pattern of river flow formation (such as the tropical, boreal and arctic zones), in regions with high elevation differences that provide large hydropower potential, as well as in dry climates.

However, nothing comes for free. Concern for the destruction of invaluable cultural, human and ecological assets has grown in the last decades. One of the least known side effects of reservoirs is their greenhouse gas (GHG) emissions to the atmosphere (Table 4.1).

The global account of greenhouse gas emissions is still largely unknown, and available estimates diverge greatly from one another. The total surface area of reservoirs is not accurately known. The estimates of emissions related to reservoirs thus differ in two basic dimensions. First, vastly different estimates of the total

---

O. Varis (✉) • M. Kummu
Water and Development Research Group, Aalto University, Espoo, Finland

S. Härkönen
Uusimaa Centre for Economic Development, Transport and the Environment, Helsinki, Finland

J.T. Huttunen
Department of Environmental Sciences, University of Kuopio, Kuopio, Finland

Table 4.1 Most important greenhouse gases with respect to reservoir emissions and their basic properties in the atmosphere

Gas	Symbol	Lifetime (years)	Global warming potential (relative to CO_2)		
			20-year	100-year	500-year
Carbon dioxide	CO_2		1	1	1
Methane	CH_4	12	72	25	7.6
Nitrous oxide	N_2O	114	289	298	153

Note: The Fourth Assessment Report of the Intergovernmental Panel on Climate Change (IPCC) lists several dozen GHGs but the others have a minor relevance in this context. The greenhouse warming potential of gases is typically expressed as CO_2 equivalents, which are different in different time perspectives since gases such as methane or nitrous oxide degrade in the atmosphere. For instance, a unit of methane has 72 times the warming potential as the same unit of CO_2 in the time scale of 20 years, but this goes down to 25 in the time perspective of 100 years. Our analysis was made using the 20-year equivalents
Source: Forster et al. (2007) (part of the IPCC Fourth Assessment report)

reservoir area are used. Second, the gas exchange fluxes as calculated per unit area, such as square metre, vary even more.

Recently, there has been a heated debate in the literature about greenhouse gas emissions from hydropower production. Traditionally hydropower was considered a clean energy source when it came to atmospheric emissions, but many recent studies challenge this view (Pearce 1996; St. Louis et al. 2000; Fearnside 2002, 2004; Rosa et al. 1997, 2004; Duchemin et al. 2002; Soumis et al. 2005; Giles 2006; Lima et al. 2008).

The goals of this chapter are twofold: first, to identify the key issues related to GHG emissions of reservoirs on a global scale, discuss the available estimates and relate them to one another; second, to present a new global estimate for GHG emissions. This is done on the basis of the available information, which is globally very sparse and so far entirely lacks data for complete climatic zones, continents, reservoir types and the like. However, a rough estimation is possible by using global databases for lakes and reservoirs as well as by making use of geographical data on ecological zoning.

Before going into the literature in detail, we first summarise the scientific basics of greenhouse gas formation in reservoirs and inundated areas. We then review what is known about the surface area of reservoirs and present the results of our database analysis, which differs substantially from many of the results presented in the literature. Three global databases are used: one for ecological zoning (EZ), one for administrative regions (AR) and another for lakes and reservoirs (Global Lakes and Wetlands Database, GLWD) (Fig. 4.1). They allow a sorting of reservoirs by ecological zones and administrative regions.

This is followed by a review of emission rates of reservoirs, and then estimates for the major ecological zones of the globe are presented. A global estimation of upstream greenhouse gas emissions from reservoirs follows. Finally, a set of conclusions and recommendations are drawn.

4 Greenhouse Gas Emissions from Reservoirs

[Flow chart figure]

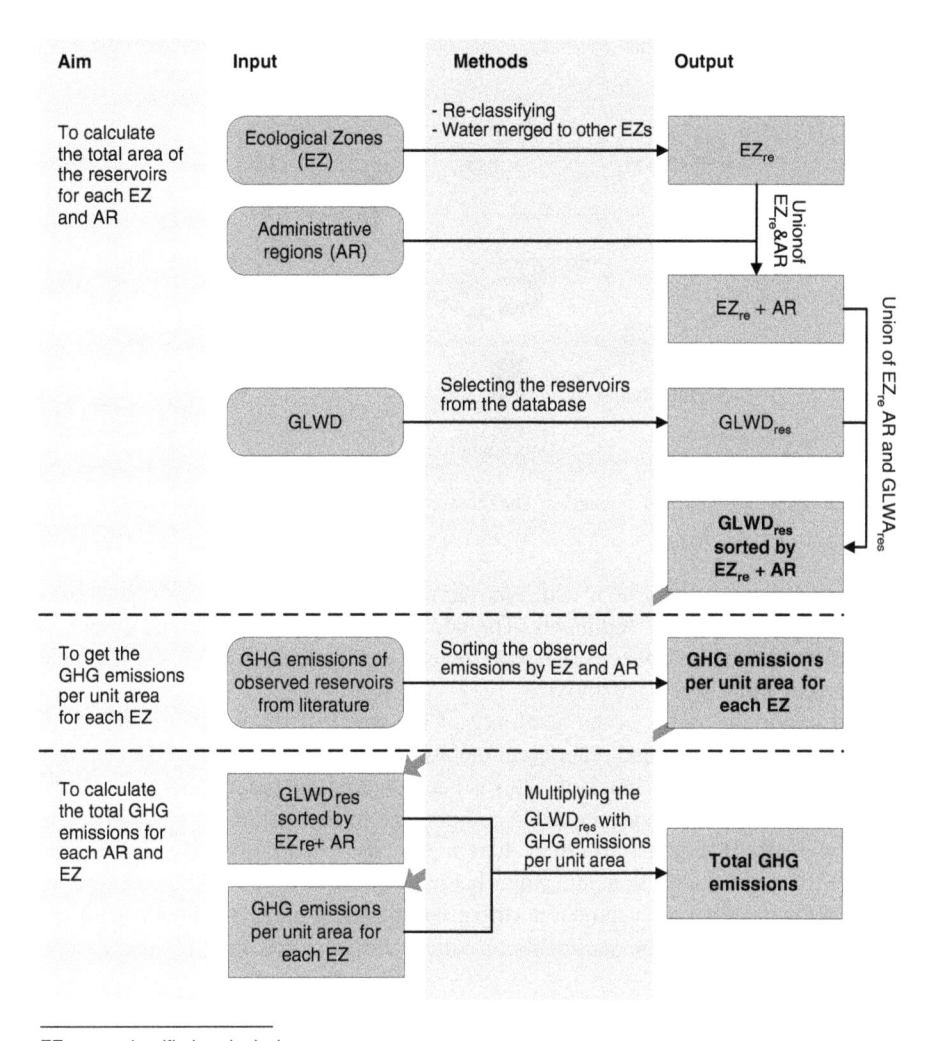

EZre = re-classified ecological zones
GLWDres = reservoirs of the GLWD database

Fig. 4.1 Flow chart of the analytic approach used in this study. *GLWD* global lakes and wetlands database, *EZ* ecological zone, *re* reservoir under study

4.2 Why Greenhouse Gases from Reservoirs?

It has been known for at least a century that natural lakes and wetlands have an active exchange of gases with the atmosphere, as summarised by Ruttner (1940). The process of photosynthesis consumes CO_2 and releases O_2 while the degradation of organic material does the opposite (Fig. 4.2). The sediment is important in this process, acting on the one hand as a sink of carbon since sedimented material is

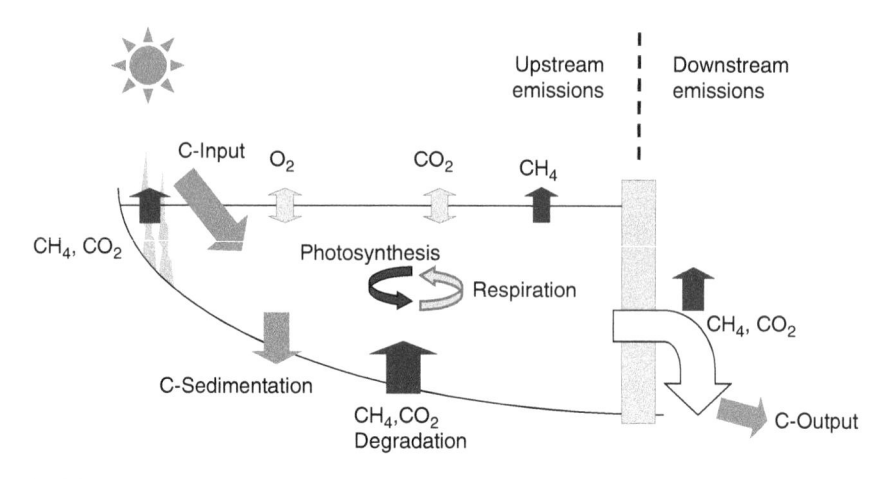

Fig. 4.2 Major gas fluxes in a reservoir. The relative role of these fluxes is highly specific

often rich in organic carbon, while on the other hand, the degradation processes release carbon from the sediments. The dominant form of this released carbon is CO_2 if the sediment surface is rich with oxygen and methane (CH_4) if the sediment and deeper water layers are anoxic.

An important player is the alkalinity of the water. If the water contains high concentrations of lime and other earth metals, the pH is high and the lake or reservoir tends to act as a carbon sink. This is because the lime precipitates to the sediment with CO_2 and forms very stable compounds. In contrast, if the water is acidic and the earth-alkaline metals are in short supply, the water body tends to emit carbon to the atmosphere. In acidic water bodies with high humic acid concentration such as the peatland reservoirs of northern Europe and Canada, the humic acids are able to precipitate to the sediments in certain amounts and thus remove carbon from circulation.

If the lake or reservoir is eutrophic, i.e. it contains plenty of nutrients and its primary production level is high, it fixes high amounts of atmospheric CO_2 and releases much of it back to the atmosphere. But a part of this carbon is sedimented, and is stored in the bottom layers. Consequently, the lakes and reservoirs act as a sink for this carbon, but in the case of allochthonous carbon (the carbon compounds that come with the inflow and are degraded in the lake), lakes act as a source of carbon for the atmosphere (Rantakari and Kortelainen 2005). Depending on the balance between these two processes, the lakes act as either sources or sinks of carbon for the atmosphere. In eutrophic conditions, the amount of settled, easily degradable organic material tends to be large enough to consume all available oxygen from the bottom layers of the water body, and the sediments release methane, phosphorus, sulphides etc., that accelerate the eutrophication process. This is usually seen as a major water quality problem in lakes and reservoirs.

The main difference between lakes and reservoirs in this regard is the following. Most reservoir bottoms, once inundated with water, contain large amounts of organic material such as wooden material, peat, other plant material and the like. This material is rich in carbon, and when it is mineralised, the carbon is released into the water body and the atmosphere. Much of this material is relatively easily degradable and leads to oxygen deficiency and methane formation, which is a common problem in the case of reservoirs. The greenhouse gas emissions of reservoirs have therefore been estimated to be in many cases far larger (as calculated per unit area) as those of natural lakes or wetlands. The estimates available vary substantially, but as a global average, the reservoirs may release 3–5 times the amount of GHGs as compared to natural lakes (Downing et al. 2006). There is enormous variation in estimates due to differences in the locations where reservoirs are being constructed, as well as due to the time that has elapsed since reservoir inundation.

The factual impact of reservoir construction on GHG emissions is the net emission, i.e. emissions in the pre-damming conditions subtracted from actual emissions. Such impact analyses are available for very few cases; they are therefore subject to rough and varying estimates and also too much argumentation.

Another major difference between natural lakes and man-made reservoirs is the following. If a reservoir has a dam and the water is released from deeper water layers, the released water might contain pressurised gases including greenhouse gases. Once the pressure drops as soon as the water exits the reservoir, large quantities of gases might be released into the atmosphere. This phenomenon, called degassing, and its contribution to total emissions of reservoirs are under heavy debate and—although the phenomenon is well known in theory—the lack of empirical studies is striking.

Even though the basic limnological and ecological mechanisms of gas balances in inland water bodies have been relatively well understood for several decades, interest in GHG emissions of reservoirs has arisen recently, as a consequence of the general growth in concern regarding global climate change. These emissions have not been studied systematically except in some specific situations, and they have not been included in the baseline monitoring, performance evaluation programmes and environmental impact analyses of reservoirs.

The GHG emissions of hydropower production have spearheaded interest in reservoir emissions. But so far not much attention has been paid to emissions of reservoirs constructed for other purposes, such as agriculture, flood protection etc.

4.3 Reservoir Surface Area

Let us first review the estimates for the number and surface area of fresh-water ecosystems. The recent study by Downing et al. (2006) came to the conclusion that the global surface area of reservoirs is around 0.26 million km^2, and that they number up to slightly over half a million (Table 4.2). This can be compared to the total number (304 million) and surface area (4.2 million km^2) of natural lakes and ponds. According to these estimates, over 3% of the earth's non-oceanic, continental area is covered by water, and 1/17 of the water-covered area consists of reservoirs and impoundments.

Table 4.2 The number, area and average size of reservoirs and impoundments of the world

Area[a] (km²)	Number	Average area (km²)	Total area (1000 km²)	Total area (%)	Comparison: area of lakes (1000 km²)
0.01–0.1	444,800	0.027	12.0	5	1294.7
0.1–1	60,740	0.271	16.4	6	523.4
1–10	8,295	2.71	22.4	9	455.1
10–100	1,133	27.1	30.6	12	392.4
100–1,000	157	271	41.9	16	329.8
1,000–10,000	21	2,706	57.1	22	257.9
>10,000	3	27,060	78.0	30	1005.8
All impoundments	515,149	0.502	258.6		4259.1

[a]Values are inclusive of lower bounds but exclusive of upper bounds
Source: Downing et al. (2006)

With regard to lake areas, these estimates are roughly twice as large as most other contemporary estimates, which range between 2.0 and 2.8 million km² in area for lakes and ponds (Meybeck 1995; Kalff 2001; Shiklomanov and Rodda 2003).

There exist far higher reservoir area estimates in the literature than those of Downing et al. (2006). The highest among these is the one by St. Louis et al. (2000), which is as high as 1,500,000 km². Shiklomanov (1993) and Dean and Gorham (1998) presented an estimate of 400,000 km², and Kelly et al. (1994) and Pearce (1996) drew figures of 500,000 and 600,000 km², respectively. Lima et al. (2008) used the ICOLD (2003) large dam database to calculate the area of large dam reservoirs at the global level to be 436,000 km². They further estimated that the area of all reservoirs, including small and large dams is 567,000 ± 48,000 km².

The differences across these estimates are an indication of the difficulty of estimating the reservoir surface area at the global level; they also make the estimation of global emissions of reservoirs more complex. Let us consider the large discrepancy between the two extremes, St. Louis et al. (2000) and Downing et al. (2006). The former estimated that the surface area of reservoirs equals that of natural lakes at 1.5 million km², whereas the latter estimated the reservoir area to be 0.26 million km² and the lake area as 4.2 million km² (Table 4.3). An obvious problem with estimates that are based on dam databases (such as that by Lima et al. 2008) is that the ICOLD database includes dams when a considerable part of dams are in natural lakes and should therefore be excluded from such an analysis. The estimate of St. Louis et al. (2000) is extremely high and unfortunately the documentation of the study is not precise enough to allow an assessment of the methodology and data used.

Although the existing estimates of reservoir surface area and unit fluxes differ dramatically, it is worthwhile to compare the approaches and definitions used in these studies. First, only Downing et al. (2006) precisely define what a 'reservoir' is, and make a distinction between impounded natural lakes and man-made reservoirs. This distinction is important since, as is the case in lake-rich Finland, the vast majority of surface water area that is dammed is in natural lakes. In such water bodies, the GHG emissions do not essentially differ from natural conditions. It is

Table 4.3 Two extreme estimates of the global reservoir and lake surface area

Reservoir area	Lake area	Reservoir–lake ratio	Source
1.5 million km^2	1.5 million km^2	1:1	St. Louis et al. (2000)
0.26 million km^2	4.2 million km^2	1:16	Downing et al. (2006)

seriously misleading to use emission estimates for such lakes along with those empirically measured from man-made reservoirs that mostly flood landscape which was originally terrestrial.

Second, the database used by Downing et al. (2006) appears far more reliable than the seemingly ad hoc approach followed by St. Louis et al. (2000) in which the earth was divided into two ecological zones, tropical and temperate. The former zone covered 40% and the latter 60% of all the world's reservoirs. No reference or data source was mentioned. Hence, we selected the analysis of Downing et al. (2006) as the starting point of our analysis. Their analysis was based on the global lake, wetland and reservoir database compiled by Lehner and Döll (2004).

4.3.1 Spatial Data

The reservoir data used in this chapter are, thus, based on the Global Lakes and Wetlands Database (GLWD) developed by Lehner and Döll (2004). The datasets GLWD-1 and GLWD-2 were used. The database includes altogether 694 reservoirs with surface areas larger than 10 km^2 and 47 smaller reservoirs with surface areas between 1 and 10 km^2. Very large reservoirs ($n=44$), with surface areas over 1000 km^2, account altogether for an area of 110,000 km^2 or 54% of the total reservoir area (205,000 km^2). The GLWD data include numerous characteristics for each reservoir. In this chapter, the following classes were used: reservoir and dam name, reservoir surface area, country and purpose. Some of the water bodies assigned as reservoirs in the first column of the datasets were then classified as open or closed lakes in later columns. These are natural lakes with a dam controlling the water flow as in Kemijärvi, Finland. These 'reservoirs' were excluded from our calculations.

The Global Ecological Zones database created by FAO (2001) and the world administrative regions (DIVA-GIS 2002) were used to identify the ecological zone and administrative region for each reservoir. The FAO (2001) database had altogether 21 EZ classes. For this study, these were collated into five EZ classes named polar, boreal, temperate, subtropical and tropical. In DIVA-GIS (2002) data, the world was divided into ten administrative regions: Antarctica, Asia, Australia, Caribbean, Europe, Latin America, North America, North Africa, Pacific (Oceania) and sub-Saharan Africa.

Each reservoir was sorted into ecological and administrative zones (Fig. 4.3). In 38 cases, a reservoir was at the border of either one or two zones. In such cases, the surface area of the reservoir was divided between these classes.

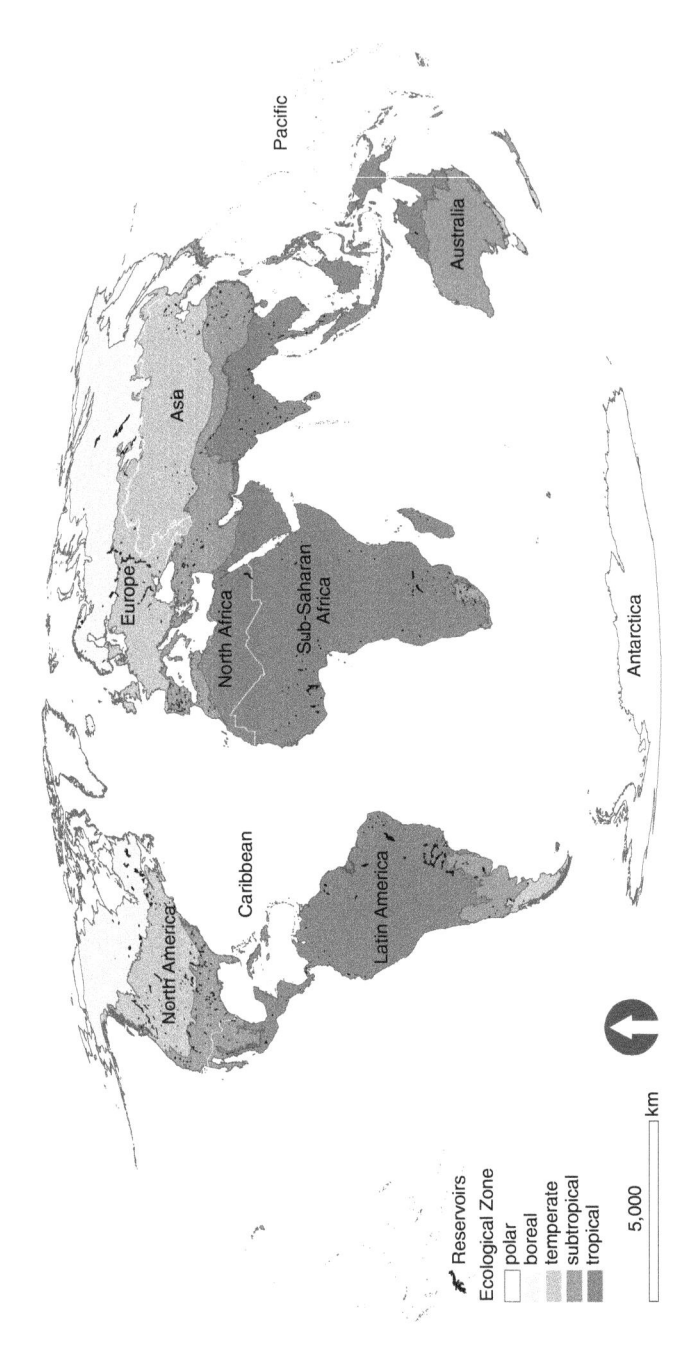

Fig. 4.3 World reservoirs plotted on top of ecological zones and administrative regions. *Source:* Lehner and Döll (2004); FAO (2001); DIVA-GIS (2002)

4 Greenhouse Gas Emissions from Reservoirs 77

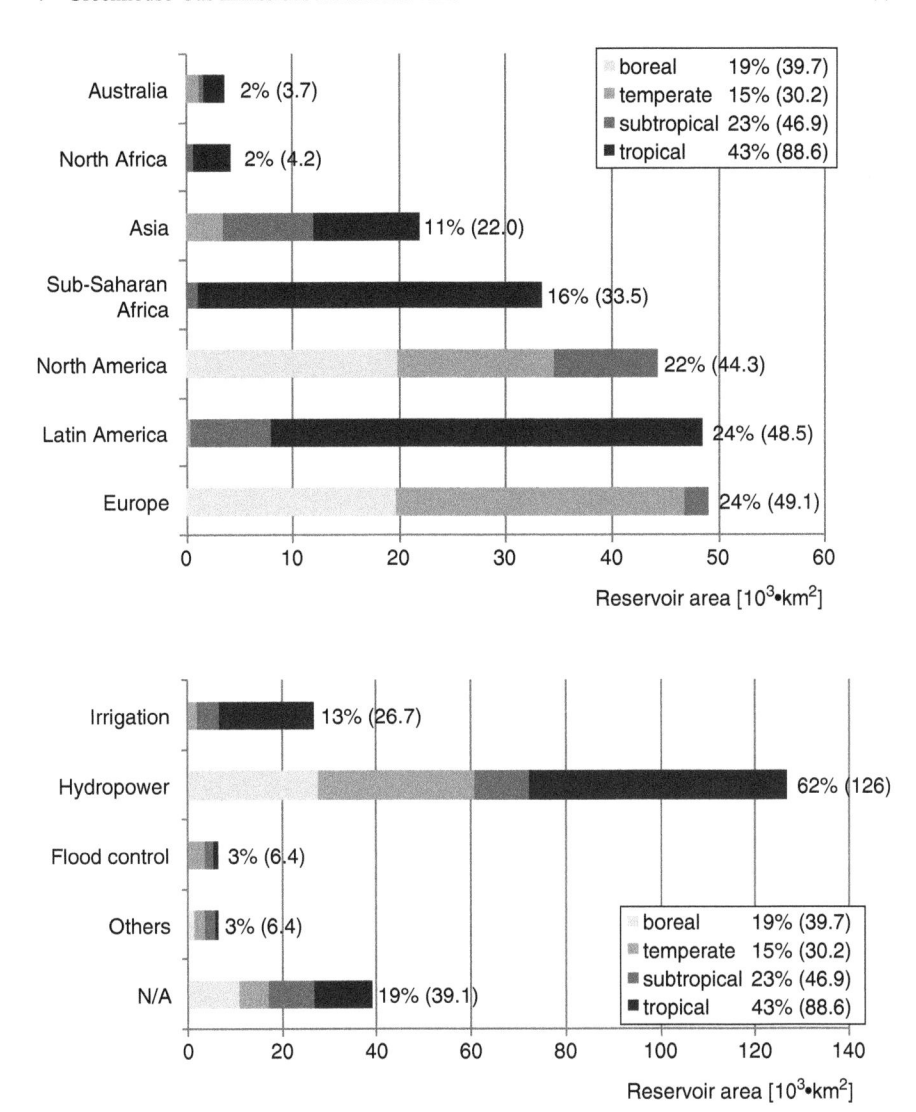

Fig. 4.4 Surface area of reservoirs sorted by administrative regions (*top*) and primary use of the reservoir (*bottom*). Each class is further divided by ecological zones. *Source*: Lehner and Döll (2004); FAO (2001); DIVA-GIS (2002)

4.3.2 Results

Europe, North America, Latin America and sub-Saharan Africa each had between 33,000 and 49,000 km² of reservoir surface area in the database (Fig. 4.4). Asia's reservoir area was a mere 22,000 km², which is astonishingly low given the large area of the continent and that Asia has altogether 39% of the world's high dams according to ICOLD (2003), whereas, for instance, Latin America has only 3%.

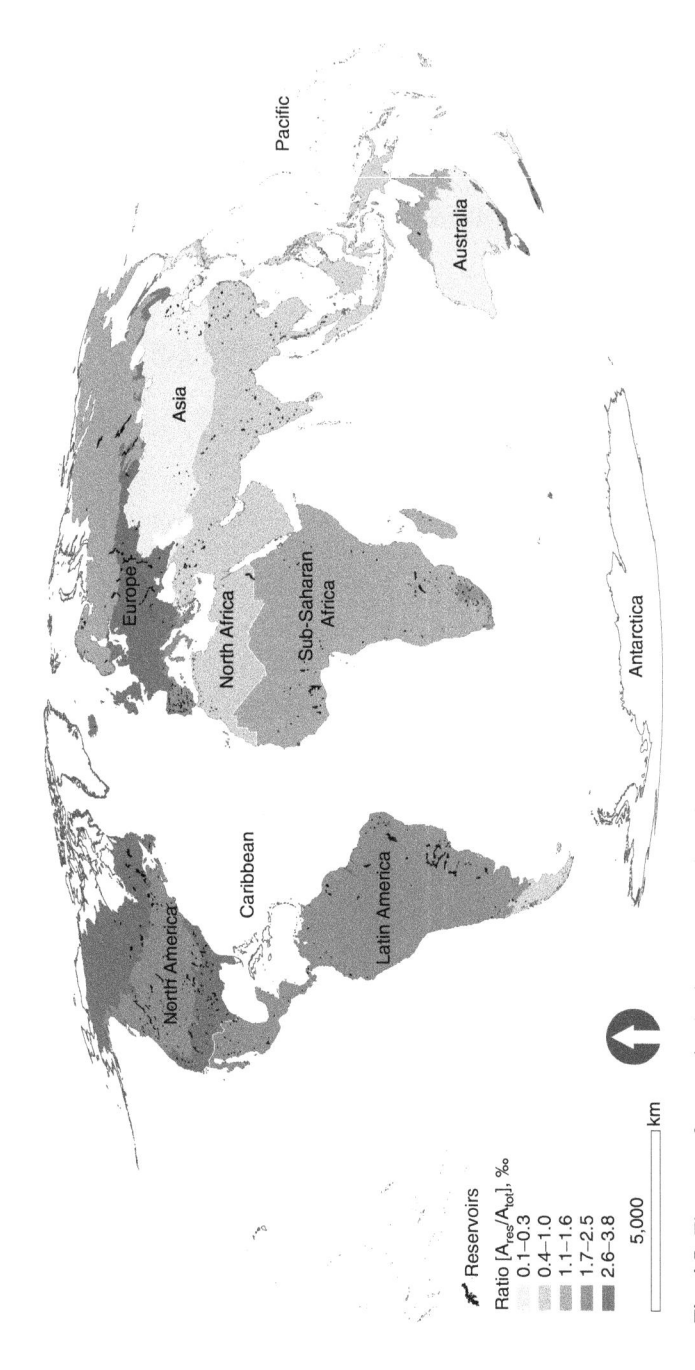

Fig. 4.5 The area of reservoirs (A_{res}) as a proportion of land surface area (A_{tot}) of administrative regions. *Source*: Lehner and Döll (2004); DIVA-GIS (2002)

It is crucial to note that the ecological zoning of reservoirs differs substantially between geographical areas. A majority of tropical reservoirs are in Latin America (40,000 km^2) and Africa (35,000 km^2), while Asia has around 10,000 km^2 of reservoir area in the tropics. The boreal reservoirs are shared between North America and Europe, with around 20,000 km^2 in each.

Hydropower reservoirs account for around 62% of the total surface area of the world's reservoirs, and irrigation reservoirs accounted for 13%. The other purposes had much smaller surface areas (Fig. 4.4). These figures were obtained by sorting the reservoirs according to their primary use. The database lists many, particularly larger, reservoirs as having several uses. However, for the sake of clarity we decided to take into account only the primary use since weighting the relative importance of each use would have been very arbitrary, given the lack of adequate information in the database.

With regard to the total land surface area of the ARs, parts of Europe and North America have the largest percentage of reservoir areas (Fig. 4.5). Latin America follows closely. Sub-Saharan Africa comes next and Asia has the lowest percentage of reservoir areas along with central Australia. Antarctica, Pacific and Caribbean do not have any reservoirs in the database.

4.4 Greenhouse Gas Emissions per Unit Area

Besides estimates of the surface area of reservoirs, we need corresponding estimates for GHG emissions per unit area. Such estimates were defined on the basis of a literature review.

GHG emission studies on 137 reservoirs were found in the literature. Table 4.6 in the Appendix lists all these reservoirs, with corresponding references. It is important to note that despite the fact that studies covered all the ecological zones relatively evenly (except the polar zone, from which there were no observations), there was a significant dominance of studies from the Americas. With the exception of two reservoirs in Finland and one in Sweden (Europe), all the studies were from either North America or Latin America. No studies were found from Africa, Asia or Australia.

The results of the literature survey are summarised in Table 4.4 and Fig. 4.6. The boreal zone was covered with 50 studies for CO_2 emissions and 39 for CH_4 emissions. The average emission rates from these studies were 1,870 mg CO_2 m^{-2} d^{-1} and 17 mg CH_4 m^{-2} d^{-1}. These values were more than twice those for the temperate and subtropical zones. The values for the tropical zone were, however, over double for CO_2 and more than six times as high for CH_4 in comparison with the boreal zone.

These results imply that special attention should be paid to methane emissions of the tropical zone. Another important feature to note is the large deviation of results in all regions, but above all in the tropical zone.

Table 4.4 Statistical summary of the emission data per unit area of the world's reservoirs (for reservoirs included in the study, and sources, see Appendix)

	mg CO_2 m^{-2} d^{-1}					mg CH_4 m^{-2} d^{-1}					mg CO_2-eq m^{-2} d^{-1}	
	Avg	Min	Max	SD	n	Avg	Min	Max	SD	n	20-year	100-year
Boreal	1870	85	5750	1190	50	17	−5.0	113	25	39	3058	2283
Temperate	550	−1190	4980	1150	43	9.0	3.0	21	6.1	10	1193	772
Subtropical	780	−1180	4790	1180	36	7.7	4.2	10	2.3	5	1331	968
Tropical	4000	−860	10400	3090	20	137	−137	1140	258	22	13862	7422

SD standard deviation, *n* number of measurements

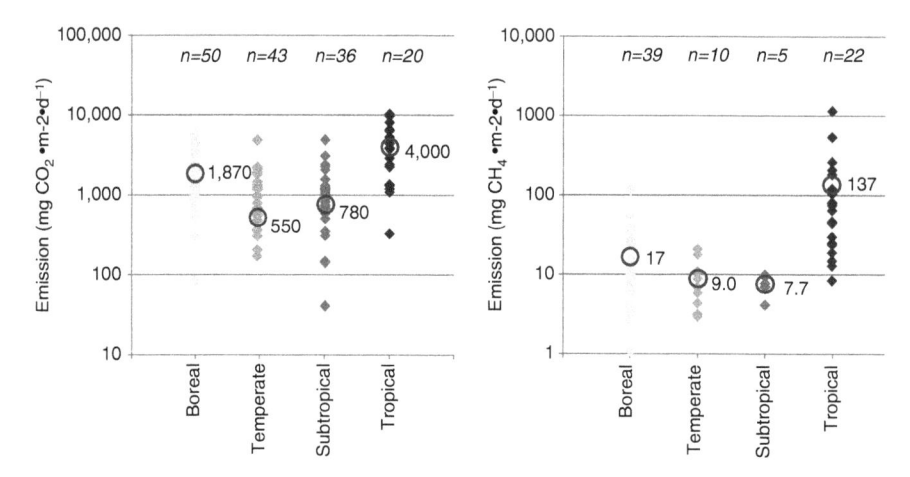

Fig. 4.6 Observed CO_2 and CH_4 emissions (for reservoirs included in the study, and sources, see Appendix). *Note*: negative observations are not shown

4.5 Total Greenhouse Gas Emissions

The method for deriving estimates for actual upstream GHG emissions from the world's reservoirs was a straightforward combination of reservoir areas and unit emissions. For the reservoirs for which we had empirical emission data, this data was used, and for the others average values documented in the previous section were used.

The gross total upstream emissions of the world's reservoirs yielded 4.8 Tg CH_4 yr^{-1} and 163 Tg CO_2 yr^{-1}. This results in 508 Tg CO_2-eq (20-year) yr^{-1} (Tg = 10^{12} g).

A number of previous estimates have been offered for global greenhouse gas emissions. To put the figures presented below in perspective, let us first mention that Houghton et al. (1996) concluded that the world's total methane emissions from all sources, either man-made or natural, would be 410–660 Tg CH_4 yr^{-1}, while the source/sink imbalance has been estimated to be about 37 Tg CH_4 yr^{-1} (IPCC 1995). Another comparative value is the estimated emission rate from natural lakes by Bastviken et al. (2004) which is 8–48 Tg CH_4 yr^{-1}. Cole et al. (2007) came to an estimated range of 0.75 to 1.65 Pg C yr^{-1} as a net contribution of inland aquatic systems to the atmosphere in the form of CO_2 (Pg = 10^{15} g).

In the analysis referred to above in the context of surface area estimates of reservoirs (Table 4.3), St Louis et al. (2000) arrived at a global estimate of 69 Tg CH_4 yr^{-1} due to reservoir emissions. This is somewhat less than the estimate of Gullenward (unpublished, preliminary results, see Giles 2006), which is represented as a range from 95 to 122 Tg CH_4 yr^{-1}. Lima et al. (2008) strikingly came to a much lower estimate of 2–4 Tg CH_4 yr^{-1}. All of these estimates were made for gross emissions from the surface of existing reservoirs (upstream emissions), thus ignoring the emission rate before construction of the dam as well as emissions after the dam (downstream emissions).

Our estimates are comparable to those estimates in both of these respects since only the gross emissions through the surface of the reservoir were estimated (with the exception of the unpublished study by Gullenward, which also includes a rough estimate for downstream emissions). When we look at the level of the estimates, our figures are close to those of Lima et al. (2008) but essentially lower than those from St. Louis et al. (2000). This is in large part due to the surface area estimates that we used which were far smaller than those used by those two analyses.

Reservoir-specific results with indications of the most remarkable emission sources are presented in Fig. 4.7. It must be noted that emissions for the reservoirs mentioned have been estimated by using unit emissions of the respective ecological zones and are thus not based on direct empirical observation from those reservoirs.

In Fig. 4.5, we presented the relative share of reservoir surface area of the total land area by administrative region. That map provides an interesting comparison to Fig. 4.8 which shows the reservoir-based GHG emissions per unit area across geographical regions of the world. The emissions from the tropics stand out clearly in these results. Latin America has the highest emissions and sub-Saharan Africa, South and Southeast Asia and northern Australia follow.

The world's reservoir-based GHG emissions are presented by geographic area in Fig. 4.9. Again, Latin America and sub-Saharan Africa dominate with combined emissions of over two-thirds of the global total from reservoirs. The shares of the other regions are considerably smaller.

As presented above, hydropower reservoirs account for about 62% of the total surface area of the world's reservoirs (Fig. 4.4). These reservoirs account for around 61% of total GHG emissions (Fig. 4.10). The GHG emissions from irrigation reservoirs are 21% of the total. When analysing the emissions for each EZ, the tropical zone dominates with 86% of total emissions while other zones account for only 4–5% each (Table 4.5).

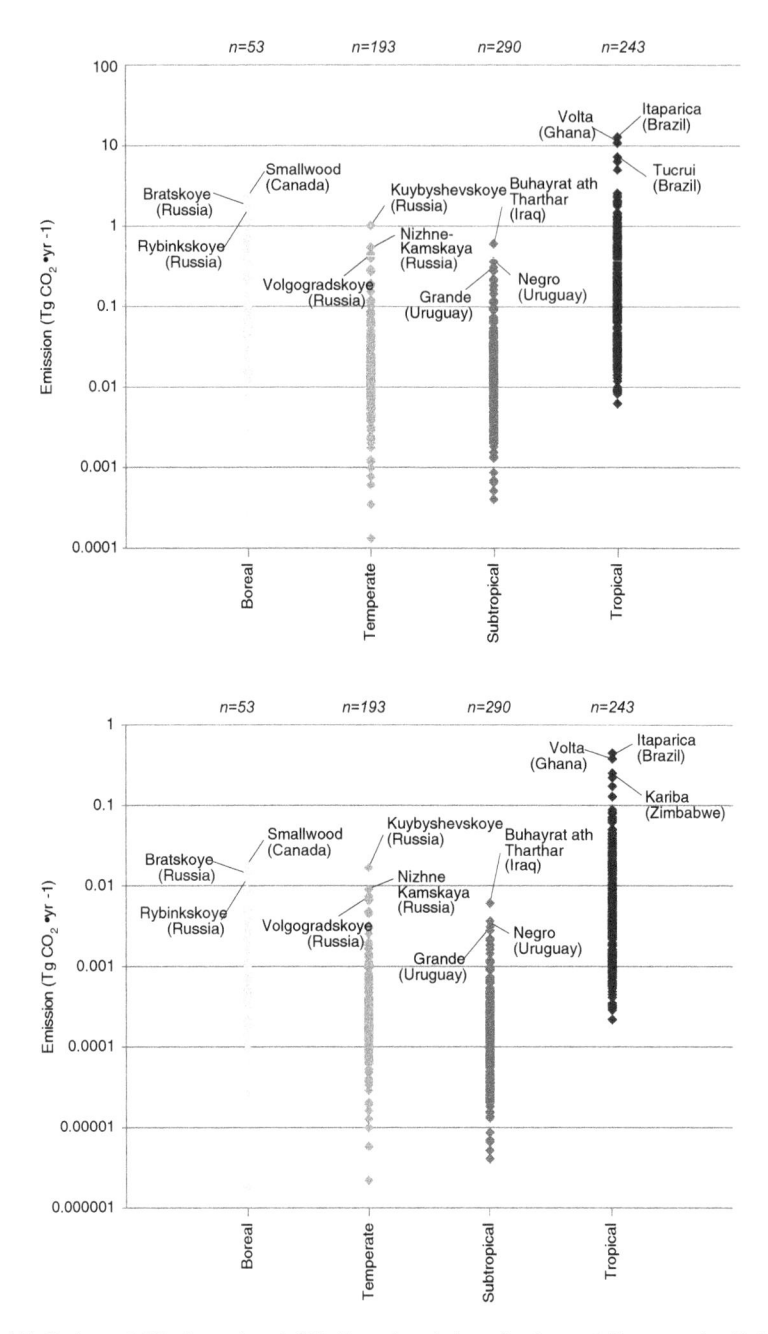

Fig. 4.7 Estimated CO_2 (*upper*) and CH_4 (*lower*) emissions for the world's reservoirs. Estimates for the largest emission sources are specifically indicated (reservoirs of the GLWD database included). *Note*: 38 reservoirs are divided between two EZs, and thus illustrated twice in the graph

4 Greenhouse Gas Emissions from Reservoirs   83

Fig. 4.8 Map of the CO_2-eq (20-year) emissions per unit area for combined EZ and AR (reservoirs of the GLWD database included)

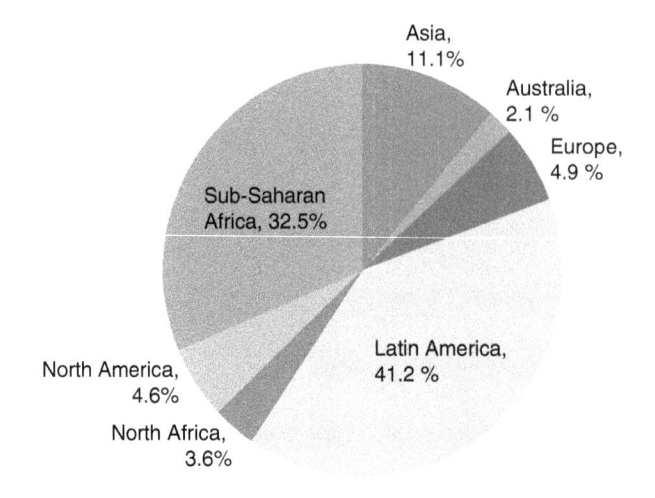

Fig. 4.9 Distribution of CO_2-eq (20-year) emissions by administrative region (reservoirs of the GLWD database included)

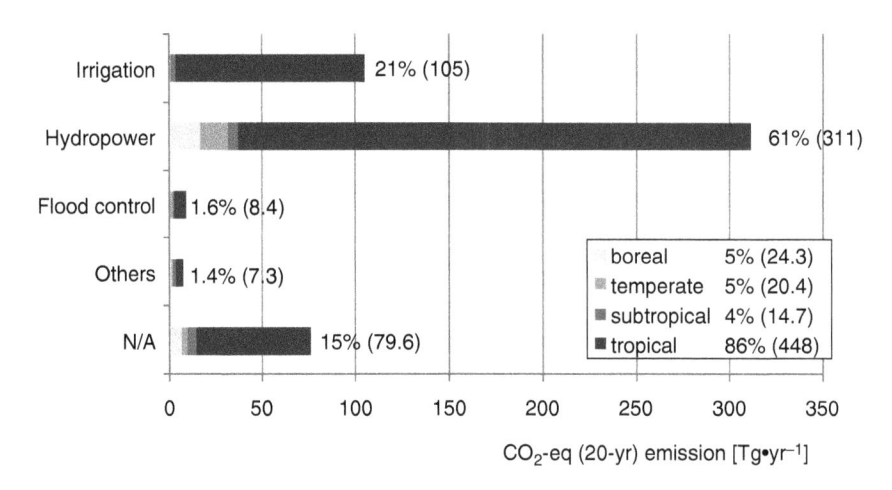

Fig. 4.10 CO_2-eq (20-year) emissions grouped by use of reservoir and EZ (reservoirs of the GLWD database included)

Table 4.5 Statistical summary of annual emission data of world's reservoirs sorted by EZ (reservoirs of the GLWD database included)

	Tg CO_2 yr^{-1}	Tg CH_4 yr^{-1}	Tg CO_2-eq (20-year) yr^{-1}	Tg CO_2-eq (100-year) yr^{-1}
Boreal	14.9	0.13	24.3	18.1
Temperate	6.0	0.15	20.4	13.2
Subtropical	13.3	0.09	14.7	10.7
Tropical	129	4.43	448	240
Total	163	4.8	508	282

4.6 Discussion and Conclusions

Man-made reservoirs are potentially an important source of greenhouse gas emissions. There is, however, little data on the topic and it is thus subject to plenty of debate and even disagreement. Better data, investigations and more unambiguous definitions are needed in order to know how significant man-made reservoirs are as a source of GHG. More importantly, they are also needed to target mitigation efforts for reservoirs, either constructed or planned, which are risky in terms of emitting large amounts of GHGs.

A number of global estimates for GHG gas emissions from reservoirs exist today, as was discussed above. The discrepancy of results is extremely wide, the most voluminous being more than tenfold the most modest one. Our estimate is at the lower end of this range, and suggests that the highest estimates might be overestimations. However, all the estimates deal with gross emissions (which are presumably far larger than net emissions) and ignore downstream emissions (which in turn ignore a part of the emissions, as pointed out by Fearnside 2002). These two issues thus contradict one another and complicate the task.

4.6.1 Paucity of Information

This analysis is based on databases that we judged as being most relevant for carrying out a global estimation of GHG emissions from reservoirs. In addition, a comprehensive literature review was carried out. The estimates that we present can only be as good as the information in these sources. It was astonishing to see how little information there was available on the entire topic. It was also very surprising to learn that studies on GHG emissions from reservoirs have been geographically concentrated on the Americas and Northern Europe. There is a burning need to extend these analyses to cover the other regions as well, particularly tropical Asia and Africa. Equally striking is the realisation that almost no studies exist on net emissions of reservoirs as well as downstream emissions. Both these aspects call for immediate scientific investigation and extensive monitoring efforts.

The studies reviewed in this context do not include measurements of nitrous oxide (N_2O) emissions of reservoirs. Therefore, we did not include that gas in our analysis. By no means does this imply that N_2O could and should be excluded from further studies. On the contrary, more information is needed on the role and importance of reservoirs as a source of N_2O.

Some of the key deficiencies in the databases that we used were the following:

- The database by Lehner and Döll (2004), as is true for all databases, needs continuous updating. The version that we used is already a few years old

and excludes a number of important new reservoirs that have been finished or are under process of construction, such as the Three Gorges Dam in China.
- Some of the definitions of reservoirs are not clear, e.g. the Finnish Porttipahta and Lokka reservoirs are listed as reservoirs in one column and as dammed lakes in another, and are thus left out from the analysis. The distinction between dammed natural lakes and man-made reservoirs should be unambiguous. Since this was not the case, for the sake of consistency we had to exclude all 'dammed lakes' although some of them are man-made reservoirs such as the ones mentioned above.
- Only the largest reservoirs are included (surface area larger than 10 km^2, plus a few of 1–10 km^2). This leaves out around 20% of the world's total reservoir area based on the global reservoir surface area estimate made by Downing et al. (2006).

4.6.2 Alert on Tropics

It seems that the tropical zone will be dominant in future increases in reservoir emissions. Therefore, paying attention to possible greenhouse gas emissions is particularly important when planning new reservoirs to the tropical zone.

Tropical reservoirs constitute the greatest future question, risk, and unknown factor for the following reasons:

- Most undeveloped potential
- Most demand
- Highest observed emissions
- High variation, very poorly known

4.6.3 What about Life-Cycle Emissions?

The construction of a reservoir is never an end in itself, but rather a means for reaching an end. Such an end is typically to gain economic benefits through improved agricultural production, energy generation, water supply for domestic and industrial purposes, control of floods and droughts etc. Therefore, an attractive approach would be to study the balances of GHGs by using the life-cycle approach, i.e. to relate balances to the total production cycle of a unit of grain, energy and so on. For instance, the global emissions of rice farming are estimated to be around 80 ± 50 Tg CH_4 yr^{-1} (Lelieveld et al. 1998). Thus, the role of irrigation dams is obviously relatively small in the entire production cycle of rice. Another example comes from hydropower. Hydropower dams are very often constructed for storing energy and providing peak power, in addition to generation of electricity. For instance, the storage capacity of reservoirs in Finland is 4,900 GWh, in Sweden 33,550 GWh and in Norway 81,489 GWh. The role of reservoirs in the entire energy distribution system should be taken into account when evaluating the environmental impacts of different energy sources.

4.6.4 Policy Recommendations

On the basis of our analysis, the key recommendations for reservoir emission investigations and monitoring programmes are the following:

- Attention should be paid to both CO_2 and particularly CH_4 emissions, but N_2O should not be neglected.
- The GHG balances should be investigated prior to the construction of a reservoir and monitored systematically after inundation so that the net emission rates can be defined. The evolution of emissions over time would also be an important issue to study.
- While emissions from the reservoir itself are an issue, an additional emission source is the water discharged from the reservoir. These downstream emissions are very scarcely investigated and there is an urgent need to track them carefully, particularly in cases in which the water is discharged from deeper water layers where the pressure is much higher than in the atmosphere and where anoxic conditions prevail.

A systematic analysis should be made of the types of reservoirs (in terms of depth, vegetation, water quality, dam type etc.) that are most risky for high greenhouse gas emissions. Equally important is a systematic analysis of the influence of water quality factors such as eutrophication level, alkalinity and concentration of humic substances to GHG balance. Special care and attention should be paid to planned reservoirs with such properties. Our results suggest that this alert is particularly important in tropical conditions. Therefore, the assessment of emissions from such reservoirs is of prime importance.

Acknowledgements This work is dedicated to the memory of our dear colleague and co-author Jari Huttunen and his ground-breaking work on greenhouse gas emissions from reservoirs. Mr. Huttunen suddenly passed away during this project.

The authors acknowledge the excellent support of the Helsinki University of Technology (now part of Aalto University), Water Resources Laboratory and its staff. Particular thanks are due to Professor Pertti Vakkilainen and Professor Asit K. Biswas for their encouragement for this project, and to Pirkko Kortelainen, Marie Thouvenot and Jukka Turunen for comments and insight on the overall topic of this work. The financial support of Maa- ja Vesitekniikan Tuki r.y. is greatly appreciated.

Appendix

Table 4.6 List of GHG observations from the reservoirs

Original reference	Reservoir name	AR[a]	EZ[b]	GHG[c]	Type of flux[d]
Åberg et al. (2004)	Skinnmuddselet	E	bo	CO_2	D
Abril et al. (2005)	Petit Saut	LA	tr	CO_2, CH_4	D, B, de and es
Duchemin et al. (2001)	Curua-Una	LA	tr	CO_2, CH_4	D and B
Duchemin et al. (1995)	Laforge 1	NA	bo	CO_2, CH_4	D and B
Duchemin (2000)	Cabonga	NA	te	CO_2, CH_4	D and B
Galy-Lacaux et al. (1997)	Petit Saut	LA	tr	CO_2, CH_4	D and B
Huttunen et al. (2002)	Lokan Tekojärvi	E	bo	CO_2, CH_4	D and B
	Porttipahta	E	bo	CO_2, CH_4	D and B
Keller and Stallard (1994)	Gatun Lake	LA	tr	CH_4	D and B
Kelly and Rudd, unpublished	Chippewa Lake	NA	te	CO_2, CH_4	D
	Day Lake	NA	te	CO_2, CH_4	D
	Moose Lake	NA	te	CO_2, CH_4	D
	Nelson Lake	NA	te	CO_2, CH_4	D
	Tigercat Lake	NA	te	CO_2, CH_4	D
Kelly et al. (1994)	Eastmain-Opinaca	NA	bo	CO_2, CH_4	D
Kelly et al. (1994), Duchemin et al. (1995), Duchemin (2000)	La Grande 2	NA	bo	CO_2, CH_4	D and B
Kelly et al. (1997)	ELARP	NA	te	CO_2	D and B
Matvienko et al. (2001)	Serra da Mesa (Sao Felix)	LA	tr	CO_2, CH_4	D and B
Rosa et al. (2002)	Tiete	LA	tr	CO_2, CH_4	D and B
	Itaipu	LA	tr	CO_2, CH_4	D and B
	Miranda	LA	tr	CO_2, CH_4	D and B
	Samuel	LA	tr	CO_2, CH_4	D and B
	Segredo	LA	st	CO_2, CH_4	D and B
	Serra da Mesa (Sao Felix)	LA	tr	CO_2, CH_4	D and B
	Repressa Tres Marias	LA	tr	CO_2, CH_4	D and B
	Tucurui	LA	tr	CO_2, CH_4	D and B
	Xingo	LA	tr	CO_2, CH_4	D and B
Schellhase et al. (1997)	Arrow (Hugh Keenleyside)	NA	te	CO_2	D
	Kinsbasket	NA	bo	CO_2	D
	Revelstoke	NA	te	CO_2	D
	Whatshan	NA	te	CO_2	D
	Dillon Lake	NA	te	CH_4	D

Table 4.6 (continued)

Original reference	Reservoir name	AR[a]	EZ[b]	GHG[c]	Type of flux[d]
Soumis et al. (2004)	Dworshak Reservoir	NA	te	CO_2, CH_4	D and de
	Franklin D. Roosevelt Lake	NA	te	CO_2, CH_4	D and de
	New Melones	NA	st	CO_2, CH_4	D and de
	Lake Oroville	NA	st	CO_2, CH_4	D and de
	Shasta Lake	NA	st	CO_2, CH_4	D and de
	Wallula	NA	te	CO_2, CH_4	D and de
Tavares et al. (1998)	Tucurui (Raul G. Lhano)	LA	tr	CH_4	D and B
Therrien et al. (2005)	Alamo Lake Reservoir	NA	st	CO_2	D
	Apache Lake Reservoir	NA	st	CO_2	D
	Arivaca Lake Reservoir	NA	st	CO_2	D
	Baker Dam Reservoir	NA	te	CO_2	D
	Bartlett Lake Reservoir	NA	st	CO_2	D
	Bill Evans Reservoir	NA	st	CO_2	D
	Brantley Lake Reservoir	NA	st	CO_2	D
	Bunch Reservoir	NA	st	CO_2	D
	Caballo Reservoir	NA	st	CO_2	D
	Cochiti Lake Reservoir	NA	st	CO_2	D
	Conchas Lake Reservoir	NA	st	CO_2	D
	Deer Creek Reservoir	NA	te	CO_2	D
	Elephant Butte Reservoir	NA	st	CO_2	D
	Green Meadow Lake Reservoir	NA	st	CO_2	D
	Gunlock Reservoir	NA	te	CO_2	D
	Havasu Lake Reservoir	NA	st	CO_2	D
	Horseshoe Lake Reservoir	NA	st	CO_2	D
	Huntingdon Lake Reservoir	NA	te	CO_2	D
	Jordanelle Reservoir	NA	te	CO_2	D
	Lake Mead Reservoir	NA	st	CO_2	D

(continued)

Table 4.6 (continued)

Original reference	Reservoir name	AR[a]	EZ[b]	GHG[c]	Type of flux[d]
	Lake Powell Bullfrog Basin Reservoir	NA	te	CO_2	D
	Lake Powell Wahweap Basin Reservoir	NA	st	CO_2	D
	Lower Charette Lake Reservoir	NA	st	CO_2	D
	Lower Enterprise Reservoir	NA	te	CO_2	D
	Lyman Reservoir	NA	st	CO_2	D
	Macallister Lake Reservoir	NA	st	CO_2	D
	Millsite Reservoir	NA	te	CO_2	D
	Minersville Reservoir	NA	te	CO_2	D
	Morphy Lake Reservoir	NA	st	CO_2	D
	Navajo Lake Reservoir	NA	st	CO_2	D
	Otter Creek Reservoir	NA	te	CO_2	D
	Parker Canyon Reservoir	NA	st	CO_2	D
	Patagonia Lake Reservoir	NA	st	CO_2	D
	Pena Blanca Lake Reservoir	NA	st	CO_2	D
	Piute Reservoir	NA	te	CO_2	D
	Pleasant Lake Reservoir	NA	st	CO_2	D
	Quail Creek Reservoir	NA	te	CO_2	D
	Robert Lake Reservoir	NA	st	CO_2	D
	Rockport Reservoir	NA	te	CO_2	D
	Roosevelt Lake Reservoir	NA	st	CO_2	D
	Upper Charette Lake Reservoir	NA	st	CO_2	D
	Upper Enterprise Reservoir	NA	te	CO_2	D
	Upperlake Mary Reservoir	NA	st	CO_2	D
	Ute Lake Reservoir	NA	st	CO_2	D
	Wallace Lake Reservoir	NA	st	CO_2	D
	Yuba Reservoir	NA	te	CO_2	D

Table 4.6 (continued)

Original reference	Reservoir name	AR[a]	EZ[b]	GHG[c]	Type of flux[d]
Tremblay et al. (2004)	Sainte-Marguerite	NA	bo	CO_2	D
	The Narrows	NA	te	CO_2	D
Tremblay et al. (2005)	Alouette	NA	te	CO_2, CH_4	D
	Arrow-Lower	NA	te	CO_2, CH_4	D
	Arrow-Narrows	NA	te	CO_2, CH_4	D
	Arrow-Upper	NA	te	CO_2	D
	Baskatong	NA	bo	CO_2, CH_4	D
	Bersimis	NA	bo	CO_2	D
	Buntzen	NA	te	CO_2	D
	Cabonga	NA	bo	CO_2, CH_4	D
	Caniapiscau	NA	bo	CO_2, CH_4	D
	Cat Arm	NA	bo	CO_2	D
	Duncan	NA	te	CO_2, CH_4	D
	EOL	NA	bo	CO_2, CH_4	D
	Gouin	NA	bo	CO_2, CH_4	D
	Great Falls	NA	bo	CO_2, CH_4	D
	Hinds	NA	bo	CO_2	D
	Jones	NA	te	CO_2	D
	Kootenay	NA	te	CO_2, CH_4	D
	La Grande 1	NA	bo	CO_2, CH_4	D
	La Grande 2	NA	bo	CO_2, CH_4	D
	La Grande 3	NA	bo	CO_2, CH_4	D
	Lac Bonnet	NA	bo	CO_2, CH_4	D
	Lac St-Jean	NA	bo	CO_2	D
	Laforge 1	NA	bo	CO_2, CH_4	D
	Laforge 2	NA	bo	CO_2, CH_4	D
	Meelpaeg	NA	bo	CO_2	D
	Meelpaeg (Granite Lake)	NA	bo	CO_2	D
	Manicouagan 1	NA	bo	CO_2, CH_4	D
	Manicouagan 2	NA	bo	CO_2, CH_4	D
	Manicouagan 3	NA	bo	CO_2, CH_4	D
	Manicouagan 5	NA	bo	CO_2, CH_4	D
	Opinaca	NA	bo	CO_2	D
	Outaouais	NA	bo	CO_2	D
	Outardes 3	NA	bo	CO_2, CH_4	D
	Outardes 4	NA	bo	CO_2, CH_4	D
	Péribonka	NA	bo	CO_2	D
	Pine Falls	NA	bo	CO_2, CH_4	D
	Pointe du Bois	NA	bo	CO_2	D
	Red Indian	NA	bo	CO_2	D
	Robert-Bourassa	NA	bo	CO_2, CH_4	D
	Robertson	NA	bo	CO_2, CH_4	D
	Sandy	NA	bo	CO_2	D

(continued)

Table 4.6 (continued)

Original reference	Reservoir name	AR[a]	EZ[b]	GHG[c]	Type of flux[d]
	Seven Mile	NA	te	CO_2, CH_4	D
	Seven Sisters Falls	NA	te	CO_2	D
	Slave Fall	NA	te	CO_2, CH_4	D
	SM 2	NA	bo	CO_2, CH_4	D
	SM 3	NA	bo	CO_2, CH_4	D
	Stave	NA	te	CO_2	D
	Toulnustouc	NA	bo	CO_2, CH_4	D
	Upper Salmon (Long Pond)	NA	bo	CO_2	D
	Upper Salmon (Cold Spring Pond)	NA	bo	CO_2	D
	Waneta	NA	te	CO_2, CH_4	D
	Whatshan	NA	te	CO_2, CH_4	D
	Williston-Finlay	NA	bo	CO_2	D
	Williston-Parsnip	NA	bo	CO_2	D
	Williston-Peace	NA	bo	CO_2	D

[a]Administrative region (*LA* Latin America, *NA* North America, *E* Europe)
[b]Ecological zone (*bo* boreal, *te* temperate, *st* subtropical, *tr* tropical)
[c]GHG measured
[d]Type of flux (*D* diffusive, *B* ebullition, *es* estuary, *de* degassing)
AR administrative regions, *EZ* ecological zones, *GHG* greenhouse gases

References

Åberg J, Bergström A, Algesten G, Söderback K, Jansson M (2004) A comparison of the carbon balances of a natural lake (L. Örträsket) and a hydroelectric reservoir (L. Skinnmuddselet) in northern Sweden. Water Res 38:531–538

Abril G, Guerin F, Richard S, Delmas R, Galy-Lacaux C, Gosse P, Tremblay A, Varfalvy L, Dos Santos M, Matvienko B (2005) Carbon dioxide and methane emissions and the carbon budget of a 10-year old tropical reservoir (Petit Saut, French Guiana). Global Biogeochem Cycles 19:GB4007

Bastviken D, Cole J, Pace M, Tranvik L (2004) Methane emissions from lakes: dependence of lake characteristics, two regional assessments, and a global estimate. Global Biochem Cycles 18:1–12

Cole JJ, Prairie YT, Caraco NF, McDowell WH, Tranvik JL, Striegl RG, Duarte CM, Kortelainen P, Downing JA, Middelburg JJ, Melack J (2007) Plumbing the global carbon cycle: integrating inland waters into the terrestrial carbon budget. Ecosystems 1:172–185

Dean WE, Gorham E (1998) Magnitude and significance of carbon burial in lakes, reservoirs and peatlands. Geology 26:535–538

DIVA-GIS (2002) Administrative boundaries for the world. Available at http://www.diva-gis.org/Data.htm. Accessed Feb 2007

Downing JA, Prairie YT, Cole JJ, Duarte CM, Tranvik LJ, Striegl RG, McDowell WH, Kortelainen P, Caraco NF, Melack JM, Middelburg JJ (2006) The global abundance and size distribution of lakes, ponds and impoundments. Limnol Oceanogr 51(5):2388–2397

Duchemin E (2000) Hydroelectricity and greenhouse gases: emission evaluation and identification of the biogeochemical processes responsible for their production. PhD Thesis, University of Quebec, Montreal

Duchemin E, Lucotte M, Canuel R, Chamberland A (1995) Production of the greenhouse gases CH_4 and CO_2 by hydroelectric reservoirs of the Boreal Region. Global Biogeochem Cycles 9:529–540

Duchemin E, Lucotte M, Queiroz A, Canuel R, Pereira H, Almeida D, Dezincourt J (2001) Comparison of greenhouse gas emissions from an old tropical reservoir with those from other reservoirs worldwide, 27th Congress SIL 1998. International Association for Applied and Theoretical Limnology, 27th Congress, pp 1391

Duchemin E, Lucotte M, St-Louis V, Canuel R (2002) Hydroelectric reservoirs as an anthropogenic source of greenhouse gases. World Resour Rev 14(3):334–353

FAO (2001) Global ecological zones. Spatial database of global ecological zones created by Food and Agricultural Organisation of the United Nations

Fearnside PM (2002) Greenhouse gas emissions from a hydroelectric reservoir (Brazil's Tucuruí Dam) and the energy policy implications. Water Air Soil Pollution 133:69–96

Fearnside PM (2004) Greenhouse gas emissions from hydroelectric dams: controversies provide a springboard for rethinking a supposedly 'clean' energy source. An editorial comment. Climate Change 66:1–8

Forster P, Ramaswamy V, Artaxo P, Berntsen T, Betts R, Fahey DW, Haywood J, Lean J, Lowe DC, Myhre G, Nganga J, Prinn R, Raga G, Schulz M, Van Dorland R (2007) Changes in atmospheric constituents and in radiative forcing. In: Solomon S, Qin D, Manning M, Chen Z, Marquis M, Averyt KB, Tignor M, Miller HL (eds) Climate change 2007: the physical science basis. Contribution of Working Group I to the Fourth Assessment Report of the Intergovernmental Panel on Climate Change Cambridge. University Press, Cambridge, UK and New York, NY, USA

Galy-Lacaux C, Delmas R, Jambert C, Dumestre JF, Labroue L, Richard S, Gosse P (1997) Gaseous emissions and oxygen consumption in hydroelectric dams: a case study in French Guyana. Global Biogeochem Cycles 11:471–483

Giles J (2006) Methane quashes green credentials of hydropower. Nature 444:524–525

Houghton JT, Meira Filho LG, Callander BA, Harris N, Kattenberg A, Maskell K (eds) (1996) Climate change 1995. The science of climate change. Contribution of Working Group I to the Second Assessment Report of the Intergovernmental Panel on Climate Change. Cambridge University Press, Cambridge

Huttunen JT, Väisänen T, Hellsten S, Heikkinen M, Nykänen H, Jungner H, Niskanen A, Virtanen M, Lindqvist O, Nenonen O, Martikainen P (2002) Fluxes of CH_4, CO_2, and N_2O in hydroelectric reservoirs Lokka and Porttipahta in the Northern Boreal zone in Finland. Global Biogeochem Cycles 16(1):3-1–3-7

ICOLD (2003) World register of dams. International Commission on Large Dams, Paris

IPCC (1995) Climate change 1994. Radiative forcing of climate change. Working Group I. Summary for Policymakers. Intergovernmental Panel on Climate Change, UNEP, Cambridge University Press

Kalff J (2001) Limnology: inland water ecosystems. Prentice-Hall, New Jersey

Keller M, Stallard R (1994) Methane emissions by bubbling from Gatun Lake. J Geophys Res 99:8307–8319

Kelly CA, Rudd JWM, St Louis V, Moore V (1994) Turning attention to reservoir surfaces, a neglected area in greenhouse studies. EOS Trans Am Geophys Union 75:332–333

Kelly CA, Rudd JWM, Bodaly RA, Roulet NP, St Louis VL, Heyes A, Moore TR, Schiff S, Aravena R, Scott KJ, Dyck B, Harris R, Warner B, Edwards G (1997) Increases in fluxes of greenhouse gases and methyl mercury following flooding of an experimental reservoir. Environ Sci Technol 31(5):1334–1344

Kelly CA, Rudd JWM, Unpublished data. Ref.: St Louis VL, Kelly CA, Duchemin E, Rudd JWM, Rosenberg DM (2000) Reservoir surfaces as sources of greenhouse gases to the atmosphere: a global estimate. Bioscience 50:766–775

Lehner B, Döll P (2004) Development and validation of a global database of lakes, reservoirs and wetlands. J Hydrol 296:1–22

Lelieveld JOS, Crutzen PJ, Dentener FJ (1998) Changing concentration, lifetime and climate forcing of atmospheric methane. Tellus Ser B 50(2):128–150

Lima I, Ramos F, Bambace L, Rosa R (2008) Methane emissions from large dams as renewable energy resources: a developing nation perspective. Mitigation Adaptation Strategy Global Change 13:1381–1386

Matvienko B, Rosa LP, Sikar E, dos Santos MA, De Fillipo R, Cimbleris A (2001) Gas release from a reservoir in the filling stage. Verh Internat Verein Limnol 27:1415–1419

Meybeck M (1995) Global distribution of lakes. In: Lerman A, Imboden DM, Gat JR (eds) Physics and chemistry of lakes. Springer, Berlin, pp 1–35

Pearce F (1996) Trouble bubbles from hydropower. New Scientist 150(2028):28–31

Rantakari M, Kortelainen P (2005) Interannual variation and climatic regulation of the CO_2 emission from Large Boreal Lakes. Global Change Biol 11:1368–1380

Rosa LP, dos Santos MA, Tundisi JG, Sikar BM (1997) Measurements of greenhouse gas emissions in Samuel, Tucuruí and Balbina Damsrsquo. In: Rosa LP, dos Santos MS (eds) Hydropower plants and greenhouse gas emissions. Coordenação dos Programas de Pós-Graduação em Engenharia (COPPE). Universidade Federal do Rio de Janeiro (UFRJ), Rio de Janeiro, Brazil, pp 41–55

Rosa L, Matvienko B, dos Santos M, Sikar E (2002) First Brazilian inventory of anthropoenic greenhouse gas emissions, background reports, carbon dioxide and methane emissions from Brazilian hydroelectric reservoirs. In: M.o.S.a. Technology (ed) Brazilian Government

Rosa LP, dos Santos MA, Matvienko B, dos Santos EO, Sikar E (2004) Greenhouse gases emissions by hydroelectric reservoirs in tropical regions. Climatic Change 66:9–21

Ruttner F (1940) Grundriss der Limnologie. Walter de Gruyter Co, Berlin

Schellhase H, MacIsaac E, Smith H (1997) Carbon budget estimates for reservoirs on the Columbia River in British Columbia. Environ Professional 19:48–57

Shiklomanov IA (1993) World's freshwater resources. In: Gleick PH (ed) Water in crisis: a guide to the world's fresh water resources. Oxford University Press, New York, pp 13–24

Shiklomanov IA, Rodda JC (2003) World water resources at the beginning of the twenty-first century. Cambridge University Press, Cambridge

Soumis N, Duchemin E, Canuel R, Lucotte M (2004) Greenhouse gas emissions from reservoirs of the Western United States. Global Biogeochem Cycles 18(13):GB2015

Soumis N, Lucotte M, Canuel R, Weissenberger S, Houel S, Larose C, Duchemin E (2005) Hydroelectric reservoirs as anthropogenic sources of greenhouse gases. In: Lehr JH, Keeley J (eds) Water encyclopedia: surface and agricultural water. Wiley Interscience, New York, pp 203–210

St Louis VL, Kelly CA, Duchemin E, Rudd JWM, Rosenberg DM (2000) Reservoir surfaces as sources of greenhouse gases to the atmosphere: a global estimate. Bioscience 50:766–775

Tavares de Lima I, Novo E, Ballester M, Ometto J (1998) Methane production, transport and emissions in Amazon hydroelectric plants, 27th Congress. International Association for Applied and Theoretical Limnology, Book of Abstracts, Dublin

Therrien J, Tremblay A, Jacques R (2005) CO_2 emissions from semi-arid reservoirs and natural aquatic ecosystems, greenhouse gas emissions — fluxes and processes: hydroelectric reservoirs and natural environments. Springer, New York, pp 233–250

Tremblay A, Lambert M, Gagnon L (2004) Do hydroelectric reservoirs emit greenhouse gases? Environ Manage 33(Suppl 1):509–517

Tremblay A, Therrien J, Hamlin B, Wichmann E, LeDrew L, Tremblay A, Therrien J, Hamlin B, Wichmann E, LeDrew L (2005) GHG emissions from Boreal Reservoirs and natural aquatic ecosystems, greenhouse gas emissions — fluxes and processes: hydroelectric reservoirs and natural environments. Springer, New York, pp 209–232

Chapter 5
Impacts of Dams in Switzerland

Walter Hauenstein and Raymond Lafitte

5.1 Introduction

In Switzerland, most dams are built to create reservoirs for the production of hydropower, some for seasonal and some for weekly or daily compensation between the pattern of run-off and that of electricity demand, respectively. Only a very small number of dams have been built for flood retention, even though, in fact, many of them are playing that role. Unlike the situation in most other countries, especially in dry areas, almost no dams in Switzerland are used for irrigation. The annual distribution of rainfall allows for agriculture without any significant irrigation and, if the necessity arises, water is taken directly from the rivers, which generally have enough run-off even in periods when there are agricultural requirements for water.

Thus, dams generate obvious benefits, such as their capacity to store water for drinking, industrial or agricultural use or for hydropower. These benefits justify their construction. But dams have further positive impacts on the environment and society which are much less obvious, as well as negative impacts. The latter are often highlighted by some environmental associations and by the media, which dwell on conflicts, while the positive aspects are thrust into the background. (Bischof et al. 2000).

In fact, very little is known about these secondary positive and negative aspects of dams. They usually do not show up in the account books of companies operating dams. Therefore, an unbiased evaluation of these aspects is only possible if external costs and benefits are included in the discussion.

In this chapter, our aim is to provide a short description of the direct and indirect or external benefits and costs of dams in our country. We are also presenting the case study of the Hinterrhein Development.

W. Hauenstein (✉)
Swiss Association for Water Resources Management, Baden, Germany

R. Lafitte
Ecole Polytechnique Fédérale de Lausanne, EPFL - ENAC - ICARE- LCH,
Laboratoire de Constructions Hydrauliques, Lausanne, Switzerland

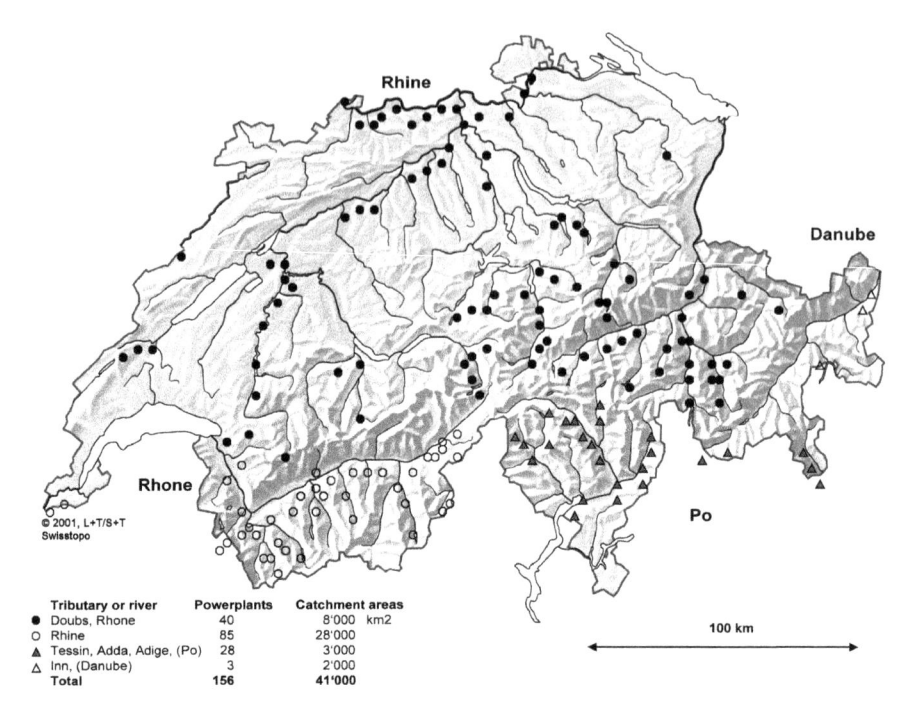

Fig. 5.1 Location of Swiss hydropower plants (>10 MW) and dams. (The Swiss dams in the catchment area of rivers Danube and Po are not located on these rivers but on their upstream affluent)

5.2 Characteristics of Swiss Dams

As mentioned before, most dams in Switzerland are built in connection with hydropower plants (Bundesamt für Energie 2007; Vischer and Sinniger 1998).

Today there are 151 large dams higher than 15 m in Switzerland (Fig. 5.1). About 80% of them are located in mountainous regions, which cover 40% of the country's surface. With Switzerland's total surface area of 41,293 km^2, the dam density in the country is about 4 dams per 1,000 km^2. The country's particular geographic position, in the centre of the European hydrographical net, explains why Switzerland does not have large rivers with important discharge, but offers a very large hydropower potential because of the high head available. The mean annual precipitation rate is about 1,470 mm, consisting of snow in winter at higher altitudes. River discharge is higher in summer than in winter. It has therefore been necessary to build large, high-altitude reservoirs to stock water for optimal use during the winter months, when the demand for electrical energy is higher.

Most of the large Swiss dams were built between 1950 and 1970 (Fig. 5.2). At the time, the construction of these dams was considered to be a major national project and a key tool to supply energy from the country's main resource, hydropower, which, along with forestry, is the only natural resource of the country. Nowadays,

5 Impacts of Dams in Switzerland

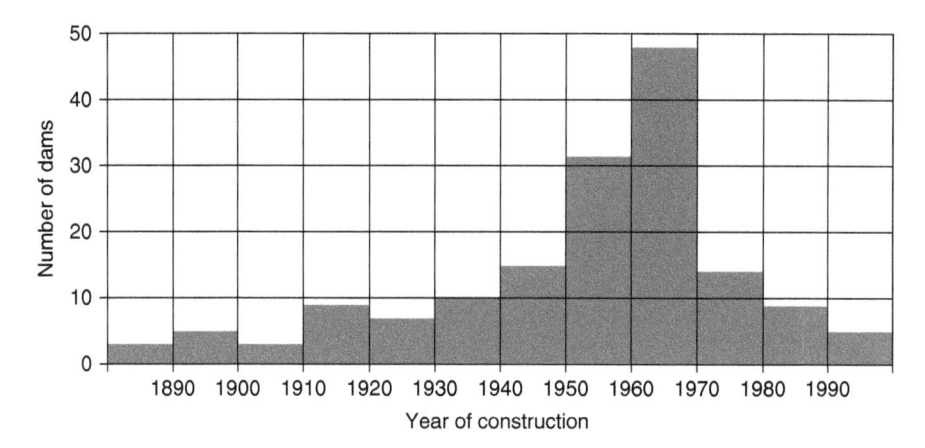

Fig. 5.2 Year of construction of major Swiss dams

Table 5.1 Top-ranking dams in Switzerland in terms of different characteristics

Dam	Year of completion	Storage volume (MCM)	Lake surface (km^2)	Height (m)	Concrete volume (MCM)	Volume of embankment (MCM)
Grande Dixence	1961	400				
Emosson	1974	225				
Sihlsee	1936/1937		10.85			
Lac Gruyère	1947		9.6			
Grande Dixence	1961			285		
Mauvoisin	1957/1990			250		
Grande Dixence	1961				6.0	
Mauvoisin	1957/1990				2.1	
Mattmark	1960					10.5
Göscheneralp	1967					9.3

Note: MCM = million m^3

about 90% of the technically feasible hydropower potential in operation is being harnessed.

Most dams in Switzerland are concrete structures. Embankment dams are mostly less than 50 m high. Due to the topographic characteristics of the Alps, the dams are relatively high and the corresponding storage volume is relatively small. Some typical figures for the top-ranking dams according to different characteristics are shown in Tables 5.1 and 5.2.

It is generally thought that the major environmental impacts of dams are linked to the reservoir area. The reservoir areas in Switzerland are less than around 10 km^2, and this is the case for 95% of large dams in the world. Indeed, we are far from very large reservoirs such as Tarbela (243 km^2), Atatürk (820 km^2), Three Gorges (1,084 km^2) and certainly from enormous ones such as High Aswan (6,700 km^2).

Table 5.2 Statistics of Swiss dams (>15 m)

Height	Number	VA	PG	TE/ER	% of total VA	PG	TE/ER
>200 m	4	3	1	0	75	25	0
150–200 m	5	4	0	1	80	0	20
100–150 m	16	12	1	3	75	6	19
50–100 m	29	16	12	1	55	41	4
15–50 m	97	18	50	29	19	51	30
Total	151	53	64	34	35	42	23

VA arch dam, *PG* gravity dam, *TE/ER* earth or rock-fill dam

To summarise, Swiss dams are mainly used for hydropower; they are high with a relatively small storage capacity and a small reservoir area; they are mainly made of concrete and are becoming old; and they are quite homogeneous in their impacts. These impacts are generally almost identical to those of power plants as a whole. Therefore, the following analyses of the impacts of Swiss power plants can be considered as representative of the impacts of Swiss dams.

5.3 Direct and Indirect Benefits and Costs

The inclusion of direct benefits and costs is relatively easy and can be done quite accurately, since the corresponding elements are part of the account books of the owner companies.

Indirect benefits and costs are more difficult to account for. They can be quantified in terms of external benefits and costs, as outlined below:

- External benefits are the advantages a company provides for a third party and for which the beneficiary does not pay any compensation.
- External costs are the damages or disadvantages a company causes to a third party and for which the sufferer does not receive any compensation.
- In the majority of cases, the sufferer or the beneficiary is the general public.

The consideration of externalities serves two purposes:

- Together with the already internalised costs, they provide a true picture of what the market price of a product, in this case the electricity, should be. As long as the external costs and benefits of the production and distribution of a company's product are not internalised, it is impossible to obtain a true picture of the performance of the product. The market price of the product is falsified.
- Since externalities are mainly related to the environmental and socio-economic fields, they can be considered as a benchmark for the sustainability of a product.

5 Impacts of Dams in Switzerland

Table 5.3 Share of different energies in Switzerland in the year 2006 (Schweizerische Energiestatistik)

Total	Fuel (heat, transportation)				Electricity		
	Oil	Gas	Coal	Other	Hydro	Nuclear	Other
Share of source in %	56.1	12.0	0.7	7.8	12.3	9.9	1.2
Value in billion CHF	17.2	2.2			8.5		

The economic estimation of external effects is particularly difficult. An attempt to do so for Swiss hydropower has been published in Hauenstein et al. (1999) and Dettli et al. (2000).

5.3.1 Direct Benefits and Costs

The direct benefits of dams in Switzerland are primarily the revenues of electricity produced and sold by the power plant associated with the dams. The role of dams is not to increase the amount of electricity produced, but to improve its quality. They cannot, of course, increase the amount of water to be turbined, but they do enable the stored water to be turbined in winter, when natural flow is low and electricity prices are high. Therefore, the benefit of dams must be measured by taking into account the surplus value of this aspect.

In order to evaluate these direct benefits, we must take a look at the economy of the Swiss electricity industry.

5.3.1.1 Swiss Electricity Economy

The total consumption of primary energy according to the different sources of energy in Switzerland is shown in Table 5.3.

The annual production of electricity (in 2006) is 62 TWh (terawatt-hour = 10^{12} watt-hour) and the consumption is 58 TWh for a population of 7.5 million inhabitants, that is to say, 7,730 kWh per person. In an average year hydropower represents 34 TWh/year (56%), nuclear 25 TWh/year (41%) and the other sources 2 TWh/year (3%). As mentioned before, 90% of the economically feasible hydropower potential has been developed. The total installed capacity is 13,200 MW. Some 513 power plants have an installed capacity of more than 0.3 MW each, and an additional 800 plants have less than 0.3 MW each.

In the year 2006 consumers in Switzerland paid CHF 8.5 billion (Swiss franc) for their electricity bill, with hydropower representing some CHF 5 billion. The gross national product of the same year was CHF 480 billion.

5.3.1.2 Investments

From the end of the nineteenth century to 2000, the total cumulative investment in production, transport and distribution of electricity in Switzerland was roughly CHF 55 billion (CHF 1 = Euro 0.64; 2000 price). The total cumulative investment in 1950 was around CHF 5 billion, corresponding to a present value of some CHF 18 billion. The value of the production plants (hydropower and nuclear) corresponds to roughly 25–35% of the total value of all installations.

The owners of Swiss hydropower plants are public corporations (80%) and private companies (20%). In a few decades, the majority of hydropower plants will come to the end of their concessions (see section on Legal Basis below). At that time, they will be financially written off, even though many parts will not be at the end of their life and thus still have a physical value. This value will be free for the governments and will thus represent a considerable gain for the public.

It is important to note that about 20,000 employees are involved in electrical energy activities.

5.3.1.3 Method of Financing

Financing of investments in hydropower in Switzerland has mainly been done within a free economy (with a competitive profit system). This means that even when public corporations are owners of power plants, they are usually only involved as shareholders of a limited company. The power plants themselves are run as limited companies, often with various partners as shareholders. The advantage of this system is the clarity of its organisation, obligations and liabilities. The flexibility of operation is comparable with other industrial and commercial activities. The system guarantees that profitability is judged in a politico-economic way (covering energy needs, creation of jobs, development of national wealth) without assigning a major priority to financial payoff and that no resources of the state are used, apart from the share capital.

The capital necessary for construction of the schemes can have two origins:

a. The owner's own funds: The owner invests his money and takes the risk of the venture.
b. Loans from third parties: These loans are usually obtained from banks or insurance companies or arranged through the issuing of bonds to the public. These loans have contractually fixed interest rates and must be reimbursed within a defined agreed period.

The more risky an operation, the higher should be the share of the owner's funds. The construction of hydropower plants in Switzerland has been considered to be a safe operation and the owner's funds are of the order of 20% of the total investments.

The possibility of raising money from the public was the result of political stability and a sound economy.

5.3.1.4 Legal Basis

According to the federal constitution of Switzerland, the right to utilise a public river is the responsibility of a canton. The federal government makes sure that a rational use of the water resources is guaranteed. The owner of a river (usually a public corporation) can use the water itself or transfer the right to a third party. In this case a concession is granted.

During the process of establishing a concession to use water, the community takes care of the public interest by ensuring that the water use is not in conflict with the laws. An environmental impact study is carried out to ensure that the effects of the scheme on the environment are within the legal limits. In a second step, based on the licence, a granting procedure concerning the authorisation to build makes it possible to ensure that the scheme will be constructed in accordance with conditions laid down in the concession. The granting procedure can be followed by a process of expropriation if it appears, according to the law, that the public interest predominates those of possible opponents.

The processes for the concession and the authorisation to build are rather complex. They contain public inquiries which allow the opponents to present their views. Agencies who think that their rights are prejudiced or who assert an infringement of public or private interests are authorised to take part in the process. The granting authority decides on the concession after considering the arguments of the opponents. The correctness of the decisions can be contested with the ordinary courts. The general result of this legal process is that the concession for and the authorisation to build hydropower plants are based on a democratic process, with an extensive involvement of the concerned public.

5.3.1.5 Profitability

An assessment of the profitability of a planned installation is the result of a comparison between the probable costs and selling prices of the product.

Production costs for hydropower—An important part of the operation expenses of a hydropower plant consists of financial charges for the invested capital (interest on loans, repayment). The initial costs mainly comprise the construction costs, as well as fees for studies, expenses for commissioning, rights and indemnities, and financing. Furthermore, fixed costs also include personnel, administrative expenses, taxes, etc. Finally, there are also variable costs to be taken into account, such as energy consumed, maintenance, etc.

Since capital costs are predominant, it is important to know the duration available for write-off. This can be as long as the duration of the concession (up to 80 years) in the case of concrete work, less for the machinery, and even less for the control system.

The actual total mean production cost for hydropower (operating and capital) in Switzerland is about CHF 0.06/kWh.

Selling price of electricity—In a free economy, the selling price varies with variations in supply and demand. In Switzerland, however, electricity supply is still monopolised (until January 2009) for customers with more than 100 MWh/y (megawatt-hour = 10^6 watt-hour) consumption. For customers with a consumption of less than 100 MWh/y, the market will be opened on an optional basis in 2014. So far, in a given area only one company was allowed to distribute electricity to consumers. Thus, a market price does not exist (yet) in Switzerland. At present, consumers pay about CHF 0.15 to 0.20/kWh. This price includes, beside production, the transportation and distribution costs. Generally, producers do not sell their product directly. They often belong to a vertically integrated utility, and are reimbursed for the costs of their electricity. Nevertheless, market prices exist in other European countries around Switzerland. Today's expectations of future price developments show that Swiss hydropower plants can be competitive at the estimated level of selling prices.

Conclusion on profitability—During the last 50 years, hydropower plants have been profitable, and the owner companies have been able to:

- Pay lenders the interest rates fixed by market rules.
- Write off part of the investment, according to the fixed duration of the concession.
- Distribute dividends to shareholders.
- Make reserves.

Shareholders and lenders, and among them, Swiss people with subscribed bonds, have made profits. The electricity consumers have also been satisfied quantitatively as well as qualitatively.

5.3.1.6 The Value of Power

Another important direct benefit of hydropower from high-head schemes with storage facilities is the availability of power. Without storage facilities, the availability of power would be much lower, since the natural run-off during winter is much smaller than in summer, and thus much smaller than the installed capacity for power production. Therefore, this capacity could not be used to its full extent in the absence of a reservoir.

As was mentioned above, an electricity market does not yet exist in Switzerland. As a result, there exists no domestic price for services of hydropower plants for the regulation of the grid: the availability of power is not yet priced. As an indicator of the value of power availability in an open market, we choose the requisite investment costs for a gas turbine power plant, which runs to some Euro 450/kW. Thus there is a considerable potential for income in these services which will be reimbursed in future.

5.3.2 Typical External Costs and Benefits of Hydropower

External costs and benefits are observed during the construction and operation of power plants. (Hauenstein 2000). The construction phase and its related effects are

5 Impacts of Dams in Switzerland

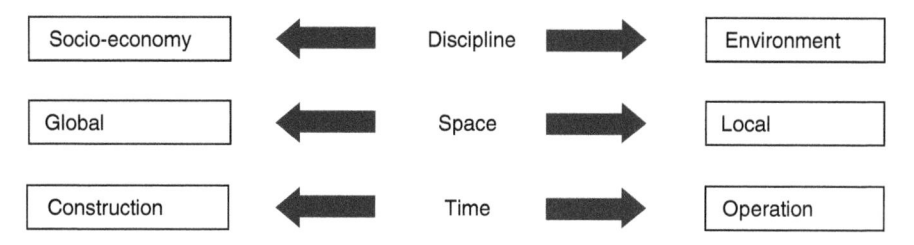

Fig. 5.3 Fields of external costs and benefits

Table 5.4a Typical externalities of hydropower during construction

Activity	Space	Field	Benefits	Costs
Construction work	Global	Environment		Atmospheric pollution, CO_2 emissions due to traffic, construction activities and provision of raw materials, noise
Material supply	Global	Environment		Atmospheric pollution, CO_2 emissions due to manufacturing of raw materials
Earthworks	Local	Environment		Change of landscape due to river diversions, quarries or deposits
Material supply	Global	Environment; Socio-economy		Loss in stock of (rare) raw materials and non-renewable energy resources
Investment activity	Local	Socio-economy	Indirect benefits for local and regional economy (restaurants, shops, etc.) due to construction activities	

of a relatively short duration, while the effects of plant operation continue for many decades. In general, the external effects impact the environmental and the socioeconomic fields. A further distinction has to be made between local and global effects. Figure 5.3 provides an overview of the different domains of external effects.

The typical externalities of hydropower are found in the local field. They concern the river and its banks within reach of the plant. The external costs of most power plants based on other sources of energy, e.g. fossil, are located in the global domain (air pollution, climate). This difference complicates the comparison of the overall sustainability of various sources of energy.

Some typical externalities of hydropower are shown in Tables 5.4a and 5.4b.

While a high number of external costs prevail in the construction phase, hydropower generates quite a large number of external benefits during the operation period.

Besides the 'ordinary' impacts mentioned above, impacts from decommissioning or caused by major accidents should also be mentioned. So far dam decommissioning

Table 5.4b Typical externalities of hydropower during the operation phase

Activity	Space	Field	Benefits	Costs
Use of renewable resource	Global	Environment	Lack of external costs inherent in non-renewable energy resources (such as CO_2 emissions, air pollution, use of primary energy resources, etc.)	
Damming up of river	Local	Environment	Creation of new habitats due to the damming of river stretches, resulting in larger water surface and reduced velocities	Loss of existing habitats due to the damming of river stretches, resulting in reduced flow velocities and changed river bed structure
Damming of river	Local	Environment	Change of landscape (positive or negative, depending on personal view)	
Interruption of flow continuum	Local	Environment		Loss of existing habitats due to the interruption of the flow continuum
River diversion	Local	Environment	Creation of new habitats due to the construction of artificial water bodies, artificial canals, etc.	Loss of existing habitats due to the diversion of water from the river to artificial canals, etc.
River diversion	Local	Environment	Change of landscape (positive or negative, depending on personal view)	
Operation of storage plants	Local	Environment		Artificial flow variations due to start and stop of plant, with short cycles, have negative effect on fish habitats
Water storage	Local	Environment	Change in run-off pattern of downstream rivers, generally a reduction in summer and increase in winter, with positive and negative effects	
Operation of storage plants	Local	Environment; Socio-economy	Flood protection due to retention of water (typically, flood peaks can be reduced by 10–20% with the existing plants in Switzerland)	
Investment activities	Local	Socio-economy	Indirect benefits for local and regional economy (employment in regions often not favoured by industry, use of new access roads for agriculture and forestry, etc.) due to presence of power plant	

Table 5.4b (continued)

Activity	Space	Field	Benefits	Costs
Operation and 331.654 maintenance of access roads	Local	Socio-economy	Potential for tourist activities in areas otherwise not easily accessible	

is not an issue in Switzerland, with the exception of the decommissioning of a few very small and old dams which had been out of operation during many years. The effects of major accidents, such as earthquakes and flood, have been taken into account in the analyses of indirect costs and benefits of dams. However, it has been found that the externalities resulting from such extremely rare incidents do not count as much as the ordinary impacts mentioned above. Therefore we will not go further into the details of these impacts in this chapter.

5.3.3 Description of Some Typical External Benefits

5.3.3.1 Renewable Character

The most evident and positive environmental effect of hydropower is its renewable character. Moreover, the generation of electricity does not require any fuel or release any pollutants or CO_2 into the air during operation. This effect cannot be considered as an external benefit, but rather as avoidance of the external costs inherent in energy produced with fossil fuels.

The energy and material balance of hydropower throughout its life cycle are extremely good. The harvesting factor, i.e. the ratio between energy gained and input of non-renewable energy is usually higher than 100. Raw material is only needed during construction and for rehabilitation work. These materials are mainly common ones, such as steel, cement, water, stone and sand.

5.3.3.2 Habitats, Landscape

The installation and operation of hydropower plants always implies modifications to habitats and landscape. In fact, these two aspects are the most relevant local effects of hydropower on the environment. The loss of existing and the creation of new habitats or the modelling of a new landscape always have a positive and a negative side. One important factor to consider when assessing these effects is how numerous and how large the lost habitats or landscapes are in the affected area. The following list indicates in qualitative terms some of the most common effects of hydropower on the habitats in or along the rivers.

Upstream of weirs, the river morphology changes: flow velocities are reduced, the river bed is covered with fine materials such as sand and silt instead of gravel, and

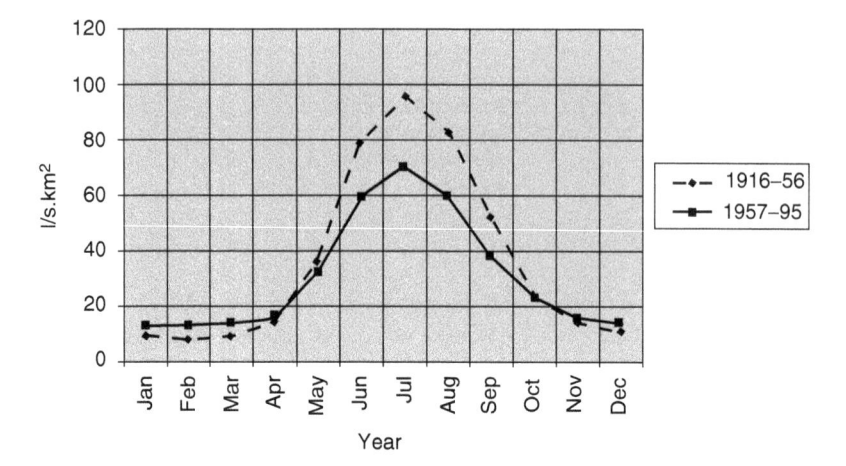

Fig. 5.4 Mean monthly flow in l/s·km² of catchment area, for the River Rhône in Sion, before and after the beginning of operation of major storage schemes

migration is more difficult for the fauna due to dams and weirs. Thus, salmon and other species of fish have disappeared from the Swiss rivers, among others, because migration to and from the sea became more difficult and the structure of river beds was no longer favourable to reproduction. The reduced dynamics of the rivers due to damming and diversion induce further changes in habitats along the rivers.

On the other hand, the new water surfaces of artificial reservoirs encourage thousands of birds to halt temporarily during their migration to the south, or for the entire winter. Some of these storage lakes have become natural reserve areas of more than local importance. Nevertheless, it cannot be denied that the negative effects in respect of the loss of habitats prevail on the positive effects.

In parallel with the modification of habitats goes the modification of landscapes. Untouched 'wild' river flow areas disappear and more fjord-like areas with calm water flow are created. Again, it is very difficult to assess the values of such changes. Are the created landscapes more valuable than the lost ones? Existing studies show that the positive effects might compensate the negative ones. The buildings themselves are generally quite well integrated in the surrounding environment or placed underground. Their influence on the scenery is therefore not considerable.

5.3.3.3 Seasonal Variation of Run-off

The flow pattern of rivers in Switzerland is marked by high discharges in summer and low flow in winter. In the mountainous regions, the difference between summer and winter is more accentuated than in the lower areas, due to the low temperatures which immobilise water in winter by freezing it to snow or ice.

What is the influence of hydropower on this flow pattern? The two curves in Fig. 5.4 show the change in annual variation of flow due to the construction of

5 Impacts of Dams in Switzerland

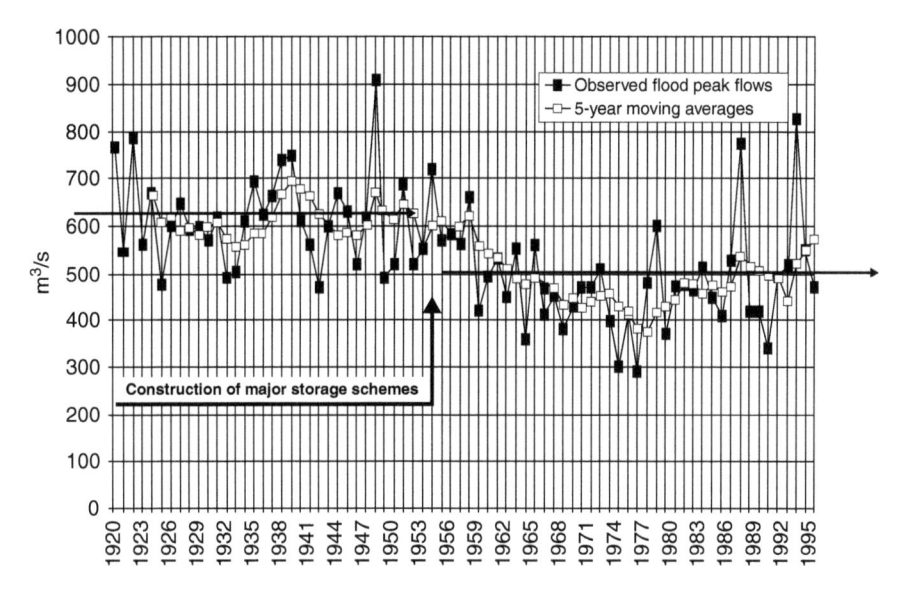

Fig. 5.5 Annual peak discharges of the Rhône River before and after the beginning of operation of major storage schemes, 1916–1995

hydropower plants with reservoirs. In the period from 1916 to 1956, before the major power plants were put into operation, winter flows were lower and summer flows higher than is the case today, as shown by the period 1957 to 1995. Thus, due to the operation on a seasonal basis of hydropower plants with storage capacity, some of the natural flow in the rivers has shifted from summer to winter. The run-of-river schemes instead have no influence at all on the seasonal variation of flow in the rivers.

This shift in flow pattern has no significant adverse effects in Switzerland. On the contrary, an increased recharge of groundwater flow due to increased river flow in winter time is welcome. In countries with a need for irrigation during the summer period of agricultural activity, such an effect might be unfavourable.

5.3.3.4 The Effect on Floods

Most major dams in Switzerland are built for the generation of hydropower, but the reservoirs also have an effect on flood protection, which is thus a secondary benefit of dams. An analysis of historical flood events in the Alps shows that the existing reservoirs are able to reduce the peak flow of a major flood by about 10–20% or more in the populated valleys some 20–30 km downstream of the dams. As an example, Fig. 5.5 shows the peak discharges of the Rhône River for the annual floods of the years 1916–1995.

The time series of annual peak flows shows a considerable decrease in the 1950s. This decrease is not primarily due to climatic changes, but to the beginning of operation of several major storage schemes. This reduction in flood peaks has the following positive effects on the downstream area:

First, as a result of retention in storage basins, the discharge of short and intense rainfall is extended over a longer period, which helps to make better use of the water in the rivers and to reduce spillway overflows.

Second, the flood risk is reduced. It can be estimated that the recurrence interval of a certain flood is roughly doubled through the influence of the existing reservoirs in the case mentioned above.

The closer a town or village is located to the dam, the more the peak flows are reduced.

The cost of damage to existing infrastructure and the agricultural losses as a result of flooding from 1972 to 1996 are estimated at Euro 2,900 per km^2/year, which corresponds to total annual damage costs of Euro 120 million per year, an amount within the range of construction costs for a large dam. This figure gives a rough indication of the value of flood protection provided by dams. If only a small percentage of this damage could be avoided through the presence of dams, the benefits would amount to several million euros each year.

5.3.3.5 The Influence of Hydropower on Water Quality

Generally speaking, the quality of the water is not affected by hydropower. Nevertheless, some minor impacts should be mentioned.

Below water intakes for diversion channels or tunnels, only a residual flow is available in rivers. In these stretches with reduced flow, an increase in the water temperature is sometimes observed, mainly in summer, due to reduced water depth and flow velocities. This temperature increase may, in extreme cases, have an adverse effect on fauna in the water. It may also cause secondary effects on water quality in cases of high concentrations of nutrients in the water, i.e. when anthropogenic effects other than hydropower are present.

The opposite effect is observed upstream of weirs, where the water body shows reduced velocities and turbulence. Here, the exchange between ambient air and water is reduced and this can cause a deficit in the oxygen content of the water. Again, it must be noted that under the conditions prevailing in Switzerland, problems only occur in the presence of extreme contamination of water by other causes than hydropower.

Furthermore, there is a second influence on the temperature regime of the water in rivers. Theoretically, the energy that is transformed in the generators into electricity is no longer available to increase the ambient temperature of the river due to flow friction. Thus the water in the river becomes cooler with hydropower. This effect plays a very minor role and has never been an issue for discussion in Switzerland. A third influence of hydropower on the temperature regime is that the exchange of water from high altitudes to low altitudes occurs much faster than would be the case with natural run-off.

5   Impacts of Dams in Switzerland

The quality of already polluted water could be affected negatively by hydropower due to reduced residual flow or reduced flow velocities. The water quality in Swiss rivers has improved considerably in past decades, and today the problems of water quality caused by hydropower, including temperature, happen very rarely. We can thus consider that hydropower has practically no negative effect on water quality.

5.3.3.6   Sediment Transport

The presence of water intakes and reservoirs changes the shape of flood waves in the rivers below such intakes or reservoirs. Below water intakes, the recession curve of a flood wave is usually steeper than would be the case in natural circumstances. Consequently, especially fine sediments are not completely flushed at the end of a flood event and often build up as deposits in the river beds. These deposits can remain over long periods since the flow in the river (and therefore the transport capacity) is reduced due to the water intake upstream. These deposits can cause damage when a new flood occurs and they are mobilised.

5.3.3.7   Socio-economic Effects

In Switzerland, hydropower plants have often been built in locations far from urban areas, mainly situated in mountainous regions, which are not considered favourable for industrial and other economic development. Consequently, migration from these areas to cities has often been observed. However, this movement could be retarded, for example, through the building and operation of power plants and dams. Thanks to the potential for secondary incomes with power plants, farming is still possible and helps to maintain a residual infrastructure in the area. The presence of hydropower plants and related financial activities also provides various forms of revenue for local enterprises such as shops, bars, restaurants and construction companies, and also employment. Today, these effects have lost their importance as these previously poor regions are now developed due to the increased mobility of people. Nevertheless, the wealth of hydropower remains in the area.

Furthermore, in the course of the construction and for plant operation, access roads have been built and are maintained by the power plant owners. These access facilities have helped develop recreational tourism and this enables visits to many sites with beautiful scenery, which would otherwise not be accessible to a great many people.

A large part of the annual costs of a hydroelectric power plant are the capital costs for interest and write-off of loans. These costs are relatively stable; they do not depend on exchange rates or price fluctuations of raw materials as would be the case for fossil fuel. Thus, the annual costs of hydropower are relatively calculable.

Furthermore, some 25% of the annual production costs are paid as taxes to the local governments, mainly as a compensation for the licence needed for hydropower operation.

With regard to technical aspects, the capabilities of hydropower to aid the regulation of the electrical grid and as a means to stockpile electricity should be mentioned.

5.4 Evaluation of External Benefits and Costs

5.4.1 Evaluation Methods

How can the qualities of hydropower be evaluated? Let us consider a list of major aspects:

- Renewable character
- Influence of flow changes on habitats
- Landscape
- Flood protection
- Water quality
- Recreation facilities
- Stable costs of energy: independent of import
- Grid regulation

We find that most of them are externalities, which do not show up in any bookkeeping either of the hydropower plants or of concerned third parties. The fact that these effects are not internalised so far indicates the difficulty of attributing monetary values to them.

Several studies have been carried out to identify, quantify, and calculate a monetary value of these effects, to assess hydropower in respect of other energy production systems. The following three studies are interesting in this context: Masuhr et al. (1993), European Commission-DGXII (1996), and Hauenstein et al. (1999). Various methods are used for monetary evaluation, such as:

a. Survey of the compensation costs which should be paid to affected people, or a survey of the cost that people should agree to pay to avoid the deterioration or loss of an environmental possession.
b. Calculation of the investment which would be needed to avoid damage. This method could also be applied for external benefits, such as those resulting from flood protection works.
c. Calculation of costs or benefits based on market prices.

Without going into the details, we provide below some examples of evaluation:

- *Atmospheric pollution and CO_2 emissions*—According to Frischknecht et al. (1994), during the construction phase, 95% of CO_2, 92% of SOx, 90% of NOx and 86% of the organic gaseous emissions result from the use of cement, steel and

diesel oil. Only these three elements are quantified, and their emissions, in kg per GWh of electricity produced per year, by the power plant. Then the cost per weight of gas emission is estimated. Various methods could be employed: for example, the amount which should be invested to avoid the damage created by the emissions, or the cost of these damages.
- *Noise pollution*—The noise intensity is measured in decibels and the pollution depends of the number of days of emission and of people, animals, or living spaces affected. The costs can be estimated by method (b) mentioned above, and the hedonist cost can also be calculated: the difference between the rent of a quiet or a noisy house or office.
- *Loss of habitats*—This impact can be calculated with method (a) listed above, the value that the people are ready to pay, per unit of area protected and per year.
- *Landscape modification*—The same calculation can be made as for the loss of habitats. Another estimation is a tourism value: The cost of travel that people would agree to pay to visit a site if it is not modified or destroyed.
- *Local or global economy*—The impact on the economy can be calculated by estimating the cost per year of the workplaces created or deleted, in relation to the construction and operation of a power plant.
- *Effects of floods* (see Sect. 5.3.3.4 above).

This method of identification, quantification, and monetary evaluation has been applied to a sample of 12 Swiss hydropower plants:

- Grande Dixence
- Cleuson-Dixence extension
- Hinterrhein
- Chippis
- Haute-Vièze
- St. Anne
- Orsières
- Beznau
- Rheinkraftwerke Schweiz-Liechtenstein (project)
- Rheinfelden
- Eglisau
- Wynau

All these schemes had previously been analysed in terms of indirect impacts. On the basis of those results, an extrapolation was carried out for all Swiss hydropower schemes, which was based on specific characteristics relevant to each impact. Thus, the impact along the river was extrapolated using the length of the affected section of the river, the number of dams or the surface of the dammed water body. Socio-economic effects were estimated using the number of employed personnel, the population or the number of tourists of an area. This procedure allows for improvements of the study in the future, by adding new case studies and repeating the extrapolation based on new samples.

Table 5.5a External benefits and costs in kCHF/y (10³ Swiss francs/year) during the construction phase ([a]including foreign share of border plants)

Construction phase	Hydropower plants with reservoirs Min	Max	Run-of-river plants High head Min	Max	Low head Min	Max
Installed capacity[a] (MW)	10,413		2,555.2		1,715.7	
Annual production[a] (GWh)	19,083		9,265.4		9,731.3	
Benefits						
Effects of employment (kCHF/y)	5,120		3,521		2,273	
Tourist attraction (kCHF/y)	170		0		105	
Costs						
Noise (kCHF/y)	97	123	49	70	310	
Air pollution (kCHF/y)	3,004	8,654	712	2,050	742	2,136
CO_2 production (kCHF/y)	24,732		3,799		3,951	
Landscape, habitats (kCHF/y)	2,390		791		187	
Loss of tourist attraction (kCHF/y)	23	67	11	32	420	

5.4.2 Results

The studies carried out so far show in the Table 5.5d (in CHF cents) that the external costs of hydropower, after deduction of external benefits, is in the range of 0.05–0.15 Euro cents/kWh for the construction phase and 0.1–0.4 Euro cents/kWh during operation. External benefits deal mainly with flood protection and recreation. Taking into account the average production costs of some 3–4 cents/kWh, it can be seen that the external costs of hydropower are very small, in the order of 1/10 of those of fossil fuel power plants.

Tables 5.5a–5.5d summarise the external costs and benefits for all the existing hydropower plants in Switzerland (Hauenstein et al. 1999).

5.5 Externalities for Electricity Produced by Other Energy Sources

Results for externalities in the hydropower sector given by Hauenstein et al. (1999) are confirmed by other studies reviewed by Dettli et al. (2000). As predicted, external costs for hydropower plants appear to be much lower than those of thermal plants. Figure 5.6 compares the estimated external costs reported in the literature for

5 Impacts of Dams in Switzerland

Table 5.5b External benefits and costs kCHF/y during operation

Operation phase (kCHF/y)	Hydropower plants with reservoirs Min	Max	Run-of-river plants High head Min	Max	Low head Min	Max
Benefits						
Effects of employment	11,450		5,559		5,839	
Tourist attraction	15,087				3,870	5,500
Flood protection	5,380					
Costs						
Landscape	21,154		19,295		2,518	
Habitats	12,768		12,768		39,840	
Hindrance of migration	2,712		2,712		2,475	
Loss of tourist attraction	0,825	2,425	0,400	1,180	4,240	

Table 5.5c External benefits and costs kCHF/y during operation

Summary in kCHF/y	Hydropower plants with reservoirs Min	Max	Run-of-river plants High head Min	Max	Low head Min	Max
Construction phase						
Total external benefits	5,290		3,521		2,378	
Total external costs	30,246	35,966	5,362	6,742	5,610	7,004
Operation phase						
Total external benefits	31,917		5,559		9,709	11,339
Total external costs	37,459	39,059	35,175	35,955	49,073	

Table 5.5d Summary, monetary values in CHF cents/kWh

Summary in cents/kWh	Hydropower plants with reservoirs Min	Max	Run-of-river plants High head Min	Max	Low head Min	Max
Construction phase						
Total external benefits	0.03		0.04		0.02	
Total external costs	0.16	0.19	0.06	0.07	0.06	0.07
Operation phase						
Total external benefits	0.17		0.06		0.1	0.12
Total external costs	0.20	0.21	0.38	0.39	0.5	

the construction and operation phase together, concerning electricity produced with various sources of energy.

It can be seen from Fig. 5.6 that hydropower, nuclear power, and wind energy have very small external costs. The externalities of photovoltaic, another renewable energy source, are higher due to the CO_2 emissions of the energy used in the construction of the solar panels. The non-renewable energies based on fossil fuel such as coal, lignite, mineral oil, etc. have much higher external costs. This significant difference is mainly due to CO_2 emissions.

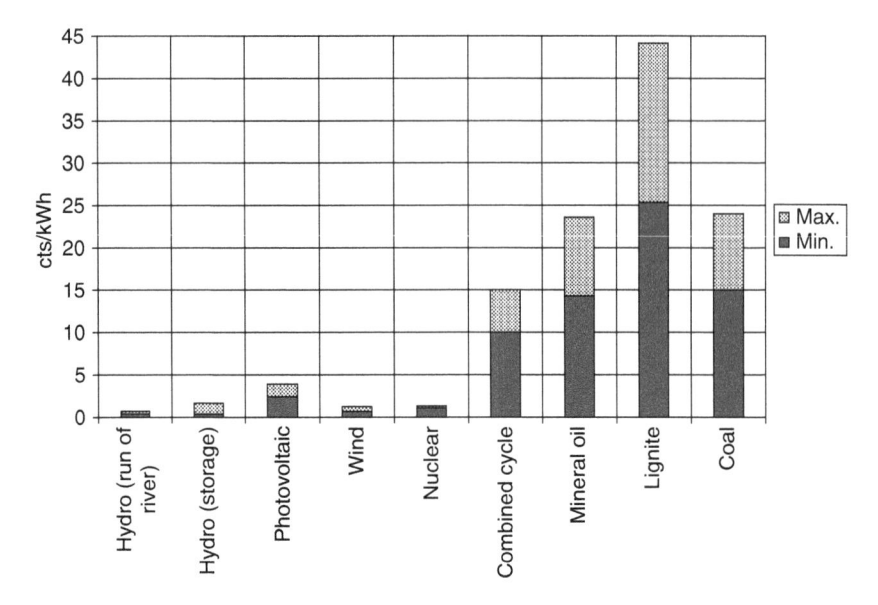

Fig. 5.6 Externalities of electricity produced based on different sources in CHF cents/kWh

5.6 Case Study of the Hinterrhein Development

5.6.1 The Existing Hinterrhein Scheme and the Valle di Lei Dam

The international Hinterrhein development on the border of Switzerland and Italy was completed in 1963. It is one of the largest power plants in Switzerland with a total installed capacity of 650 MW, an annual production of 1,420 GWh, and a total storage volume of 216 MCM Its main features are typical for a Swiss high-head scheme (see Fig. 5.7).

Some 80% of the shares are held by Swiss utilities and communities. An Italian utility holds the remaining 20%. Most of the catchment areas and installations are located in Switzerland. Only the major reservoir and its direct catchment area lie in Italy. For the construction of the dam itself, an exchange of territory has been executed, so that this dam is now on Swiss territory.

The main characteristics of the three steps with regard to storage facilities and power stations are given in Tables 5.6–5.8.

5.6.1.1 Direct Benefits and Costs

The direct costs and benefits are evaluated for the Valle di Lei Dam alone. Therefore, construction and operation costs for the dam are only compared to the added value

5 Impacts of Dams in Switzerland 115

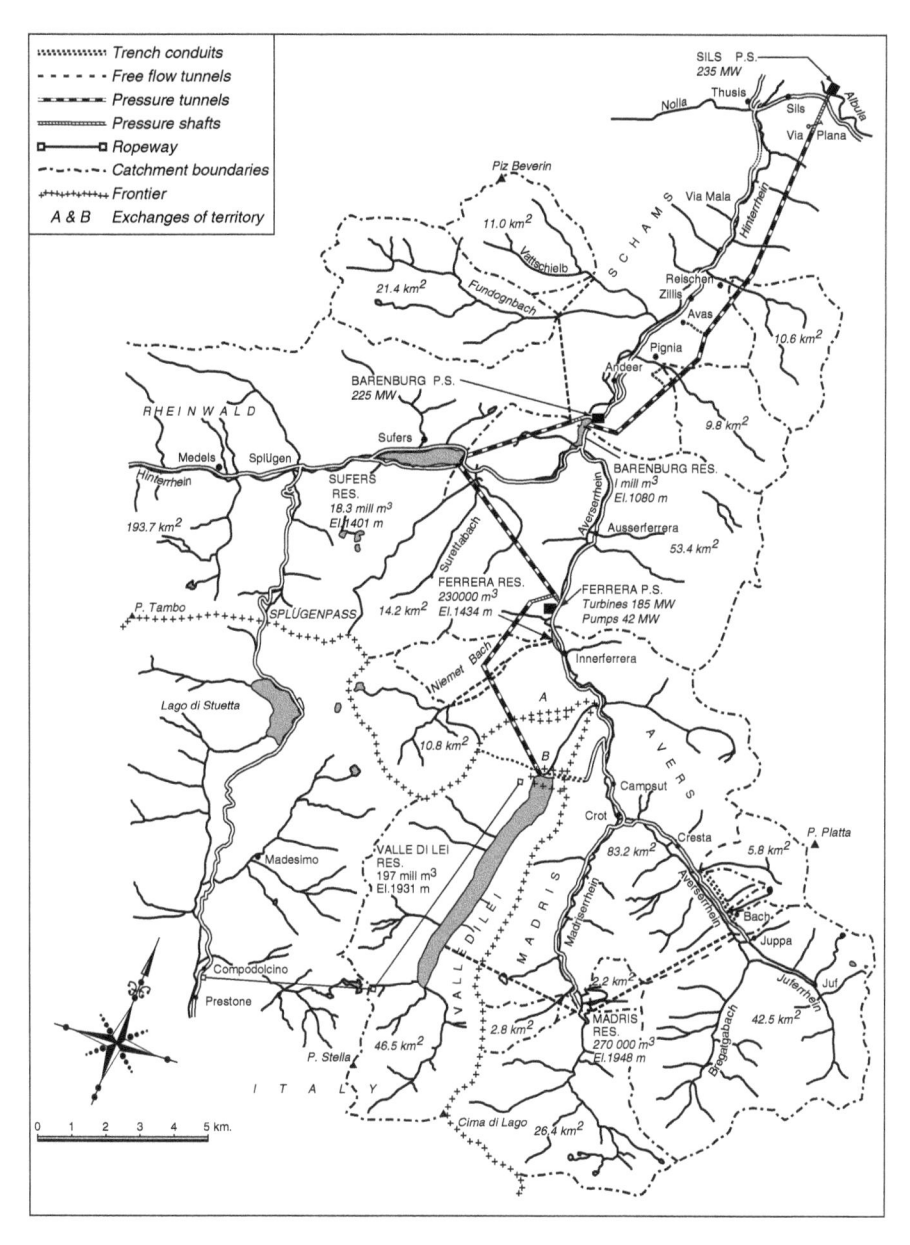

Fig. 5.7 Overview of the Hinterrhein Development

of the energy output resulting from the storage and not from other equipment such as water intakes, tunnels and shafts, as well as the powerhouses.

Table 5.6 Main characteristics of Step 1 of the Hinterrhein scheme

Step 1	Storage facility		Powerhouse	
	Valle di Lei		*Innerferrera*	
	Storage capacity	197 Mm^3	Equipment	3 horizontal Francis turbines, two of them equipped with storage pumps
	Storage height above sea level	1,931 m	Installed capacity turbines/pumps	185 MW/80 MW
	Type of dam	Arch dam	Discharge turbines/pumps	45 m^3/s/16 m^3/s
	Height of dam	143 m	Gross head	524 m
	Concrete volume	0.862 Mm^3	Average annual output	Winter 310 GWh Summer 0 GWh
	Ratio storage volume/dam volume	228	Average annual pump energy	Winter 0 GWh Summer 110 GWh
	Catchment area	137 km^2		
	Preda			
	Storage capacity	0.27 Mm^3		
	Storage height above sea level	1,948 m		

Table 5.7 Main characteristics of Step 2 of the Hinterrhein scheme

Step 2	Storage facility		Powerhouse	
	Sufers		*Bärenburg*	
	Storage capacity	18.3 Mm^3	Equipment	4 vertical Francis turbines
	Storage height above sea level	1,401 m	Installed capacity	220 MW
	Type of dam	Arch dam	Discharge	80 m^3/s
	Height of dam	58 m	Gross head	321 m
	Concrete volume	0.058 Mm^3	Average annual output	Winter 230 GWh Summer 250 GWh
	Ratio storage volume/dam volume	315		
	Catchment area	460.5 km^2		
	Ferrera			
	Storage capacity	0.23 Mm^3		
	Storage height above sea level	1,443 m		

Direct Benefits

The electricity output of the total scheme is 800 GWh during winter months and 620 GWh during summer adding up to a total of roughly 1,420 GWh. (The figures give the total of energy production without deduction of the energy for pumping: 110 GWh.)

5 Impacts of Dams in Switzerland

Table 5.8 Main characteristics of Step 3 of the Hinterrhein scheme

Step 3	Bärenburg		Sils	
	Storage capacity	1 Mm³	Equipment	4 vertical Francis plus 2 vertical Pelton turbines
	Storage height above sea level	1,080 m	Installed capacity	240 MW plus 5 MW
	Type of dam	Gravity dam	Discharge	73 m³/s
	Height of dam	64 m	Gross head	413 m
	Catchment area	534 km²	Average annual output	Winter 260 GWh Summer 370 GWh

a. *Value of peak load energy*—These 1,420 GWh are of a much better quality than the output of a run-of-river scheme. The value of this energy is therefore higher than that of base load energy coming from a run-of-river power plant. The surplus value of the electricity produced in schemes with reservoirs such as the Hinterrhein scheme is due to the fact that it can be produced when market prices are high and not only when the natural inflow occurs. This is in wintertime and during peak hours. This surplus value amounts to about Euro 30 per MWh (price forward curves for 3 years in winter months). It can be assumed that of the total winter production, some 75%, is possible thanks to the Valle di Lei Dam. Thus, the yearly benefit of this dam in winter is: 800,000 MWh × 0.75 × Euro 30/MWh = Euro 18 million per year.

b. *Value of power availability*—In addition to the increased quality of the electricity produced, the availability of power must be taken into consideration. Without storage capacity, the installed power would not be available during wintertime, since natural run-off is very limited at this time of year. This power would have to be purchased elsewhere.

If again it is assumed that the installed capacity of the three steps, i.e. 490 MW (which represents 75% of the total capacity of 650 MW, with Innerferrera: 185, Bärenburg: 220, Sils: 245) had to be procured elsewhere, this would lead to additional costs of some Euro 220 million (equivalent to an investment in a gas turbine scheme of equal power with specific installation costs of Euro 450/kW). This amount is equivalent to another Euro 11 million per year, given an annuity of 5% of the investment costs.

Thus the total direct benefits of the Valle di Lei Dam run to some Euro 30 million per year.

Direct Costs

Construction costs for a concrete dam of the size and shape of Valle di Lei Dam at 2004 prices amount to roughly Euro 2/m³ storage volume. The investment costs for Valle di Lei Dam at 2004 prices can therefore be estimated at Euro 394 million (only Valle di Lei dam with 197 million m³). With a discount rate of 5%, annual costs of Valle di Lei Dam amount to some Euro 20 million.

This estimation can be confirmed if we look at the annual profit and loss account of the hydrological year 2002–2003, where the total costs for the operation of the plant amounted to some Euro 40 million. Thus, the dam itself would account for some 50% of the total costs, which seems to be a reasonable assumption.

5.6.1.2 Conclusion on Direct Effects

Direct benefits by far outweigh direct costs of a dam such as the Valle di Lei, located at a high altitude and with relatively advantageous topographical conditions.

5.6.1.3 Indirect Benefits and Costs

Indirect benefits and costs are evaluated according to the results given by Hauenstein et al. (1999). This evaluation is not restricted to the dam, but includes other components of the development as well (see Tables 5.9 and 5.10).

5.6.1.4 Conclusion on Indirect Effects

Indirect effects amount to roughly Euro 2 million per year for the benefits, and Euro 4.5 to 5 million per year for the indirect costs. Thus, the net indirect costs are in the order of Euro 2.5 to 3 million per year or Euro cents 0.18 to 0.2/kWh. These indirect costs of the total hydropower plant (not only the dam) are much smaller than the direct costs for hydropower, which amount on average to some Euro cents 3.5/kWh in Switzerland.

Table 5.9 Indirect benefits of the Hinterrhein development

	Cents (Euro)/kWh	10^6 Euro (annual production: 1,420 GWh)
Construction phase		
Visitors	0.001	
Indirect economic profit	0.02	
Total	0.02	0.28
Operation phase		
Flood protection	0.02	
Indirect economic gain	0.04	
Attraction landscape, visitors	0.05	
Total	0.11	1.56
Both phases	0.13	1.84

5 Impacts of Dams in Switzerland

Table 5.10 Indirect costs of the Hinterrhein development

	Cents (Euro)/kWh	10^6 Euro (annual production: 1,420 GWh)
Construction phase		
Noise	0	
Air pollution	0.01–0.03	
CO_2	0.09	
Habitats and landscape	0.01	
Loss of tourist attractions	0	
Total	0.11–0.13	1.56–1.85
Operation phase		
Landscape, infrastructure	0.01	
Landscape, storage reservoirs	0.09	
Habitats, reduced wet area	0.09	
Habitats, damming up of river	0	
Habitats, loss of flow dynamics	0.003	
Migration opportunities	0.013	
Reduced tourist attractions	0.003–0.007	
Total	0.21	2.98
Both phases	0.32–0.34	4.54–4.83

5.6.2 Extension Project for the Hinterrhein Development

5.6.2.1 Description of the Extension Project

Towards the end of the twentieth century, an extension of the existing scheme was studied, including a new dam with a reservoir of 400 MCM in place of the existing dam and reservoir and two steps parallel to the existing Steps 1 and 2 with installed turbine capacities of 366 MW and 122 MW respectively, as well as pumping equipment of approximately similar electrical capacity.

Table 5.11 shows the details of this extension.

5.6.2.2 Indirect Benefits and Indirect Costs

The indirect benefits and costs have been analysed based on an environmental impact study carried out. Tables 5.12 and 5.13 give estimates of these evaluations.

Table 5.11 Characteristics of the extension project

	Storage facility		Powerhouse	
Step 1	*Valle di Lei*		*Valle di Lei – Sufers*	
	Storage capacity	400×10^6 m^3	Equipment	Open
	Storage height above sea level	1969 m	Installed capacity (turbines/pumps)	366 MW/366 MW
	Type of dam	Arch dam	Discharge	79 m^3/s/65 m^3/s
	Height of dam	193 m	Gross head	533 m
Step 2			*Sufers – Bärenburg*	
			Equipment	Open
			Installed capacity (turbines/pumps)	122 MW/122 MW
			Discharge	45 m^3/s/33 m^3/s
			Gross head	326 m
Step 3	No extension planned		No extension planned	

Table 5.12 External benefits of the extension project

	Cents (Euro)/kWh	10^6 Euro (annual production: 1,420 GWh)
Construction phase		
Visitors		0
Indirect economic profit	0.02	0.26
Total	0.02	0.26
Operation phase		
Flood protection	0.03	0.44
Indirect economic gain	0.04	0.56
Landscape attractions, visitors	0.0	0.02
Total	0.07	1.02
Total construction and operation	0.09	1.28

5.6.2.3 Conclusion on Indirect Effects

Again, indirect costs are higher than indirect benefits. Nevertheless, even this new project leads to net externalities of only Euro 1 million per year or, given an annual production that is similar to the existing project, of 0.07 Euro cents/kWh.

5.7 Conclusions

The balance of direct benefits and costs of dams and hydropower plants in Switzerland is largely positive. This small country provides an example of how the only natural wealth of a nation—in this case, hydropower (apart from forest

Table 5.13 External costs of the extension project

	Cents (Euro)/kWh	10^6 Euro (annual production: 1,420 GWh)
Construction phase		
Noise	0.0	0.005–0.006
Air pollution	0.01–0.02	0.13–0.38
CO_2	0.07	0.98
Habitats and landscape	0.007	0.10
Loss of tourist attraction		0.0
Total	0.09–0.1	1.21–1.47
Operation phase		
Landscape, infrastructure	0.0	0.01
Landscape, storage reservoirs	0.05	0.76
Habitats, reduced wet area	0.006	0.091
Habitats damming up of river	0	0
Habitats, loss of flow dynamics	0	0
Migration opportunities	0	0
Reduced tourist attraction	0	0
Total	0.056	0.86
Both phases	0.146–0.156	2.07–2.33

resources)—has been nearly completely developed since the end of the nineteenth century. Dams play a major role because they help to considerably increase the quality of the electricity produced and have a major effect on flood protection.

External benefits and costs are not easy to quantify, since often their monetary evaluation cannot be based on widely accepted methods and values. However, existing estimations show that hydropower generates external benefits and costs that are very low compared with the internalised price of the product (1/10 at a maximum).

The quantification of the external benefits and costs is a very adequate tool to compare the overall performance of different energy systems. It appears that the value of the externalities of hydropower, which is comparable with the value of the externalities of nuclear power, are considerably smaller than those of other renewable sources and much smaller than the externalities of fossil-fuel-driven plants.

An evaluation of the indirect benefits and costs has been carried out for the Hinterrhein development in Switzerland and an extension project, studied in the 1990s. The evaluation shows that these externalities are very small compared with average electricity prices and that external benefits are of the order of 50% of external costs.

Although it is widely accepted that the externalities of hydropower are very small, the opposition to new projects remains strong due to unilateral views which focus on negative local ecological effects instead of considering an overall evaluation of all external effects. That was one of the reasons why the extension project was not implemented. Only if the externalities of energy sources other than hydropower could be internalised and the price for these products be increased accordingly, would better conditions for new hydropower projects be achieved.

Acknowledgements The authors of the present chapter are very grateful and express their thanks to Jean Remondeulaz, former president of Energie de l'Ouest Suisse SA (EOS), for his review of this chapter and his many valuable comments. The authors also acknowledge Alison Bartle of Aqua-Media International for her review of the final English text.

References

Bischof R, Hagin B, Hauenstein W, Lafitte R, Mouvet L (2000) 205 dams in Switzerland for the Welfare of Population (R49, Q77). 20th ICOLD Congress, Beijing, China

Bundesamt für Energie (2007) Statistik der Wasserkraftanlagen der Schweiz/Statistique des Aménagements Hydroélectriques de la Suisse/Statistica degli Impianti Idroelettrici della Svizzera, Bundesamt für Energie, 3003 Bern

Dettli R et al (2000) Ökologische (Teil A) und technisch/ökonomische (Teil B) Qualitäten der Wasserkraft, Vebandsschrift Nr. 64, Schweizerischer Wasserwirtschaftsverband, CH-5401 Baden, Switzerland

European Commission-DGXII (1996), ExternE – Externalities of Energy, Volume 6, Wind and Hydro, Part 2, Hydro Fuel Cycle, pp 127–249, European Commission, EUR 16525 EN

Frischknecht R, Hofstetter P, Knoepfel I et al (1994) Ökoinventare für Energiesysteme. Laboratorium für Energiesysteme, ETHZ, Zürich

Hauenstein W (2000) Role and evaluation of external costs and benefits of hydropower. Congress Proceedings: Hydro 2000, making hydro more competitive. International Hydropower Association, Sutton, Surrey, UK

Hauenstein W, Bonvin JM, Vouillamoz, Wiederkehr B (1999) Externe Effekte der Wasserkraftnutzung in der Schweiz, Verbandsschrift 60, Schweizerischer Wasserwirtschafsverband, CH-5401 Baden, Switzerland

Masuhr K, Weidig I, Tautschnig W (1993) Die externen Kosten der Stromerzeugung aus Wasserkraft. Prognos, Basel

Vischer D, Sinniger R (1998) Wasserkraft in der Schweiz. Gesellschaft für Ingenieurbaukunst, Stäubli AG, Zürich. ISBN 37266 00337

Chapter 6
Hydrodevelopment and Population Displacement in Argentina

Leopoldo J. Bartolome and Christine M. Danklmaier

6.1 Introduction

6.1.1 Large Dams in Argentina

6.1.1.1 Population Displacement and Resettlement

Until quite recently population displacement due to hydrodevelopment was very rare in Argentina and not a very important topic for public discussion. However, the construction in the last decades of some large projects as well as the increasing sophistication and power of ecologist movements led to a new prominence of this topic in public conscience and the mass media. The result was a strong movement 'against' large dams and its plethora of ecological and social consequences.

For this discussion of the Argentine experience with population displacement and dams, we have selected three projects which are the only ones that caused a considerable displacement of population. The Salto Grande Project, finished in 1979, resulted in the displacement of 12,000 persons on its Argentine side and posed the problem of the need to resettle an entire town of more than 5,000 inhabitants.

This is an updated and revised version of a report originally commissioned by the World Commission on Dams.

L.J. Bartolome (✉)
Graduate Programme on Social Anthropology, National University of Misiones, UNAM, Posadas, Argentina

C.M. Danklmaier
National University of Misiones, El Bolson, Rio Negro, Argentina

The Piedra del Águila Project is slightly unusual, since a little more than 400 persons were resettled, but it is included here because all those persons were Mapuche, an Amerindian minority group within Argentina, and the project therefore presents some important aspects. Finally, the still incomplete but already operative Yacyreta Project relocated by 2006 more than 5,725 families in Argentina and 2,366 families in Paraguay (this resettlement is still in progress) and most of them urban dwellers. These cases account for the bulk of the Argentine experience with population resettlement associated with the construction of dams.

6.1.1.2 The Normative and Policy Framework

There is no general policy framework concerning displacement and resettlement of populations affected by large-scale development projects. Each project develops its own policy according to the circumstances, the private or public nature of the endeavour and—this is very influential—the origin of the funds. Until the 1970s, it was only Argentina's so-called Public Interest Appropriation Law that was applicable to such cases. This law covered monetary compensations for proprietors of the land and goods affected, with no provisions for resettlement or other kinds of compensations or mitigation measures. The increasing intervention of international banks (such as the World Bank and the Inter-American Development Bank) led to the adoption—reluctantly in many cases—of the resettlement and compensation policies developed by those institutions. In the early 1970s there was an attempt by the then National Secretary for Environmental Regulation to promote the establishment of a general set of norms and a handbook of procedures to be applied in the implementation of large-scale development projects. This endeavour received little support from the national authorities. Furthermore, the agencies managing such projects refused to cooperate and the attempt was finally abandoned.

Despite the absence of an official national policy framework, some kind of resettlement and rehabilitation component is becoming required for all projects that may cause population displacement. Provincial governments as well as private firms are more and more conscious of social impacts and consider the implementation of mitigation measures, though in some cases they attend more to form than to substance. Although Article 41 of the New Argentine Constitution (1994) establishes the need to protect the environment and the obligation of mitigating anthropic impacts, the article has not been yet developed into specific norms and is open to diverse interpretations. The same is the case with Article 75, Clause 17, that grants rights to aboriginal (Amerindian) communities to participate in the management of their natural resources. However, some provincial governments are more advanced in this respect. For instance, Decree 1741/96 (Annex 6) of the Province of Buenos Aires obliges all new industries to obtain a Certificate of Environmental Aptitude prior to their installation. These cases notwithstanding, the absence of a national legislation allows for disorder and multiple violations of constitutional principles.

Fig. 6.1 Map of the Salto Grande Project. *Source*: www.saltogrande.org/saltogrande/Principal.htm

6.2 The Salto Grande Project and the Resettlement of the City of Federación

6.2.1 Description

The Salto Grande Hydroelectric Dam (1973–1979) is an Argentine-Uruguayan endeavour located on the Uruguay River, which serves as a border between both countries. The dyke is located 15 km upriver from the cities of Concordia (Argentina) and Salto (Uruguay), and 500 km north from Buenos Aires and Montevideo, the capital cities of Argentina and Uruguay, respectively. Built between 1973 and 1979, the dam has a generating capacity of 1,890 MW and formed a reservoir of 783 km^2, which flooded 45,500 ha of agricultural and forested lands on the Uruguayan side and 30,000 in Argentina. This impoundment forced the displacement and resettlement of 8,000 persons living in the towns of Constitución and Belen (Department of Salto, Uruguay) and of 12,000 persons who lived in the city of Federación and in the township of Santa Ana (Province of Entre Rios, Argentina). The works were financed with loans from the Inter-American Development Bank and with special funds contributed by the partner countries. An ad hoc binational outfit, the Joint Technical Commission for Salto Grande (Comisión Técnico Mixta de Salto Grande, or CTM), was in charge of construction and operation. Although construction took place in the 1970s, the project actually originated many years before and suffered repeated delays because of political and economic reasons (Fig. 6.1).

6.2.2 A Brief History of the Project

The Salto Grande Treaty between Uruguay and Argentina was signed in 1946. This treaty created the CTM and established objectives for the project. These were: (1) improvement of local environmental and health conditions, (2) improvement of navigation conditions, (3) production of energy and (4) provision of water for irrigation. However, the first technical studies were conducted after 1966, under a military dictatorship, and the actual works were begun in 1974, under a democratically elected populist government. The first turbine was installed in 1979, again under a military dictatorship, and the whole complex was officially inaugurated in 1983, a few months after Argentina's return to democracy. After all this tumult in the political arena, the objectives of the project had been reduced to energy production.

6.2.3 Population Affected by the Project

Most of the land impounded on the Argentine side corresponded to the Province of Entre Rios, a rich agricultural and beef-producing area that was the site of a large European immigration in the nineteenth century. The main impact was in the city of Federación, which had to be resettled almost entirely. Federación (population 6,162 in 1970) was an administrative and services centre for a rich hinterland dedicated to agriculture and forestry, and one of the oldest urban settlements in the province (founded during the period of the independence wars with Spain). The city concentrated on commercial activities and also diverse agro-industries, such as plants producing oil and fruit juice, sawmills and tanneries, etc. This city and the neighbouring village of Santa Ana account for most of the 12,000 people that had to be displaced on the Argentine side.

The urban centre of Federación was occupied by the upper echelons of local society, composed of merchants, industrialists (of small- and medium-sized enterprises), public officers of high rank and professionals. The areas surrounding the city were inhabited by the lower strata: blue collar workers, rural labourers, lower level public servants and occasional workers. Seventy per cent of downtown Federación, i.e. the residential area with the best service infrastructure and housing for the local upper class, was directly impounded. Among sites and buildings lost to the waters were the central square or 'plaza', the city hall, the cathedral, police headquarters, banks, etc. The so-called industrial quarters, where sawmills and tanneries were located, were also flooded. Only 30% of the city was not submerged, with most of this area corresponding to newer and peripheral settlements, housing the lower classes.

6.2.4 Main Social Impacts

The construction of the Salto Grande Project meant not only the forced displacement of more than 20,000 people in Argentina and Uruguay, but also affected roads, railways and urban and rural infrastructure. Besides, the project disrupted the hierarchy among urban centres in the region and modified political and economic relationships. Social impacts were not limited to the resettlement period but continued to be felt many years after completion. Some of these impacts can be traced to defective planning and implementation of resettlement, such as the failure to build enough housing for all inhabitants of the city of Federación, which led to the concentration of the excluded (mostly the poor) in the remnants of the old city, where they live under very deprived conditions. On the other hand, the construction of the new city attracted large numbers of workers from other parts of the province, many of whom also settled in the city remnants once the building works were finished, generating a large shanty town. This remnant remained isolated from the new city until the construction of a bridge several years later. The result of faulty policies and lack of provisions is that New Federación exhibits a degree of socio-economic residential segregation unparalleled among Argentine provincial towns.

6.2.5 Compensation and Resettlement Policies

Affected houses were evaluated and received a market price. Proprietors or owners could apply for money to buy a resettlement house and, should that amount not be enough, apply for a 30-year loan to pay for the rest. Although the original project (1973) did not consider compensation for tenants, they were included in September 1976 under a similar provision of loans and repayment if they were able to demonstrate a stable occupation and livelihood. Merchants and industrialists were also entitled to resettlement or to receive monetary compensation; in some cases they were granted lots in the remnants of Old Federación. Those Federeños that took loans had to make payments until 1993, when the Entre Rios government annulled outstanding debts. No rehabilitation and social assistance plans were implemented, besides physical removal.

6.2.6 Implementation of Compensation and Resettlement

New Federación was built on the shores of the Salto Grande Lake, 5 km to the north of its old site. The new location involved an entire redefinition of the city's spatial relationships with neighbouring settlements. National Route 14 and the railways were removed to a distance of 18 km from the new site, increasing the distance to agricultural colonies, and some of these factors changed commercial links with other nearby towns and cities (Chajari and Concordia, for instance). The resettlement project considered transforming the remnant of Old Federación into an industrial area and, consequently, it promoted the establishment of productive activities there. The so-called remnant became a sort of satellite town of New Federación, composed of un-demolished old residential neighbourhoods (quickly occupied by squatters), shacks housing the new settlers attracted by job opportunities in the construction of the new city, and some buildings utilised by an elementary school, a high school, a police station, a health centre and a church. This remnant remained isolated from the new city for several years due to the presence of an inlet of the Salto Grande Lake that forced residents to make a detour of 36 km. Those dwelling in the remnant had to wait until 1985 for the construction of a bridge that reduced the distance to 5 km.

6.2.7 Participation of Stakeholders

The Salto Grande Project originally received great support from the population of both countries and particularly inspired the enthusiasm of 'entrerrianos' (people of Entre Rios), who saw this project as an opportunity for addressing the relative isolation of their province and for promoting local economic development. However, they had to wait 25 years (1946–1971) before the first effective steps were taken. Only in 1971 did the CTM complete the feasibility studies and presented to the Argentine government a request to pronounce all land and goods required to implement the project of public interest and subject to appropriation.

Although there was no official information about the compensation policy to be applied, rumours circulated stating that only those whose properties would be impounded would be compensated. In the face of such rumours, the inhabitants of Federación—actually the industrial and business community—created an association under the name of Commission for the Promotion of Federación's Interests and in Favour of the Hydroelectric Salto Grande Complex (Comisión Pro-Defensa de los Intereses de Federación y Apoyo al Complejo Hidroeléctrico de Salto Grande), with the objectives of securing the realisation of the endeavour and of advancing the interests of the city.

Population resettlement as well as the replacement of affected infrastructure were categorised as 'not shared works', which meant that each country had exclusive responsibility for their implementation. These components became the subject of a bitter dispute over the definition of policies to be applied, including who was going to be resettled, where and how. This debate involved political parties, political and administrative authorities, the Argentine officers of CTM, the national government and even several ad hoc associations that were supposed to represent affected people. A first confrontation took place between the government of the province (and all other local interest groups) and the CTM, which insisted on resettling only those directly affected by the reservoir. By mid-1972, the Entre Rios government commissioned its Housing Institute to undertake studies oriented to defining a resettlement policy for Federación. These studies advised the resettlement of the city as a whole, arguing that it should be considered as a 'working system' and not a simple sum of parts. In spite of this recommendation, in February 1973 the national government issued an Appropriation Law that adhered to the CTM position in this respect. These standards meant the exclusion from resettlement of approximately

1,400 people residing on the outskirts of the city, most of them belonging to the lower classes. These norms were actively resisted by the provincial authorities and by the people of Federación. In May 1973, taking the opportunity of a change in the national government—a populist party came to power—the Entre Rios authorities requested that the resettlement of Federación be decided by democratic vote. This criterion was finally accepted by the national government and the CTM and in April 1974, ENFYSA (Studies for New Federación and Santa Ana) was created, an outfit endorsed by CTM and the government of the Province of Entre Rios that was charged with planning the resettlement of these urban centres. ENFYSA conducted various feasibility studies and prepared a set of alternative locations for the new city. It also arranged, with the provincial government, a poll in which the inhabitants of Federación would chose the new emplacement of the city.

Still another organisation was created for this purpose: the so-called Commission for Popular Participation for the City of New Federación and Township of Santa Ana. This commission was presided over by the mayor of Federación, and included representatives of the industrial and business community, trade unions, provincial legislators and officers of ENFYSA. Although it was supposed to represent all local interests, in practice its operations were strongly coloured by internal divisions of the hegemonic Justitialist Party, then in power at the national level and also in the Province of Entre Rios.

The Commission for Popular Participation started to work by the end of May 1974 and although the organisation of the poll was its main objective, it tried to obtain a postponement, alleging the lack of enough information and the existence of the Appropriation Law. However, given that work on the dam had already begun, they had to accept that this was organised the same year in October. The people of Federación received scarce and confusing information about the alternative sites and about the resettlement policy that was to be applied. The results of the poll were revealing of the social, economic and political cleavages dividing the inhabitants of Federación. Of the five alternative sites offered, two received 92.5% of the vote. One of these received the vote of the upper and middle classes and of the orthodox leadership of the Justitialist Party, while the other (the winner) was supported in large numbers (56.6% of the vote) by the lower classes, including those not entitled to resettlement, and the left-wing faction of the Justitialist Party. The site chosen was the one closest to the old location of the city and also the one best suited to the interests of the poor and of those excluded from the resettlement programme. In December 1974, once the resettlement site had been decided, the Entre Rios government and the CTM celebrated a new agreement, after which the government assumed responsibility for developing a new resettlement plan and the CTM was put in charge of obtaining the necessary funds. In September 1975, the existing Appropriation Law was derogated and a new one was issued, incorporating into the resettlement project all the people living in Federación and its surroundings.

The armed forces took power in Argentina in March 1976 and suspended all civil liberties. This event modified the whole political context of the Salto Grande Project and the implementation of a strict free-market economic policy led to attempts to reduce the funds required for the resettlement project. Thus, the national government promoted a plan that proposed the payment of individual compensation and the resettlement of people in exiting urban centres nearby, such as Chajari and Concordia. Despite the highly repressive nature of the regime, this proposal was met with active opposition from all sectors of the local population. This opposition was successful and, in September that year, the national government decided to go ahead with the project, including construction of the new city. However, the government dissolved the existing organisations and created a new one: Council for the Salto Grande Region (COPRESAG) that initiated work at the site of New Federación in early 1977. It also created, within COPRESAG, the so-called Council for New Federación (CONFED)—composed mainly of local representatives—that was charged with responsibility for urban planning and for carrying out the resettlement plan. By this time the resettlement programme was already lagging behind work on the dam. Thus, when the first turbine was put into operation in 1979 there was a shortage of 500 resettlement houses. As a consequence, many affected people could not move to New Federación and had to remain in the non-flooded areas of the old site, living in precarious conditions.

A new confrontation took place during the early 1990s, when the CTM decided to raise the height of the reservoir by 1 m. In response, the inhabitants of New Federación organised a mobilisation that also raised the banner of the numerous families still to be resettled, and insisted on the fulfilment of many of the original benefits promised by Salto Grande (irrigation works, port facilities, a navigation lock for the dyke, etc.). The Entre Rios government, in turn, requested to take over the administration and management of the dam. These requests have been unsuccessful so far. All these confrontations caused by the Salto Grande Project contributed greatly to a change in the mood of the Entre Rios population towards large-scale development projects, and particularly towards large dams. Thus, when in 1996 the so-called Parana Medio Project was brought into public discussion, the popular vote showed massive support for the law approved by the Entre Rios legislature that prohibited the construction of new dams on the Parana and Uruguay Rivers.

6.2.8 *Evaluation of Consequences*

The Salto Grande Project illustrates the real priorities in large-scale development projects that include the production of energy as one objective. When it was announced it was propagandised as a multi-purpose regional development endeavour, with energy production in third place among its main objectives.

Throughout its implementation and particularly once the power plant began energy production, all other purported objectives faded into the background and eventually vanished from official discourse. From one point of view, of the hidden but real agenda, the need to displace people and even whole towns was seen as an obstacle to be overcome and not as a set of tasks rightfully belonging to the project. Even today, the region's people are demanding the fulfilment of promises made at the beginning, when the project was being 'sold'.

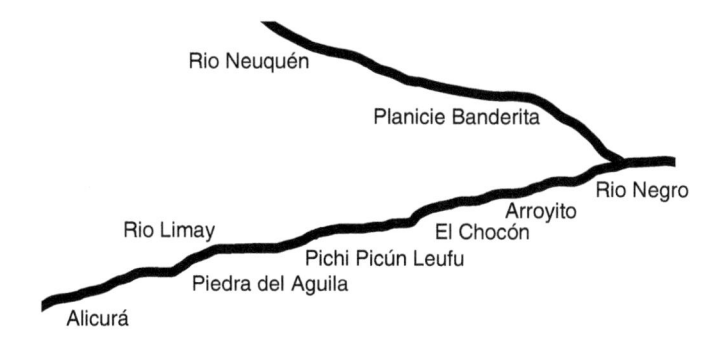

Fig. 6.2 Map of dams of the Limay River. *Source*: www.argentinaturismo.com.ar/piedradelaguila

The influence of social actors with some degree of 'social agency'—power by another name—helps to explain the rather unusual, by Argentine standards, participation of stakeholders in the resettlement of the city of Federación, and their relative ability to discuss, reject and even fight back against government decisions under different kinds of political regimes. In other respects, the Federación resettlement scheme reveals poor planning and defective implementation that placed an undue and extra burden on the shoulders of the deprived and powerless, who were excluded almost entirely from the benefits associated with the resettlement project. Resettlement was limited to the construction and adjudication of houses and buildings, with little consideration paid to the mitigation of social and economic impacts.

6.3 The Piedra del Águila Dam and the Mapuche Indians

6.3.1 Description

The Piedra del Águila hydroelectric plant is located at the limits of the Argentine provinces of Neuquen and Rio Negro, both in the region known as Comahue, about 24 km from the small town of the same name. Its construction (1983–1990) was carried out by the state-owned company Hidronor, though it was later sold to private firms. Fed by the Limay and Collon Cure Rivers, the artificial lake formed by the dam covers an area of 292 km^2 and accumulates a volume of 12,400 hm^3 of water (Fig. 6.2).

Its concrete dyke, 173 m in height, categorises it as a high dam. The generating equipment of the plant is composed of four turbines of 350 MW each. This production capacity makes it one of the largest privately held hydroelectric complexes in the country. It produces 5,500 GW/h per year, which amounted in 1995 to 10% of national power generation, 32% of the energy produced by the Comahue region and 24% of hydroelectric energy produced in Argentina. In spite of the dam's privatisation, the national government holds 26% of its shares and the Province of Neuquen

13%. The construction of the complex was partially financed by loans from the World Bank and the Inter-American Development Bank and its final cost is estimated to be about US$1,400 million.

6.3.2 Population Affected by the Project

Most of the population affected by the project belongs to the Mapuche Indian group. The Mapuches (people of the land) are one of the more numerous aboriginal groups in Argentina and also one of the most successful in maintaining their ethnic and cultural identity, in spite of pressure from mainstream society. They are found living in scattered groups throughout the southern and Andean piedmont regions of Argentina. Besides constituting an important part of the rural and urban proletariat of the region, many of them are settled in 'reservations' and make a living from extensive cattle raising, and are subject to diverse forms of social, cultural and economic discrimination. In the Province of Neuquen most of the 'reservations' date from 1964, when the provincial government decided to put into effect provisions contained in the provincial Constitution of 1957.

Article 239'd' of that Constitution established the following: "The existing (aboriginal) reservations must be maintained and enlarged. Its inhabitants must receive technical and economic assistance in order to promote the education and training of its youth and the rational utilisation of the received land, improving their living conditions and working toward the objective of the progressive elimination of this form of 'de facto' segregation [the reservations]" (Constitution of the Province of Rio Negro, p. 15). The creation of those reservations was a partial response to the struggles of the Mapuche people, for whom the land constitutes not simply a means of production but also a symbolic expression of their ethnicity (ñuke mapu=mother land), and possesses a value that transcends its materiality, giving meaning to their daily life.

The community that was affected by the construction of Piedra del Águila was the indigenous reservation Pilquiniyeu del Limay, composed of about 500 people, of which 150 were finally resettled, grouped into 23 domestic groups. This reservation is located 230 km from the city of Bariloche, one of the most important tourist resorts in the country.

It is a dispersed settlement of smallholders (by Argentine standards) served by a small community centre and characterised by precarious housing and an almost complete lack of basic services. Economic activity was centred around goat husbandry (with an average of 300 heads per domestic group) and complemented with orchards of vegetables for domestic consumption. The main commodity produced for the market was goat wool that was sold to non-aboriginal intermediaries. A significant proportion of the monetary income was obtained through salaried work on neighbouring sheep ranches. Since the young people of the community have historically emigrated, it was common to resort to various forms of reciprocal labour exchange among neighbours and relatives to compensate for the labour shortages that occurred at certain points in the productive cycle. Group solidarity and a sense of belonging were maintained not only by existing kinship ties, but also by ethnic consciousness and common use of the language (mapudungun), in spite of the fact that its utilisation was strongly discouraged until recently.

Public education was available at an elementary school located in the 'community centre' of Pilquiniyeu. The infrastructure of this school was somewhat deficient and lacked, for instance, capacity for boarding, which could be a great help for those students who lived some distance away. Primary health assistance was provided through a 'sanitary post' periodically visited by physicians and paramedics who came by ambulance from the hospital in the locality of Comallo. Cases of more serious illness had to be sent directly to the aforementioned hospital. The reservation was served by dirt roads and, in the absence of any public transportation system, the mobility of its inhabitants depended exclusively on horses and on the goodwill of occasional travellers.

The impoundment directly affected grasslands, freshwater sources, orchards, springs, forested areas, 23 houses and—in the old community centre—the school building, the multiple use salon, the office of the Local Promotion Commission (Comisión de Fomento) and the sanitary post. Total land impacted included 2,422 ha in the Paso Flores Colony (inhabited by farmers of German descent), 3,995 ha in the Paso Limay Ranch, a small area of the aboriginal reservation of Ancatruz, and 9,600 in the aboriginal reservation of Pilquiniyeu del Limay. To this should be added 5,586 ha of the María Sofía Ranch, expropriated for resettling the people who had to be displaced in Pilquiniyeu.

6.3.3 Compensation and Resettlement Policies

The compensation and resettlement policy finally adopted was the result of a process that included long discussions and political manoeuvres and in which diverse private and public interest groups participated. Of particular importance was the intervention of formal organisations and non-governmental organisations (NGOs), both Mapuche and winka (white people), which actively lobbied in favour of the recognition of indigenous rights. A crucial step in this respect was the establishment of the Joint Commission (Comisión Mixta) with representatives from the community, the provincial government and the agency responsible for the construction of the dam, Hidronor.

The units in charge of implementing the programme were the Office for Regional Interests of Hidronor and the Joint Commission. They elaborated a resettlement and rehabilitation plan the main objective of which was to carry out the displacement of people affected in the indigenous reserve, trying to preserve the unity of the community, maintaining the rights acquired by its inhabitants, and resettling them in an ecologically equivalent area. It was understood that the new settlement should constitute a way to historically compensate the Mapuche people for their long sufferings, as well as to give them the opportunity to recover their practices on land and natural resources management.

This programme was initially designed as an integrated and participative development plan that would include not only the Mapuche but also other inhabitants of the surrounding region. The programme envisaged the replacement and improvement of existing infrastructure and services, the physical movement of persons and goods to the new site, the implementation of rehabilitation through productive, health and educational projects, and the provision of direct assistance to the community during the stage of transition. The site chosen for resettlement was the María Sofía Ranch, located close to the reservation. This ranch was expropriated by Provincial Law 2180/87 and was incorporated into the Pilquiniyeu Reservation by Disposition 714/72.

6.3.4 Implementation of Compensation and Resettlement

Although the resettlement and rehabilitation plan considered some of the negative impacts to be mitigated, it did not take into account some others produced by the expectations raised among inhabitants during the 4 years between the announcement and the effective implementation of displacement (1986–1990/1). Among these impacts, those that affected the productive system negatively can be mentioned:

1. The settlers stopped planting new trees and abandoned existing plantations.
2. They ceased maintenance works on irrigation channels and other facilities needed for horticultural labour.
3. They did not carry out needed repairs and improvements to their houses.
4. Pasture fields were overgrazed.
5. Merchants and intermediaries withdrew credit from producers facing the prospect of displacement.

Another problem derived from the topographic characteristics of the resettlement site. Although it was chosen, among other reasons, because of its proximity to the original reserve, it was not as 'ecologically equivalent' as was intended. Its lands were higher and colder than those of Pilquiniyeu, had less water, and lacked any native or planted tree cover. These factors represented serious problems for the development of economic activities when settlers were moved to the new locations between February and August in 1991. The land was distributed among the 23 domestic groups according to the size of their previous holdings and that of their herd, regardless of household size and composition. However, those who were resettled benefited from a substantial improvement both in quality and quantity of their herds. Not only did the project provide for the replacement of any animal that could be lost or injured during the move, but resettlers also received enough sheep according to the sustainable capacity of the new land free of cost.

The new houses had substantial improvements over the existing ones. They also took into account the preferences expressed by beneficiaries for 'modern' houses. Houses were allocated according to the size and composition of the resettled household up to a maximum of five rooms. Although there was general satisfaction with the new houses, some resettlers complained about the quality of building materials. The problem most frequently voiced pertained to the difficulties experienced in replacing domestic orchards, whose contribution to the subsistence of the family was important. Problems arose because of both lack of adequate care in the original site (those orchards were soon supposed to be abandoned) and scarcity of water for irrigation in the new locations. Nevertheless, such problems were mitigated by the delivery of assistance in the form of food and even clothes during the transition stage.

As mentioned above, although the original programme contemplated an ambitious gamut of actions, oriented by the notion of 'multidimensional resettlement stress' and the need to assist the resettled population in rebuilding their subsistence strategies and rehabilitating their production system, none of these actions was thoroughly implemented, once the construction of the replacement infrastructure and the physical movement were completed. Failure in carrying out the complete plan was due, among other factors, to misunderstandings between the technicians and the population, and to insufficiency of the financial and human resources assigned to post-movement actions.

6.3.5 Participation of Stakeholders

The return of democracy to the country in 1983 made possible a degree of participation by the affected population and other stakeholders in the design of the resettlement plan that would have been unthinkable during the military period. In the first place, in the time between the public announcement of the project and the formation of the Provincial Mixed Commission (Hidronor) and the adoption of the final resettlement plan, the entire Mapuche community in the region mobilised itself to defend the rights of those affected in Pilquiniyeu. It is important to note that participation was not limited to leaders and/or family heads, but involved women and the youth. Through this process the Mapuche were able to develop social practices that replaced individual strategies with collective ones.

The resettlement programme finally adopted allowed for the participation of the affected population in different instances, though the 'public discourse' did not always corresponded with reality. Thus, for example, future recipients intervened in the design of the houses—into the original design of which they incorporated elements from their traditional dwelling patterns—and also chose the location of the new community centre. Participation of those affected in the construction of the new houses not only permitted control over the quality of building materials, but for a time was also an important source of income. However, the social distance between technicians and the affected population frequently resulted in highly unequal degrees of participation and particularly of influence upon decisions.

6.3.6 Evaluation of Overall Impacts

Attempting, as a summary, an assessment of the resettlement process, a first conclusion is that it was basically sound, particularly from the point of view of those effectively resettled. These resettlers received more and better lands, increased and improved their herds, and also received better houses, services and infrastructure. Although there were various inconveniences during the move and occupation of the new sites—mainly the shortage of water to irrigate the new orchards and delays in providing energy to the houses—these do not seem to have been very serious and resettlers were compensated with additional assistance.

However, the situation was quite different for the population that remained in their old homes. They were left out of any benefits from the project. This difference of treatment was not perceived as totally just by the non-displaced, and provoked some degree of resentment towards the project. The failure to fully implement the so-called Integrated Development Plan meant not only a lack of any possible improvement, but real hindrances such as increased distance to the new community centre and consequently, to the school, the student lodging facility, the health centre, etc. On the other hand, the reorganisation of roads and paths increased the distances separating them from their neighbours and other populated centres. People settled in some locations became isolated due to the irregularity of the lake shore, making difficult both the marketing of produce as well as the supply of needed goods.

The few professionals from the social sciences who participated in the process could not convince Hidronor's authorities to accept the concept of 'indirect impact' (those not directly flooded), nor to recognise the existence of 'hidden social costs', which frequently cause greater impacts than the evident ones. However, a crucial result was that Hidronor accepted the notion of 'equivalent ecological area' as a compensation modality. It must be noted that the Pilquiniyeu Project did not consider the alternative of cash compensation for the affected population, as was provided, for instance, for the people living in Paso Flores, a German settlement nearby. Those working for the settlers, both Mapuche and non-Mapuche labourers, received cash compensation, and ended up living in barracks in the urban centres of the region.

6.4 The Yacyreta Project

6.4.1 Description

The Yacyreta Hydroelectric Project was in the 1980s one of the world's largest hydroelectric developments then under construction. It is located on the Parana River, between Argentina and Paraguay, at a site near the localities of Ituzaingo (Argentina) and Ayolas (Paraguay), and located approximately 100 km downstream from the cities of Posadas, the capital city of the Argentine Province of Misiones, and Encarnación (Paraguay).[1] The works include a 69.6 km long dam, a 2,700 MW

---

[1] The dam site is located 1,470 km upstream from Buenos Aires (Argentina) and 310 km southeast from Asunción (Paraguay). In terms of latitude, this location is roughly equivalent to that of Miami in the northern hemisphere.

6 Hydrodevelopment and Population Displacement in Argentina

power plant, a navigation lock, fish-passage facilities and irrigation outlets on both riverbanks. At its maximum, the reservoir will cover an area of 1,650 km^2, with a capacity of 21×10^9 m^3. Here are some further figures:

- An annual power generation capacity of 17,500 GWh (Gigawatt hour), representing 30% of current Argentine supply including all sources of generation.
- A flow through each turbine of 2,630 million l/h, equivalent to the consumption of drinking water in the city of Buenos Aires over 2 days, or Asunción city over 13 days.
- A reservoir surface of 1,650 km^2: eight times the area of Buenos Aires and 13 times the area of Asunción.
- A navigation lock allowing vessels to overcome a maximum difference in elevation of 23 m.

The project has 20 generators with an aggregate capacity of 2,700 MW and will generate about 17,500 GWh of electricity each year once the reservoir is operating at its ultimate water level. In 2008, the level of the reservoir was raised to 80 m, at which the generators already produced close to 90% of their rated output. So far the level of the dam could not be set to its ultimate design height of 83 m due to the lack of progress in various so-called complementary works, including the resettlement process.

The construction of the dam and the consequent creation of the reservoir have had and will continue to have a significant impact on the environment, both social and natural. The artificial lake will extend itself more than 200 km upstream the Parana River, with a maximum width estimated at 21 km, and will flood large areas in Argentina (29,900 ha) and in Paraguay (78,900 ha). The area to be flooded includes prime agricultural lands as well as substantial portions of the cities of Posadas and Encarnación, industrial and commercial facilities, roads, bridges, railways, etc.

The Yacyreta Project is regulated by the Entidad Binacional Yacyreta (EBY), a binational entity with legal, financial and administrative capacity and technical responsibility for the study, design, supervision and execution of the hydroelectric project.

6.4.2 A Brief Historical Description

Even though the Yacyreta Treaty was signed in December 1973—under the constitutional but troubled government of General Perón—actual work on the 'complementary works' (access roads, housing for technicians and workers, etc.) began only in 1979 (under a military Junta), and those corresponding to the dam were initiated in March 1984 (under a democratic government). Consequently, the project has already had a long history of 32 years (1978–2010), and its implementation has been affected by successive problems. It has been plagued by increasing economical and financial problems, as well as by political and institutional turbulence.

Presently, work on the structures of the dam is completed, while the environmental and resettlement components are still delayed (60% progress), and there persist serious uncertainties and doubts about their implementation.

The resettlement programme of the EBY was launched in 1979— more than 30 years ago!—with social teams working on both sides. The resettlement office for the Argentine side included a substantial number of social scientists drawn from the faculty of the National University of Misiones, a recruitment encouraged by the World Bank. The first action was to conduct exhaustive censuses in the affected areas. These censuses still constitute one of the main criteria for defining resettlement rights, although the EBY has recognised relocation rights also for those registered by the 1989/90 census updates, known as RAU/89. Increasing economic and financial problems, lack of administrative organisation, changes and confusion in resettlement policies, and the drastic and indiscriminate reduction in staff working on social issues, resulted in the dismantling of professional resettlement teams, serious delays in the construction of resettlement houses, and postponement of rural programmes. The implementation of this first Social Action Plan was discontinued in 1990, in conjunction with the rise to power of a new national government and a thorough change in EBY's directive and professional staff.[2]

In 1991, the EBY initiated new negotiations with the World Bank and the Inter-American Development Bank, seeking the concession of additional loans to finance not only the construction of the dam, but also the resettlement and environmental components, so far financed exclusively by the Argentine government. This was the beginning of negotiations of the loan identified as Yacyreta II by both Banks. Because of delays in executing the required resettlements and in implementing measures to protect the environment, and also because of the pressing need to minimise immediate investments, the EBY devised a scheme according to which the reservoir would be filled in three stages. In 1992, and with technical assistance from the World Bank, EBY developed the first version of the Action Plan for Resettlement and Rehabilitation or PARR (after its Spanish initials), which received initial approval from the Banks and, after some public consultations, started its implementation. Nevertheless, despite efforts to supervise and evaluate its implementation, delays and under-financing have continued to plague the resettlement component up to the present.

6.4.3 Population Affected by the Project

The area affected by the Yacyreta Project encompasses regions in Argentina (Provinces of Corrientes and Misiones) and of Paraguay (Itapua and Misiones Departments) where the main economic activities are agriculture and cattle ranching. The main crops are rice, soybean, corn and manioc. In the precolonial period,

---

[2] For an assessment of Yacyreta's experience with urban resettlement on the Argentine side, see Bartolome (1993).

this area was inhabited mainly by Amerindians of Guarani origin and developed a traditional Creole and mestizo society in colonial and postcolonial times. From the end of the nineteenth century, it saw European immigration in significant numbers (Germans, Slavs, etc.), and more recently has received Japanese immigrants. Settlers of immigrant origin predominate among farmers, locally known as 'colonos'. The most important urban centres in the region are Posadas (population 500,000) in Argentina, and Encarnación (population 80,000) in Paraguay. Both cities are commercial and service centres, with few industries not directly related to agricultural production.

More than 92.6% of the population affected in Posadas lack legal rights over the land and another high percentage are very underprivileged, with poor housing and living conditions. According to EBY's estimates, based on data from the RAU/89 census update conducted in 1989/90, 72% of the economically active population works in the tertiary sector, with domestic services, construction and infrequent services accounting for 50.2%. There are no available updated data on family incomes, but the historical average has oscillated around the legal minimum salary.

According to data provided by EBY, a full 77% of families living in the affected areas were composed of relatively recent rural immigrants. These urban poor tend to inhabit the low-lying coastal areas and the abrupt slopes of river cliffs along the river front of the city. Most of that land has little or no value as real estate because it is subject to periodic flooding. Its owners have traditionally kept it for speculation or as their patrimony. Rural immigrants settled in these areas, either joining the older settlements inhabited by Paraguayan immigrants or creating new ones.

Although somewhat slowed now, the past growth of the city originated considerable demand for unskilled labour in the construction industry and in the 'personal services' sectors of the economy. This labour market operates to a large extent within the informal economic sphere and is characterised by low wages, occupational instability and the absence of any social security benefits. Those who make a living in this market try different survival mechanisms in order to compensate for these unfavourable conditions. Such circumstances have also contributed to the emergence of a disproportionately large sector of self-employed workers. According to the 1979 census, about 20% of the employed males in the affected population worked in the construction industry, while another 13% made a living from different temporary jobs (locally known as changas). Also, more than 79% of the employed males were unskilled workers, low-rank employees in public administration and commerce, or occasional workers (changarines).

Working women accounted for 31% of the total labour force, with an overwhelming concentration in domestic services (80%). Only half of the paid workers held permanent positions of any kind, while almost 50% of those receiving cash incomes were declared as self-employed. When only squatters were considered, the percentage of those holding temporary or very low paying jobs reached 77%. In summary, more than two thirds of the population affected by the Yacyreta Project in Posadas is in the category of 'urban poor'. They work in activities related to construction, transportation and merchandising, and in the provision of diverse low-cost services to the middle and upper social classes.

The final number of persons to be displaced by the project is rather difficult to estimate. According to estimates in August 2008,[3] the total number of persons that have been resettled or have yet to be resettled in both countries amount to 68,000 (15,033 families), a figure that is likely to increase if the resettlement process suffers further delays. Out of this figure, 37,000 persons (8,108 families) are in Argentine and more than 31,000 persons (6,925 families) in Paraguay. A majority of these project-affected people are urban, particularly in Argentine, and almost 92% of the affected population lives in Posadas. In Paraguay, the urban component is also important (69.2%)—most of these in the city of Encarnación—but the rural affected population is more than 10,000 persons.

6.4.4 Main Social Impacts

Besides affecting large agricultural areas in both countries, the Yacyreta Project affects urban residential areas as well as infrastructural works and services. A majority of the residential dwellings to be impounded in Posadas are of low quality or little more than precarious huts. Only 8% of the affected dwellings are brick and masonry constructions, with tile or zinc roofs; 18% are built using refuse materials, and the rest are wooden structures with zinc, fibre or board roofs. The reservoir will also affect several urban infrastructure, community services, and commercial and industrial facilities. Among the affected structures are 10 elementary schools, two churches, two health centres, police and river guard facilities, nine sawmills, one boat factory and approximately 120 'olerias' or small artisan brick factories. The water intake of Posadas is also affected, as well as the port installations, the railway station, 30 km of railways, 10 km of paved roads, the power plant and parts of the electricity, telephone and water distribution networks.

The urban impacts in Paraguay are mainly in Encarnación and, to a much lesser extent, in Carmen del Parana. In contrast to the Argentine case, a large percentage (67%) of affected urban families belongs to the middle and upper social classes of the area. Hence, there are many constructions of good quality among the affected houses. The Yacyreta Reservoir will flood the main commercial district in Encarnación, including about 600 shops. It will also affect 280 agro-industries, tanning plants, sawmills, grain storages, etc. Of particular importance is the impact of a large number of brick-producing factories, ranging from small artisan firms to medium-sized mechanised factories. Among the infrastructure affected are government buildings, three elementary schools, two churches, military headquarters, the city port, 106 km of railways, the power plant, approximately 70% of the electricity distribution network, 80% of telephone lines, the water intake of the city, and a major part of the water distribution pipes.

---

[3] Data provided by EBY.

6.4.5 Compensation and Resettlement Policies

As stated earlier, the resettlement programme of the EBY was launched in 1979—when the country was ruled by a military Junta—with social teams working on both sides. The first action it undertook was the implementation of exhaustive censuses of the affected areas. These censuses still constitute one of the main criteria for defining resettlement rights, although the EBY has also agreed to provide relocation rights to those registered in the 1989/90 census updates.

That first resettlement programme was based on the following laws and international treaties: (1) the Yacyreta Treaty and its Annexes (Law No. 20.646 R.A. (Argentina) and Law No. 433 R.P. (Paraguay)); (2) Additional Protocols and Reciprocal Notes, N.R. No. 10 of 15.09.83, approving Resolution N^a 141/83 of the Administrative Board of EBY, List of Resettlement Works; (3) expropriation and relocation laws: Law No. 21.499 R.A. and Law No. 944 R.P. (expropriations), Law No. 22,313 R.A. (resettlements); (4) national decrees approving the delimitation of affected areas (R.A.), Decree No. 1,585 of 21.12.82; (5) loan agreement with the World Bank and execution agreement with the Inter-American Development Bank—Loan Contracts; and (6) Norms for Resettlements in the Yacyreta Project—Resolution No. 48/78 of 7.4.78 of the Administrative Board of EBY.

The resettlement policy adopted was based on the following guidelines: (a) Yacyreta was to resettle all population and infrastructure in the affected areas, at a reasonable cost, in such a way as to avoid the violation of personal rights and undue privileges; (b) The displaced population would be offered compensation or resettlement in houses to be built by Yacyreta, maintaining appropriate standards of comfort and health, and endowed with basic services (electricity, water, sewage, etc.); (c) Population resettlement would aim to avoid social inequality. It would also seek to maintain the affected population's existing access to working sites and service and educational centres; (d) Small-scale shops were to be resettled in commercial areas, depending on their activities; (e) Large rural properties would receive compensation, but small farms would be offered resettlement into equivalent land; and (f) all resettlement planning and action would be based upon precise censuses and inventories of affected persons and property, and would be coordinated following the rules and regulations of the government institutions of each country.

On the basis of these guidelines, Yacyreta prepared manuals for the implementation of resettlement of rural, urban and suburban populations, commercial and industrial facilities, community services and equipment. Those procedures were approved by the Administrative Board of EBY in its ordinary meeting No. 46/83 (Item 6) of 13 and 14.07.83. The criteria for the assignment of resettlement houses were also approved by Resolution No. 665/84 of 17.10.84. Almost in tandem with the census operations, Yacyreta launched a Social Action Plan (PAS). This plan was more intensively implemented in Posadas, where the first resettlements were to take place. It included the following programmes: (1) Programme for Labour Training and Adult Education; (2) Programme on Public Health for Resettlement; (3) Programme on Resettlement Planning; (4) Programme for Resettlement and Housing

Adaptation; (5) Programme on Documentation and Social Security Normalisation; (6) Programme on Social Promotion; (7) Programme on Communication for Resettlement; and (8) Programme on Institutional Coordination. With the exception of programmes 3, 4 and 8 which were implemented by EBY staff, the rest were executed through agreements with federal and provincial organisations having competence in the relevant area.

A similar plan was implemented in Paraguay, although the programmes had slightly different names. Rural plans in both countries included technical assistance programmes for small farmers.

Yacyreta built several housing developments for resettlement purposes. These were composed of single-story family houses with all necessary urban services. In compliance with the approved operational procedures, owners of urban and suburban houses received a resettlement house. Non-owners (tenants, occupants, etc.) received resettlement houses under a 30-year sale plan, requiring the payment of monthly instalments that would not exceed 20% of the family income. This implied a 65% subsidy on the real cost of the houses, including urban infrastructure. The EBY Administrative Board approved this repayment plan via Resolution No. 181/85 of 11.12.85.

Regarding impacts in the rural areas, the operational guidelines gave the option of resettlement to small farmers whose plots were up to 20 ha. Non-owners who chose resettlement had to buy the land in instalments. Progress was very slow and different for each country: minimal on the Argentine side and of some magnitude on the Paraguayan side. Until the second semester of 1989, Yacyreta resettled or paid compensation to approximately 1,281 urban and rural families. Increasing economic and financing problems, lack of administrative organisation, changes and confusion in resettlement policies, and the implementation of a drastic and indiscriminate reduction in staff resulted in the dismantling of professional resettlement teams, delays and cancellations in the construction of resettlement houses, and delays in implementing rural programmes.

Resettlement actions were paralysed for all of 1990 and part of 1991. In the latter year, Yacyreta again initiated negotiations with the World Bank and the Inter-American Development Bank in order to obtain an additional loan to continue the resettlement process. In 1992, and with technical assistance from the World Bank, Yacyreta elaborated the first version of the Action Plan for Resettlement and Rehabilitation. This action plan received initial approval from the Banks and, after some public consultations, Yacyreta began its implementation.

The new and currently valid resettlement policy is based on the following principles: (a) A definition of 'affected population' which includes all those who are forced to move because of the Yacyreta Project, whatever their tenancy condition, provided they have been registered in the census updates of 1989/90; (b) The recognition of resettlement rights to all persons registered in the 1989/90 censuses, including the right to receive either houses or land (known as Basic Productive Units in the rural areas), with infrastructure enabling resettlers to raise their standards of living, and the rehabilitation of the rural and suburban economic activities. This proposal makes the solutions offered to those affected in urban and rural areas more equitable, since the old procedures required rural resettlers to pay for the land,

6 Hydrodevelopment and Population Displacement in Argentina

and they were not entitled to receive replacement houses; (c) Population settled in affected areas after 1989/90 are not counted as beneficiaries of the resettlement programme, and must be resettled by the respective governments, albeit with assistance from EBY.

The current policy also proposes that non-owners receive houses or land, free of cost and without incurring any debt. In contrast, owners are to pay up to 35% of the value of the houses or land along a 30-year term, plus interests at an annual rate of 6% when their pre-existing properties are less valuable than the new ones. It must be pointed out that the new batch of resettlement houses are of lower quality and are poorly finished as compared to the existing ones, and are no longer assigned according to the household size. Since there is a considerable number of resettled families that have contracted payment obligations due to their houses, they are granted the option of continuing with the payments, or moving to a new, cheaper house that will be provided without further payments. The policy also states that all legal and administrative expenses will be paid by Yacyreta. For affected rural plots, the alternative of resettlement is offered provided that the area does not exceed 20 ha. Larger units are entitled only to compensation. Small farmers, both owners and non-owners, are entitled to receive land of equal size as their affected units, but within a minimum of 7.5 ha.

The Action Plan for Resettlement and Rehabilitation also contains a series of actions geared at providing social support for the resettlement process and to help the socio-economic rehabilitation of the resettled population. Included in the former is a programme of social communication, and in the latter are programmes for health, education, community development, agricultural technical assistance, economic growth and assistance to Amerindian communities. All these programmes require the active participation of governmental organisations and NGOs. There is also a monitoring programme to be implemented by Yacyreta, and one for evaluation of the resettlement programme that was contracted with a Brazilian consultancy firm.

After long negotiations, the World Bank and the EBY agreed on launching the Yacyreta II Project, which assigned responsibility for resettlement implementation to the Department of Complementary Works. At the time of appraisal, the department had a staff of only five experts, clearly an inadequate number for managing the resettlement of approximately 50,000 persons. The agreement between the World Bank and EBY established that this department should (a) recruit an internationally renowned consultant to work as an experienced resettlement coordinator; (b) hire approximately 50 additional professionals to deal with resettlement and environmental issues; (c) involve existing agencies to deliver key services such as health and education, and fund the additional consultants; (d) design an integrated training programmes for all resettlement staff; (e) add two resettlement specialists to the project's panel of experts; and (f) establish through the Secretariat of Natural Resources and Human Environment (Argentine) a national forum for governmental organisations and NGOs to review the annual implementation reports. It was also agreed that EBY should monitor progress in resettlement and environmental plans, and hire a third institution, from neither Argentina nor Paraguay, to carry out an evaluation of the programme.

6.4.6 Implementation of Compensation and Resettlement

From early 1993 to the present, EBY has been implementing the Action Plan for Resettlement and Rehabilitation and trying to meet the Bank's requirements, although at a slow and half-hearted pace. An experienced foreign resettlement specialist was appointed as Resettlement Coordinator for both countries and began working in January 1993, but resigned after 2 years. The number of field officers was increased but other experts mentioned in the agreement with the World Bank were not appointed. The EBY also met the requirement of appointing two resettlement specialists to the project's panel of experts, although no effort has so far been made to fulfil the conditions in point (f) (see above). Yacyreta has resisted the requirement of hiring a third-country institution to carry out monitoring and evaluation of the resettlement programme. Instead, there are signed agreements with the Catholic University of Asunción, Encarnación Campus, and with the National University of Misiones to monitor the resettlement programme. It must be mentioned that the budget assigned to this task was clearly insufficient and that the quality of the proposals put forward by these institutions was not taken seriously in the process of selection. Not surprisingly, the monitoring reports from the Catholic University of Asunción, Encarnación Campus, were very poor and their involvement had to be cancelled, without any further developments till date. However, the participation of the National University of Misiones was effective until 2000, although EBY appeared unwilling to accept any criticism.

Little progress has been made in relation to point (c), regarding the delivery of key services such as health and education by existing agencies, and the financing of expert consultants. An agreement with the Provincial Secretariat for Welfare, Women and Youth of Argentine was soon suspended because of its unsatisfactory results. The agreement with the Provincial Secretariat for Health (Argentine side) is being executed but proving difficult to manage it efficiently. The only external agreement that seems to have resulted into a clear output is one with the Provincial Sub-secretariat for Mining and Geology (Province of Misiones), which produced a resettlement and compensation plan for affected artisans (brick-makers). Even so, EBY accepted the report and promised to implement it only under heavy pressure from the provincial government of Misiones and from the affected people, who demonstrated on several occasions in front of EBY's offices in Posadas. The continuous delay on key decisions by EBY authorities and their reluctance in providing clear information to the community resulted in a severe deterioration of relationships between the Yacyreta Project, the provincial government and the public.

One of the arguments presented by Yacyreta's authorities for reducing resettlement benefits was that any additional investment in this area would threaten the profitability of the entire project. It is therefore interesting to consider the issue of investment in resettlement from a wider perspective. The World Bank's worldwide review of projects (1994) compared per capita resettlement allocations with per capita GNP figures, and found a close correlation between investment levels and project capabilities. None of the projects with a ratio of 3.5 or higher has reported major resettlement difficulties. On the contrary, virtually all projects with a ratio lower than 2.0 are experiencing serious implementation problems. It is worth noting that the Yacyreta Project's figures are slightly below this threshold. It is also interesting to note that the same international financing institution (World Bank 1994: Table 5.5) estimates that as of 1994 resettlement represented 2.2% of the cost of the Yacyreta Project. To some degree, this answers the frequently raised argument that larger investments in resettlement cannot be undertaken without jeopardising a project's feasibility. Since the rate of return on the project was estimated at 24–28% at the time of appraisal, it would be necessary to increase resettlement costs 571% to lower the project's return below (an acceptable) 12%!

Although the construction work is practically complete and the reservoir almost full, only 4,927 families (32.8%) have been resettled so far, 62.7% of these on the Argentine side. This means that it took the project almost 20 years to relocate little more than 30% of the affected population, while there is talk of completing the filling of the reservoir in a year or two!

6.4.7 Participation of Stakeholders

Among the conditions set by the international banks was the implementation of public consultations and discussions regarding the 1992 Action Plan for Resettlement and Rehabilitation. These consultations were to include governmental organisations as well as NGOs. Yacyreta accepted this suggestion and initiated a series of consultations and meetings that were supposed to culminate in the approval of the final version of the above plan. Yacyreta sent copies of the April 1992 version of the Action Plan for Resettlement and Rehabilitation to several governmental organisations and NGOs at the national, provincial (departmental) and municipal levels.

The first of this series of consultations took place in Ayolas, Paraguay, from 15 to 17 July 1992, and was devoted to discussion of the Action Plan for Resettlement and Rehabilitation and the Environment Programme (Plan de Manejo de Medio Ambiente, PMMA in Spanish). Several important NGOs did not attend the meeting, arguing that they were not given sufficient advance notice or that they lacked the funds to cover the participation of their representatives. Others expressed direct opposition and refused to participate. Such was the case with the Foundation for the Defence of the Environment and 200 other Argentine NGOs, including the Southern Cone chapter of Greenpeace, which not only rejected the call but also issued an international press statement, denouncing the meeting as a fraud organised by Yacyreta simply to obtain new loans from international banks. In July 1994, the Municipal Council of the City of Posadas pronounced EBY's Executive Director of Argentine a *persona non grata*.

The final meeting took place in Posadas on 3 August 1992, and it was intended to close the consultation series with a talk on the proposals accruing from the Ayolas meeting as well as Yacyreta's answers, and a presentation of the final version of the Plan for Resettlement and Rehabilitation. It was also expected that the meeting would close with the signature of an agreement between Yacyreta, governmental organisations and NGOs, formally approving the Plan for Resettlement and Rehabilitation. Events did not follow this plan. Although no serious observations were made on the Plan, the Environment Programme was met with strong criticism. The Inter-Sectorial Commission of the Province of Misiones presented a harsh statement. It was signed by the Ministries of Public Health, Ecology, and Economy, the Mayor of Posadas, the Undersecretary for Hydroelectric Projects, the Rector of the National University of Misiones, the Coordinating Commission of Neighbourhood Commissions of Level 84 and the Yguazu Foundation.

This statement questioned the motives of Yacyreta in calling the meetings, pointed out the contradiction between the apparent purpose of the meetings and the fact that the authorities of Yacyreta were continuing dismantling its Department for Resettlement and the Environment. It also stated their intention of not attending further meetings or collaborating with Yacyreta, until the EBY presented proof of the seriousness of its intentions in relation to resettlement and environmental considerations. Representatives of the Province of Corrientes agreed with this statement and announced that the government of the province would issue a similar document. In the face of such opposition, the meeting was closed without even a mention of the possibility of signing the expected agreements.

6.4.8 Evaluation of Overall Impacts

It is still difficult to evaluate the full impacts of the Yacyreta Project but after more than 30 years of an unfinished process, it is certainly possible to conclude that the people have paid a high cost, both in terms of money and suffering, because of the rambling course of this project, its chronic financial problems and the continuous policy changes. In fact, Yacyreta represents a good example of the consequences of assigning secondary, or even marginal, importance to the social and environmental components of large-scale development projects. Even the term given to these components—'complementary works'—clearly shows what the project's managers and technicians consider to be the 'core' of the endeavour.

It must be noted that the fact that a project designed and started under an unpopular dictatorship included a resettlement and rehabilitation plan was the result of pressure exerted by the World Bank. As a matter of fact, most of the EBY authorities regarded considered these requirements to be undue and arbitrary. However, the World Bank was not consistent in its pressures and demands, and accepted again and again EBY's shortcomings and changes in policies. In other words, threats without consequent actions resulted in a pattern predictable for the EBY authorities and repeatedly used to either confirm their decisions or to postpone even crucial ones. The monitoring of resettlement was ineffective and performed by diverse specialists who did not always share the same point of view, not even in respect to those policies that the World Bank has already considered as 'good practices'.

When the EBY changed its personnel, greatly reduced the number of staff working in social issues and cancelled the social plan in 1990, the World Bank granted a new loan and accepted the establishment of a Resettlement Office staffed by poorly qualified personnel and without any autonomy. It also approved the downgrading of resettlement-related decisions (for instance, houses of poor quality and of only one size for all families), the execution of poorly planned social assistance programmes, etc. Of course, the bank is not the only party responsible for the failures and shortcomings of the project, but it can at least be said that it was weak in enforcing its own guidelines and policies. The result is that almost three generations have lived under the threat of displacement and the uncertainty of changing policies and resettlement solutions. Even now many of the affected people do not know when and under what conditions they are going to be resettled.

6.5 Lessons and Conclusions

6.5.1 Assessment of the Argentine Experience of Compensation and Resettlement

Although not numerous, the Argentine cases considered here present a wide variety of institutional contexts and specific problems. In this sense they allow us to draw some conclusions that are in line with worldwide experience in these matters. These lessons can be stated as follows:

a. The production of energy is the overwhelming objective of all hydropower projects, even when they are frequently presented as 'multi-purpose' development projects. Once the production of energy is secured all other objectives fade away and, consequently, also the funds for these components.
b. Under various planning and financing schemes, funds become scarce when the time comes to implement resettlement and rehabilitation components and these are either under-financed or wholly abandoned. The conclusion we can draw is that it is not enough to count on a good and well-structured resettlement and rehabilitation programme if the funding is not guaranteed. International financial organisations and banks must place particular emphasis on this aspect when negotiating loans for a project that involves population displacement, particularly when the loan is intended specifically for resettlement and rehabilitation.
c. When responsibility for resettlement is assigned to agencies other than the one implementing the main project, the outcome is not better than otherwise, and the risk for asynchronies is greatly increased. We think that the best alternative is to have a centralised organisation, provided that the sector in charge of resettlement and rehabilitation is adequately funded and empowered from the administrative viewpoint.

d. Except at the beginning of the Yacyreta Project—and then too, only because of the pressure of the World Bank—social science knowledge has not been deemed necessary for planning or executing resettlement and rehabilitation plans. Social science professionals are seldom involved in resettlement and rehabilitation schemes, although they are sometimes called in as consultants or to provide specific inputs. Resettlement is often left in the hands of architects and social workers, with the latter excluded from formulating policy suggestions and reduced to the status of 'field' personnel in charge of the implementation of policies designed by others frequently without any training in the social sciences. Although nobody ever dreams of leaving the design of the dams to an amateur, they do not hesitate to do so with social matters. Social science inputs should be included from the earliest stages of planning, and trained professionals must participate in the execution and evaluation of resettlement and rehabilitation activities. Particular attention should be paid to the qualifications of such professionals.

6.5.2 Stakeholders and Participation

Participation of stakeholders is often a formal matter or treated as such by those responsible of a development project. When such participation is allowed and included in the process, it is frequently because of imposition (by lending institutions, for instance) or due to political circumstances that make some degree of participation unavoidable. The fact is that participation is at its core a matter of 'power' and that stakeholders are rarely homogeneous in this respect. In Orwellian terms: all stakeholders are equal but some are more equal than others. Real participation implies the capacity to influence or even modify decisions, and the poor and powerless are seldom 'granted' that capacity. Instead, they are manipulated, co-opted or directly excluded. The case of those holding some degree of power is different; for instance, the upper and middle sectors of Federación were able to change the decisions of the national government even under a military dictatorship.

Ignoring this political reality leads to naiveté and even to becoming an unwilling accomplice in the make-believe scenarios set up by the powerful. The zero-conflict models frequently accepted by planners unrealistically assume that wolves and sheep may have common interests and the same capacity to lobby for them. The answer, we believe, is to encourage the self-organisation of the powerless and to accept that conflict may be a part of the process of negotiating a solution. The cost of a lengthier negotiation stage may be repaid through swifter implementation and better (more just) results.

6.5.3 Normative Framework and Policy Implementation

The existence of nationwide norms and of an officially approved resettlement and rehabilitation policy can be of great help in securing compliance. However, in the absence of such norms and policies, the role of the international lending organisations becomes crucial. The fact is that no resettlement and rehabilitation policies would exist in Argentina without the requirements imposed by the World Bank or the Inter-American Development Bank. This acknowledgement notwithstanding, it must be also noted that the banks have not been systematic and congruent in applying these requirements. Often monitoring missions were not consistent in their appraisals or accepted undue delays and changes from the agreed actions, which were already a compromise. The case of the Yacyreta Project illustrates this point. We are aware of the difficulties in striking a balance between 'guidance' and 'undue intrusion', but we are also convinced that the development and enforcement of clear rules throughout the process can be very effective in securing 'good practices'.

Bibliography

Asociación Internacional de Fomento (1994) Invertir en el Futuro. Banco Internacional de Reconstrucción y Fomento, Washington DC

Balazote A (1987) Tenencia de la Tierra, Producción y Organización Social en Dos Comunidades Mapuche (Pcía de Río Negro). Proyecto de investigación, UBACyT, Buenos Aires

Balazote A (1993) ¿Nuevas Propuestas o Viejas Constricciones? Cuadernos de Antropología, No. 4, UNLu, Luján

Balazote A (1994) Relocalización de la Comunidad Mapuche de Pilquiniyeu del Limay en el Marco del Proyecto Piedra del Águila. Universidad de Buenos Aires (disertación doctoral), Buenos Aires

Balazote A, Radovich J (1988) Economía Doméstica en la Comunidad de Naupa Huen. Cuadernos de Antropología, No.1, EUDEBA-Universidad Nacional de Lujan

Balazote A, Radovich J (1990) Reproducción Social y Migraciones en Naupa Huen, Provincia de Río Negro. Revista de Antropología, Año V, No. 9. Buenos Aires

Balazote A, Radovich J (1991a) Piedra del Águila y el Impacto Social de las Grandes Represas. Ciencia Hoy 2: 11 (enero-febrero), Buenos Aires

Balazote A, Radovich J (1991b) La Represa de Piedra del Águila: La Etnicidad Mapuche en un Contexto de Relocalización. América Indígena, Ciudad de México

Balazote A, Radovich J (1992) El Contexto Político-administrativo en la Relocalización de Pilquiniyeu del Limay. Papeles de Trabajo No. 2, Centro de Estudios Interdisciplinarios en Etnolingüística y Antropología Socio-cultural, Universidad Nacional de Rosario

Balazote A, Radovich J (1993a) Estudio Comparativo del Proceso Migratorio en Dos Comunidades Indígenas de Río Negro y Neuquén. Cuadernos del Instituto Nacional de Antropología y Pensamiento Latinoamericano, No. 14, INAPL

Balazote A, Radovich J (1993b) Gran Obra e Impacto Social en Pilquiniyeu. CEAL, Buenos Aires

Baranger D, Dieringer A (1982) Estudio sobre los Medios de Subsistencia y la Capacidad de Pago de la Población no-propietaria de la Etapa I de Relocalizaciones. Instituto de Investigación de la Facultad de Humanidades y Ciencias Sociales para la Entidad Binacional Yacyretá, Posadas

Barrios L (1991) El Impacto de la Construcción y Operación de la Represa Hidroeléctrica de Salto Grande. Instituto de la República. Universidad de la Republica, Regional Norte, Fundación de Cultura Universitaria, Salto, Uruguay

Bartolome LJ (1982) Financiamiento de Viviendas Económicas para no Propietarios a ser Relocalizados por el Proyecto Yacyretá en Posadas. Posadas: Programa de Relocalización y Acción Social (PRAS), Entidad Binacional Yacyretá, Posadas

Bartolome LJ (1983) El Papel de los Programas de Acción Social en los Procesos de Relocalización Compulsiva de Población. Revista Interamericana de Planificación [SIAP, México] XVII(68): 115–131

Bartolome LJ (1984a) Aspectos Sociales de la Relocalización de la Población afectada por la Construcción de Grandes Represas. In: Suárez F, Franco R, Cohen E (eds) Efectos Sociales de las Grandes Represas en América Latina. Fundación Cultura Universitaria, Montevideo, pp 115–144

Bartolome LJ (1984b) Forced resettlement and the survival system of the urban poor. Ethnology XXIII(3):177–192

Bartolome LJ (1985a) La Familia Matrifocal en los Sectores Marginados: Desarrollo y Estrategias Adaptativas. Runa XIV:23–49, correspondiente a 1984

Bartolome LJ (1985b) Estrategias Adaptativas de los Pobres Urbanos: el Efecto Entrópico de la Relocalización Compulsiva. In: Bartolome LJ (ed) Relocalizados: Antropología Social de las Poblaciones Desplazadas. Ediciones del IDES N° 3, Buenos Aires, pp 67–115

Bartolome LJ (1993) The Yacyreta experience with urban resettlement: some lessons and insights. In: Cernea MM, Scott G (eds) Anthropological approaches to resettlement: policy, practice, and theory. Westview Press, Boulder, pp 109–132

Bartolome LJ (1994) Theoretical and operational issues in resettlement processes: The Yacyreta Project and urban relocations in Posadas (Argentina). Environmental and Social Dimensions of Reservoir Development and Management in the La Plata River Basin, UNCRD Research Report Series 4, Nagoya, pp 43–57

Bartolome LJ, Baranger D, Herrán C (1978a) Vivienda y Trabajo en las Zonas Urbanas y Periurbanas de Posadas a ser Afectadas por la Represa de Yacyretá. Centro de Investigación Social de la Facultad de Ciencias Sociales para Entidad Binacional Yacyretá, Posadas (unpublished report)

Bartolome LJ, Baranger D, Herrán C (1978b) Análisis Socioeconómico y Tipificación de la Población de las Zonas de Inundación (Posadas): Planteo de Opciones y Factores de Decisión. Centro de Investigación Social de la Facultad de Ciencias Sociales para Entidad Binacional Yacyretá, Posadas (unpublished report)

Bartolome LJ, Baranger D, Herrán C (1978c) Caracterización y Diagnóstico Socioeconómico de la Población a ser Relocalizada por la Construcción del Obrador y de la cabecera del Puente Internacional Posadas-Encarnación. Centro de Investigación Social de la Facultad de Ciencias Sociales para Entidad Binacional Yacyretá, Posadas (unpublished report)

Catullo MR (1987) Identidad Comunitaria e Identidad Barrial en un Proceso de Relocalización Compulsivo de Población (ciudad Nueva Federación), Entre Ríos. In: Ringuelet R (org.) Procesos de Contacto Interétnico. Editorial Búsqueda, Buenos Aires, pp 113–136

Catullo MR (1992) Reconstrucción de la Identidad y Proyectos de Gran Escala: Ciudad Nueva Federación, Provincia de Entre Ríos, Argentina. Serie Antropología, Departamento de Antropología, Universidad de Brasilia, Brasilia

Catullo MR (1993) La Antropología y los Proyectos de Gran Escala: los Estudios sobre Represas Hidroeléctricas en Brasil. Anuario Antropológico/90, Universidad de Brasilia, Brasilia

Catullo MR (1996a) A Antropologia e as Represas Hidroelétricas no Brasil. CEPPAC- Universidad de Brasilia, Brasilia, Cuadernos de América latina

Catullo MR (1996b) Poder y Participación en Proyectos de Gran Escala. Análisis Comparativo de los Procesos de Relocalización por la Construcción de la Represa Binacional Argentino-Uruguaya de Salto Grande. Doctorado FLACSO-Universidad de Brasilia (UnB) en Estudios Comparativos sobre América Latina y el Caribe, Brasilia

Catullo MR (1997) Poder, Participación y Niveles de Integración en Proyectos de Gran Escala. Actas del V Congreso Argentino de Antropología Social, La Plata

Catullo MR (1999) Proyectos de Gran Escala en el Marco del Mercosur. Clases Sociales, Intereses Sectoriales y Brokers en Procesos de Relocalización. Cuadernos del INAPL, No. 18, Buenos Aires

Choucri N (1982) Energy and development in Latin America. Massachusetts Institute of Technology, Lexington

Cluigt J, Francioni M, Poggiese H (1987) Implicancias del Emprendimiento Hidroeléctrico de Piedra del Águila sobre el área Pilquiniyeu del Limay. Secretaría de Planificación de la provincia de Río Negro, Viedma (MS)

6 Hydrodevelopment and Population Displacement in Argentina 151

Cluigt J, Francioni M, Poggiese H (1990) Planificación-gestión y Autodesarrollo Rural en la Relocalización de la Comunidad Indígena Pilquiniyeu del Limay. III Congreso Latinoamericano de Sociología Rural (10–14 Octubre 1990), Neuquén. Comisión Mixta Provincial-Hidronor

Dibble S (1992) Paraguay plotting a new course. National Geographic 182(2):88–113, Washington DC

Entidad Binacional Yacyreta (1979) Programa de Relocalización y Acción Social Urbano. Margen Argentina. Programa de Relocalización y Acción Social (PRAS), Entidad Binacional Yacyretá, Posadas

Entidad Binacional Yacyreta (1981) Informe Analítico del Censo 1979 de Viviendas y Hogares Afectados por el Proyecto Yacyretá. Dirección de Coordinación, Programa de Relocalizaciones Urbanas y Acción Social, Posadas

Entidad Binacional Yacyreta (1988) Memoria Anual del Programa Urbano de Relocalización y Acción Social. PRUAS, Posadas

Entidad Binacional Yacyreta (1992) PARR-Programa de Acciones de Reasentamiento y Rehabilitación. EBY, Posadas

Equipo de Antropología Social (UBA) (1987) Documento Síntesis del Seminario-Taller. Secretaría de Planificación-C.F.I, Viedma, Río Negro (unpublished report)

Escay J (1988) Summary data sheets of 1987 Power and Commercial Energy Statistics for 100 developing countries. International Bank for Reconstruction and Development, Washington DC

Ferradas CA (1989) Communication processes in a development project: the Yacyreta Hydroelectric Dam, Misiones, Argentina. Dissertation in Anthropology, City University of New York, New York (unpublished Ph.D.)

Grenfell, Morgan and Co. Ltd. (1993) Evaluation of options for the private sector participation in EBY, Posadas (final report)

Grubb M (1993) The Earth Summit Agreements: a guide and assessment; an analysis of the Rio '92 UN Conference on Environment and Development. Brookings Institution, Washington DC

Hamilton S (1991) Finding the forgotten: researching displaced urban families in Argentina. Syracuse University, Albany (unpublished manuscript)

Hidronor SA (1981) Informe de Impacto Ambiental, Neuquen (unpublished manuscript)

Hidronor SA (1987) Relevamiento Antropológico y Propuestas para la Relocalización en la Reserva Indígena de Pilquiniyeu del Limay. Buenos Aires (unpublished manuscript)

Integracion Latinoamericana (1991) Hidrovia Paraguay Parana: Viejo Cauce para Una Integración Renovada, Buenos Aires

International Energy Agency (1992) Hydropower, energy and the environment: options for increasing output and enhancing benefits, Washington DC

Mills R, Toke A (1985) Energy, economics, and the environment. Vermont Technical College, Englewood Cliffs

Moore E (1991) Capital expenditures for electric power in the developing countries in the 1990's. Interamerican Bank for Reconstruction and Development, Washington DC

National Geographic Special Edition (1993) Water: the power, promise, and turmoil of North America's fresh water. National Geographic, Washington DC

National Research Council (1989) Global change and our common future. National Academy Press, Washington DC

Olivera M, Briones C (1986) Luces y Penumbras: Impacto de la Represa de Piedra del Águila en la Comunidad Mapuche Neuquina de Ancatruz. II Congreso Argentino de Antropología Social, Buenos Aires (unpublished manuscript)

Olivera M, Briones C (1987) Proceso y Estructura. Transformaciones Asociadas al Régimen de "Reserva de Tierras" en una Comunidad Mapuche. Cuadernos de Historia Regional, No. 10, Vol. IV, UNLu, Luján

Peralta C (1989) Informe de Campo (Pilquiniyeu del Limay), Neuquen (unpublished manuscript)

Poggiese H, Francioni M, Tassara J (1987) Documento Síntesis del Primer Seminario-taller: Proyecto Integrado Pilquiniyeu. SEPLA/CFI, Viedma

Price C (1994) World Bank Environmental Guidelines: the new international standard. Public Utilities Fortnightly, Washington DC
Radovich J (1987) El Proceso Migratorio entre los Mapuche del Neuquén. Informe al CONICET, Buenos Aires (unpublished manuscript)
Radovich J (1993) Política Indígena y Movimientos Étnicos: El Caso Mapuche. Cuadernos de Antropología No. 4. UNLu, Lujan
Radovich J, Balazote A (1989) Mercachifles y Cooperativas: un Análisis del Intercambio. Runa F.F.y L. UBA, Buenos Aires
Radovich J, Balazote A. (1990a) Formas de Discriminación hacia el Pueblo Mapuche de la República Argentina. Fundación Hebraica Argentina, Buenos Aires
Radovich J, Balazote A (1990b) Trabajo Doméstico y Trabajo Asalariado en la Unidad de Explotación Campesina. III Congreso Argentino de Antropología Social, Rosario
Ribeiro GL (1985) Proyectos de Gran Escala: hacia un Marco Conceptual para el Análisis de una Forma de Producción Temporaria. In: Bartolome LJ (ed) Relocalizados: Antropología Social de las Poblaciones Desplazadas. Ediciones del IDES N° 3, Buenos Aires, pp 23–47
Ribeiro GL (1991) Capitalismo Transnacional na Terra da Lua. Poder y desenvolvimento enum grande projeto. ANPOCS/Marco Zero, São Paulo
Steen N (1990) Sustainable development and the energy industries: implementation and impacts of environmental legislation. Brookings Institution, Washington DC
United Nations, Economic and Social Commission (1992) Energy, environment and sustainable development. United Nations, New York
Veiga D (1989) El Proyecto de Salto Grande en la Perspectiva de la Población Local: la Evaluación de Impactos Socioeconómicos. In: Brunstein F (ed) Grandes Inversiones Públicas y Espacio Regional. Experiencias en América Latina. Ediciones CEUR, Buenos Aires, pp 199–222
Virgolini M (1980) Efectos del Transplante Poblacional en Relación a Grandes Obras de Infraestructura: Federación y Santa Ana, Informe Parcial, Comisión de Investigaciones Científicas, Universidad Nacional de La Plata, (mimeo), La Plata
World Bank (1992) World Bank development report. Development and the Environment, Washington DC
World Bank (1994) Annual report for the Yacyreta Project. World Bank, Washington DC
World Bank (1995a) Mainstreaming the environment. World Bank, Washington DC
World Bank (1995b) Striking the balance. The environmental challenge of development. World Bank, Washington DC
World Bank (1995c) Environmental bulletin 7(3). World Bank, Washington DC
Wu K (1995) Energy in Latin America: production, consumption, and future growth. Westport, Connecticut, London

Chapter 7
Impacts of Sobradinho Dam, Brazil

Benedito P.F. Braga, Joaquim Guedes Correa Gondim Filho, Martha Regina von Borstel Sugai, Sandra Vaz da Costa, and Virginia Rodrigues

7.1 Introduction

Brazil is a large country with an area of 8.5 million km^2 and a population of approximately 184 million in the year 2008. The country is rich in water resources, with the impressive rank of holding 14% of the world's waters. This large amount of water is unevenly distributed in time and space across the territory. The northern region of the country accounts for 68% of the total available water in the country while the semi-arid northeast region faces severe water shortages. The population distribution is also diverse. The northern region, where most water is available, is home to only 7.6% of the population. The northeast, with only 3% of the country's water, is inhabited by 30% of the population. The semi-arid climate of this region has very irregular annual average rainfall, ranging from 200 to 700 mm, and its population is the poorest in Brazil, facing serious social problems. The few humid areas of this region are limited to those bordering the northern region and the coastal strip. Its high climatic variability has led to the construction of dams to regulate flows, thus providing water supply and irrigation for food production. The construction

---

B.P.F. Braga (✉)
Department of Hydraulic and Sanitary Engineering, EPUSP, São Paulo, Brazil

J.G.C.G. Filho
National Water Agency of Brazil, Brasilia, Brazil

M.R. von Borstel Sugai
Water Resources Consultant of COPEL, Parana State Power Utility, Curitiba, Brazil

S.V. da Costa
Kaerl Lake Project, Fluor Canada, Calgary, Canada

V. Rodrigues
National Water Agency of Brazil, Brasilia, Brazil

of these dams and reservoirs was responsible for a significant increase in the social and economic development of this region.

Besides flow regulation, dams in Brazil play an important role in providing hydropower. Brazil is rich in freshwater and waterfalls, and hydropower accounts for about 42% of the national energy matrix, contributing approximately 77% of the total electricity produced in the country (excluding the Paraguayan production portion of the ITAIPU hydropower plant imported by Brazil). This matrix is being diversified but it is anticipated that hydropower will continue to be the main source of electric energy in Brazil for many years to come. Brazilian hydropower potential is estimated at 260 GW, 40% of which is located in the Amazon River Basin. Other basins of importance and their respective potentials are: Parana River Basin (23%), Tocantins River Basin (11%) and São Francisco River Basin (10%). In 2008, the National Interconnected Electric System had an installed capacity of the order of 100 GW, with 76.5 GW from hydropower sources. This figure includes Brazil's share of 6.3 GW from the ITAIPU hydropower plant, a joint venture with Paraguay. It is important to note that only 30% of the country's hydropower potential is currently exploited.

Construction of large dams for power generation resulted in the formation of large reservoirs that flooded vast areas. In several cases, such areas were productive and/or had considerable biological diversity, thus requiring interventions such as relocation of local residents/workers and rescue of wild animals. This chapter presents an objective evaluation of the impacts (positive and negative) associated with the construction of one such reservoir in the São Francisco River Basin, the Sobradinho Reservoir, one of the largest in the world.

7.2 Energy for Northeastern Brazil

Northeastern Brazil is characterised by scarce water resources with periodic droughts that in the past have induced large migrations in the country. In order to increase the socio-economic standards in this region, the federal government realised that energy and water sustainability were essential. Since these resources were not available in the region, nearby basins were used to supply energy through the development of hydropower plants. The most important one is the São Francisco River Basin, responsible for 80% of the power supply in the northeast. Figure 7.1 shows: (a) the location of the Brazil's northeast region, (b) the areas of low precipitation and high evapotranspiration rates that characterise a large semi-arid region and (c) the location of the São Francisco River Basin. It can be appreciated that a large portion of the northeast territory and of the São Francisco River Basin are under semi-arid conditions.

The nearby São Francisco River Basin extends from the humid southeast region to the northeast of Brazil with an area of 638,324 km² covering six states (Minas

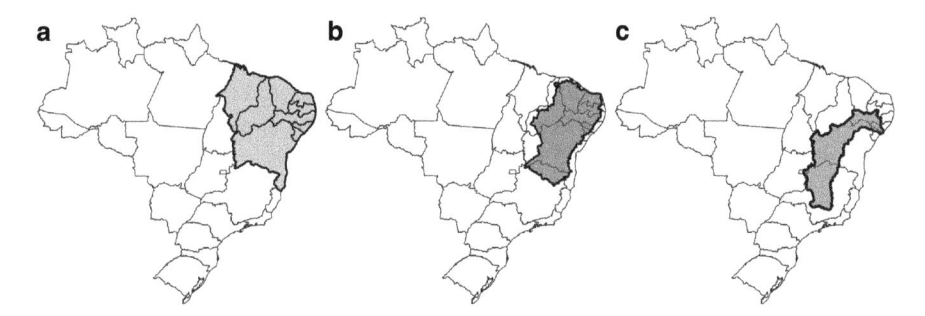

Fig. 7.1 Location relationship between the northeast region (**a**), the semi-arid conditions region (**b**) and the São Francisco River Basin (**c**) in Brazil

Gerais, Bahia, Sergipe, Pernambuco, Alagoas and Goiás) and the Federal District, and with a population of 13 million people (8% of the country). The São Francisco River is the third largest river of Brazil, flowing over 2,700 km and discharging an annual mean discharge of 3,037 m^3/s into the Atlantic Ocean.

The pioneering efforts to develop the hydropower potential of the São Francisco River go back to 1913 and the establishment of the Angiquinho Hydropower Plant (HPP), later named Delmiro Gouveia HPP. The plant was located between the river's left margin and a river island, using only a portion of the Paulo Afonso Falls hydraulic head. This plant was inaugurated in January 1913 with a total installed capacity of 1,120 KW. The energy generated provided power supply for a thread-making industry and to the village of Pedra, where the factory workers lived.

Following the creation of the São Francisco River Hydropower Company, CHESF, in 1948, a series of hydropower plants were built along the São Francisco River to supply electricity to the northeast region. The first hydropower plant built by CHESF was a pilot plant named PA, located in the municipality of Paulo Afonso in Bahia state, which started operations in 1949. It was built on the left margin of Gangorra Stream and has one generating unit with an installed capacity of 2 MW. In 1954, the first hydropower plant of the Paulo Afonso Complex, Paulo Afonso I was inaugurated. It comprised three generating units of 60 MW each, with a total capacity of 180 MW. In 1955 the Paulo Afonso Complex started expanding with the construction of Paulo Afonso II, inaugurated in 1961 with a total installed capacity of 445 MW. The construction of Paulo Afonso III was initiated in 1967 and completed by 1971. At that time, the Paulo Afonso Complex was considered the largest in the country, generating enough power to supply most of the northeast region's demand for electricity for nearly three decades.

Construction of the Três Marias HPP began in May 1957 and the plant was finalised in January 1961. The main purposes of this HPP were flow regulation of the São Francisco River, flood control, improvement of navigation, generation of electric energy and irrigation. Três Marias HPP has an installed capacity of 396 MW.

In 1971, construction work began on the Apolônio Sales (Moxotó) HPP, part of the Paulo Afonso Complex. This hydropower plant is located 3 km upstream of the first dam built in the Complex, and the water flowing through its turbines supplies power for the operation of the Paulo Afonso I, II and III plants. It started operations in 1977, supplying an extra 400 MW of energy to the northeast subsystem.

Building of the Sobradinho HPP started in 1973 and the plant began operations in 1979 with an installed capacity of 1,050 MW. The main objective of this reservoir is flow regulation; it was planned to increase the minimum flows of the São Francisco River during periods of drought and to enable an improvement at the operational capacity of the Paulo Afonso IV HPP. Construction of the Paulo Afonso IV plant started in 1972, and it was inaugurated in 1979. The water that flows to Paulo Afonso IV HPP comes from a diversion channel of the Apolônio Sales Reservoir, excavated from its right margin. The hydropower plant has six generating units with a nominal capacity of 410 MW each, with a total of 2,460 MW.

The construction works for the Luiz Gonzaga HPP, also known as Itaparica HPP, were initiated in the year 1979. It is located 50 km upstream of the Paulo Afonso Complex. Besides generating energy, the plant allows a more efficient operation of the Complex, contributing directly and decisively to the control and regulation of the daily and weekly discharges flowing to the Paulo Afonso plants. The Luiz Gonzaga plant comprises six generating units of 250 MW each, making up a total installed capacity of 1,500 MW. It can be expanded by the addition of four generating units similar to those currently in operation. Construction of the Xingó HPP started in 1987 and operations commenced in 1994. It is located between the states of Alagoas and Sergipe some 65 km downstream of the Paulo Afonso Complex, and its reservoir is located in a natural canyon. Boating in the reservoir is a major tourist attraction in the region. It provides water supply for the city of Canindé in Sergipe and for downstream irrigation projects. The Xingó HPP has an installed capacity of 3,000 MW (Centro da Memória da Eletricidade no Brasil 2001). Figure 7.2 is a schematic illustration of all hydropower plants located in the São Francisco River, indicating their respective installed capacities. Volumes are depicted only for storage reservoirs.

7.3 Sobradinho Dam and Reservoir

The Sobradinho Reservoir was built to achieve multiple uses through: (i) multi-annual flow regulation for hydropower, navigation and irrigation and (ii) flood control management for the riverine communities of the São Francisco River Basin. At the time of its completion, the Sobradinho Reservoir was the largest man-made lake in the world, and today it remains one of the largest. The total storage of the reservoir is 34.1 billion cubic meters (BCM), and it extends over 350 km with an artificial lake surface of 4,214 km^2.

Três Marias
TSC = 19 billion m³
LS = 15 billion m³
396 MW

Sobradinho
TSC = 34 billion m³
LS = 28 billion m³
1,050 MW

Moxotó
400 MW

Paulo Afonso
I -180 MW
II -443 MW
III -864 MW

Itaparica
TSC = 10 billion m³
LS = 3 billion m³
1,500 MW

Xingó
3,000 MW

Paulo Afonso IV
2,460 MW

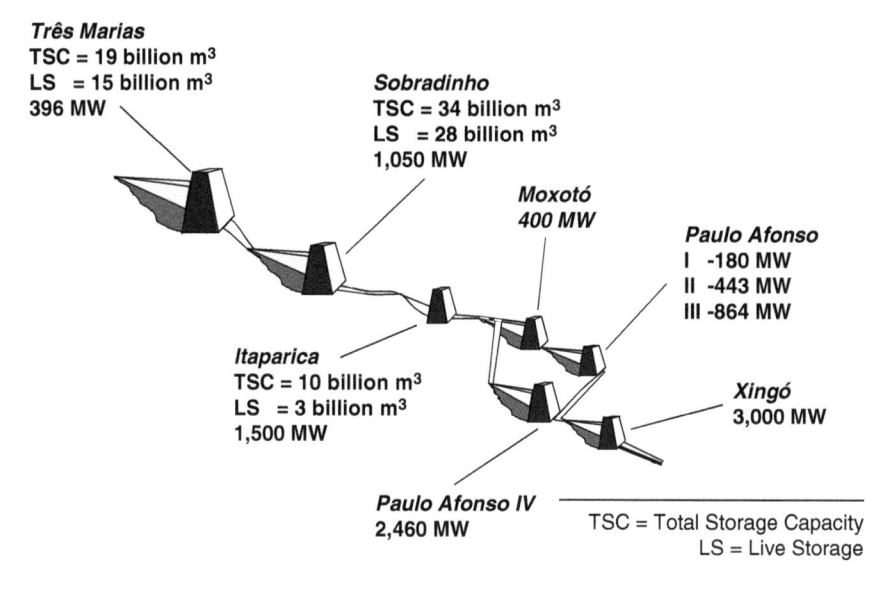

TSC = Total Storage Capacity
LS = Live Storage

Fig. 7.2 Main reservoirs located in the São Francisco River

The main features of Sobradinho Hydropower Plant are as follows:

- Dams and dikes: Two main dams and four earth dikes allow for the formation of the Sobradinho Reservoir and a maximum water depth of 32.5 m. Operation water levels are: (i) upstream maximum normal 392.5 m, and (ii) upstream minimum normal 380.50 m. At its maximum normal level, the reservoir extends for 350 km, covers an area of 4,214 km² and has a volume of 34,116 km³.
- Overflow system: Consisting of four spillway gates and 12 sluice gates, with a total capacity of 22,080 m³/s.
- Water intake and irrigation channels: Concrete structure located in one of the left margin dikes, used for irrigation of areas downstream of Sobradinho.
- Navigation lock and canals: Capable of dealing with a maximum head of water of 32.5 m, consisting of: (i) 1 chamber (120× 17 m) with a maximum filling time of 16 min, (ii) navigation canals and (iii) emptying flushing canals.
- Powerhouse: Comprises (i) water intake, (ii) outlet works and (iii) powerhouse with six generating units, and a total installed capacity of 1,050 MW.
- Substation and Transmission System: The substation consists of three single-phase units of 13.8/500 kV each, while the transmission system has two transmission lines in 500 kV for the Luiz Gonzaga HPP and a third interconnecting CHESF and ELETRONORTE, the North of Brazil Electric Utility energy lines. The substation also has a 500/230 kV transformation area (Portal do São Francisco 2004).

Other details of the construction works are provided in Table 7.1.

Table 7.1 Main characteristics of Sobradinho Dam

Name: SOBRADINHO	Reservoir at max normal water level Area: 4,214 km² Volume: 34,116 km³ Length: 350 km
Power Utility: CHESF	
Main purpose: hydropower	
Other uses: irrigation and navigation	
Construction period	*Main dam*
Start of civil works: June, 1973	Type: Rock-fill
Start of operations: November, 1979	
Volumes	Length: 8,532 m
Concrete: 1,400,000 m³	Height: 398 m
Soil: 12,100,000 m³	Foundation: 43 m
Rock-fill: 4,000,000 m³	Volume: 14,060,700 m³
Excavation: 21,000,000 m³	
Lock	*Powerhouse*
Length: 120 m	Type of turbine: Kaplan
Width: 17 m	Number of units: 6
Filling time: 16 min	Unitary power: 175 MW
	Total head: 27 m
Water depths	*Spillway*
Max normal: 32.5 m	Number of reaches: 16
Maximum maximorum: 33.5 m	Dimensions (width × height): 9.80 × 7.50 m
Wall height: 41.5 m	Design flow: 22,080 m³/s
Annual traffic capacity: 8,000,000 tons/year	
Navigation canal	*Operation levels*
Length: 1,770 m	W.L. Max normal upstream: 392 m
Width: 50 m	W.L. Max normal downstream: 371.9 m
Guide walls	
Length: 110 m	
Upstream crest: 394	
Downstream crest: 374	

7.4 Impacts of Sobradinho Dam and Reservoir

Hydropower generation in Brazil is produced by large power plants. Twenty-three hydropower plants of 1,000 MW or higher represent 71.4% of the total installed capacity in the country, while 337 plants with 30 MW or less represent 2.4% of the installed capacity (ANEEL 2002). The São Francisco River Basin is no exception to this rule. As shown above, the basin hosts the largest reservoir in the country. The implementation of large infrastructure certainly produces both positive and negative impacts. The science and art of balancing these effects is the task of the responsible water resources decision maker. The impacts of the Sobradinho Dam and Reservoir are described in this section.

7.4.1 Social Impacts

Construction of reservoirs usually intensifies migration processes. People are attracted by work opportunities that the plant brings during construction, and later, the population living in areas flooded by the reservoir are relocated. There is currently much discussion around the loss of urban centres and the relocation of populations affected by construction of dams. The creation of new urban centres involves complex social and cultural aspects including attachment to land and loss of identity in relocation to new areas. These people leave behind urban, historical and cultural references which only have meaning in the physical space that is lost to flooding. There are important questions which must be taken into consideration when constructing artificial lakes for hydropower generation, such as: (i) the resistance of long-term residents in leaving their homes; (ii) the occurrence of spontaneous communities and (iii) the logistics of resettlement (ANEEL 2002).

Upon construction of the Sobradinho Reservoir, both urban and rural areas of Casa Nova, Remanso, Sento Sé and Pilão Arcado were inundated. Flooding also affected the rural areas of the municipalities of Juazeiro and Xique-Xique. By the end of 1977, the inhabitants of four cities and 30 rural villages had left their homes, a total of 11,853 families (over 70 thousand persons). Of those families, 5,806 remained in rural lots around the Sobradinho Reservoir and 3,851 families moved to newly built municipalities (Portal do São Francisco 2004). The construction of the workers residence village took place in an older district of the municipality of Juazeiro, about 46 km from the city centre. Soon, the old district became a large urban centre inhabited mainly by construction workers of the hydropower plant. The Planning Department of CHESF, following their experience in the construction of other dams, designed temporary housing for the workers as part of the plant's infrastructure. These settlements were expected to last only for the duration of the works. However, settlers strongly resisted moving and Sobradinho became an established municipality by 1989 with about 38,000 inhabitants.

Positive social impacts can be observed with the construction of dams and reservoirs. In the case of the Sobradinho Reservoir, a most important social indicator showed improvement in the northeast region: the infant mortality index. Table 7.2 presents the trend of infant mortality in Brazil and in the northeast region from 1930 to 1990. There is an important decrease in these rates from the 1970s onwards.

7.4.2 Economic Impacts

As a component of the São Francisco River hydropower cascade, the Sobradinho power plant has brought important positive economic impacts to the northeast region of Brazil. It has enabled multiple water use through irrigation and navigation, which in turn produces economic benefits for the population in the river basin.

Table 7.2 Infant mortality rates (per 1,000 live births) in Brazil and in the northeast region

Year	Brazil	Northeast
1930	162.4	193.2
1935	152.7	188.0
1940	150.0	187.0
1945	144.0	185.0
1950	135.0	175.0
1955	128.2	169.6
1960	124.0	164.1
1965	116.0	153.5
1970	115.0	146.4
1975	100.0	128.0
1980	82.8	117.6
1985	62.9	93.6
1990	48.3	74.3

Source: Censuses Historical Data (IBGE 1999)

7.4.2.1 Hydropower

The São Francisco River hydroelectric power plants are part of the National Interconnected Electric System (SIN), a hydrothermal power production system strongly dominated by hydropower plants. It is divided into subsystems, as follows: South, Southeast/Centre-West, Northeast and North. Only 3.4% of the power produced in Brazil is generated outside of SIN, in small isolated systems located mainly in the Amazon region.

ONS (the national operator of the Electric System) is a private entity in charge of the national dispatching of SIN. It was charged in 1998 by the Brazilian federal government with the responsibility of operating the National Interconnected Electric System as well as managing the main transmission network in Brazil. The company's institutional mission is to assure a continuous, qualitative and economic supply of electricity to all users of the interconnected system. ONS is also responsible for guaranteeing the maintenance of the synergic gains resulting from a coordinated operation and creating conditions for fair competition among the public and private agents of the sector. ONS follows rules, criteria and methodologies established by a code called the Grid Procedures, a collection of governance manuals approved by the member agents and ratified by ANEEL (Brazilian Electricity Regulatory Agency).

The northeast (NE) subsystem is basically supplied by:

- The hydropower plants located in the São Francisco River and in other basins of the NE region.
- Thermal energy plants distributed all over the NE region.
- Power imported from other subsystems of SIN through transmission lines.

Currently, the São Francisco HPP's supplies 80% of the energy in SIN's NE subsystem. The electrical infrastructure of the NE region, particularly that of the

São Francisco River Basin, contributes to ensuring regional economic growth compatible with the country as a whole, and occasionally higher. Between 1960 and 2000, the NE economy registered an expansion of its annual GNP (Gross National Product) in the order of 4.6%, much the same as the rest of the country.

Moreover, economic changes took place in the production structure of the region, with a noticeable growth in urban activities, such as industry and services, as opposed to agriculture or ranching. The latter comprised 30% of the region's GNP in the 1960s, and today it represents less than 10%. Industrial activities represented 22% of the GNP in the 1960s, and have increased to 26%, while the services sector has increased from 47% to 64% in the same period. This growth was followed by the consolidation of important urban economies in the region followed, with the emergence of large metropolitan centres such as Fortaleza, Recife and Salvador, the results of the merging of several urban centres.

7.4.2.2 Navigation

Traditional navigation in the São Francisco River can be divided into four stretches (Fig. 7.3):

- The section of free surface flow located between Pirapora (state of Minas Gerais) and the entry point to Sobradinho Lake in Xique-Xique (state of Bahia).
- The section corresponding to the entrance to the lake, located between Xique-Xique and Pilão Arcado.
- Sobradinho Reservoir.
- The section of free surface flow located between the Sobradinho Dam's lock and the cities of Juazeiro/Petrolina.

The São Francisco River waterway is 1,371 km long, and is located between Pirapora and Juazeiro/Petrolina. It is one of the most important transportation corridors that link the Southeast (SE) and Northeast (NE) regions of the country. In this waterway, multimodal transportation is the safest and cheapest alternative for the transport of agricultural goods from the regions of Western Bahia and Unaí to the NE region for local consumption and for export, using the NE ports.

Commercial navigation is currently limited to the section downstream of Ibotirama. The increase in grain production in Western Bahia, Barreiras region to supply the demand of NE cities via the distribution centre Juazeiro/Petrolina, has stimulated navigation activities. Table 7.3 presents cargo demands for this reach downstream of Ibotirama.

During the construction of the Sobradinho Dam, between 1972 and 1978, access to the fluvial ports of Juazeiro and Petrolina, located 40 km downstream of the reservoir, was interrupted. This waterway was reopened in 1978 with the construction of the Sobradinho lock and dam, connecting the reservoir to the river.

Sobradinho Dam was built to store water at a maximum elevation of 392.5 m, and a minimum operational elevation of 380.5 m. As a consequence, the water level may vary for about 120 km, between Pilão Arcado and Xique-Xique. Water levels fluctuate between 388.0 m and 383.5 m for 95% of the time (ANA 2004a).

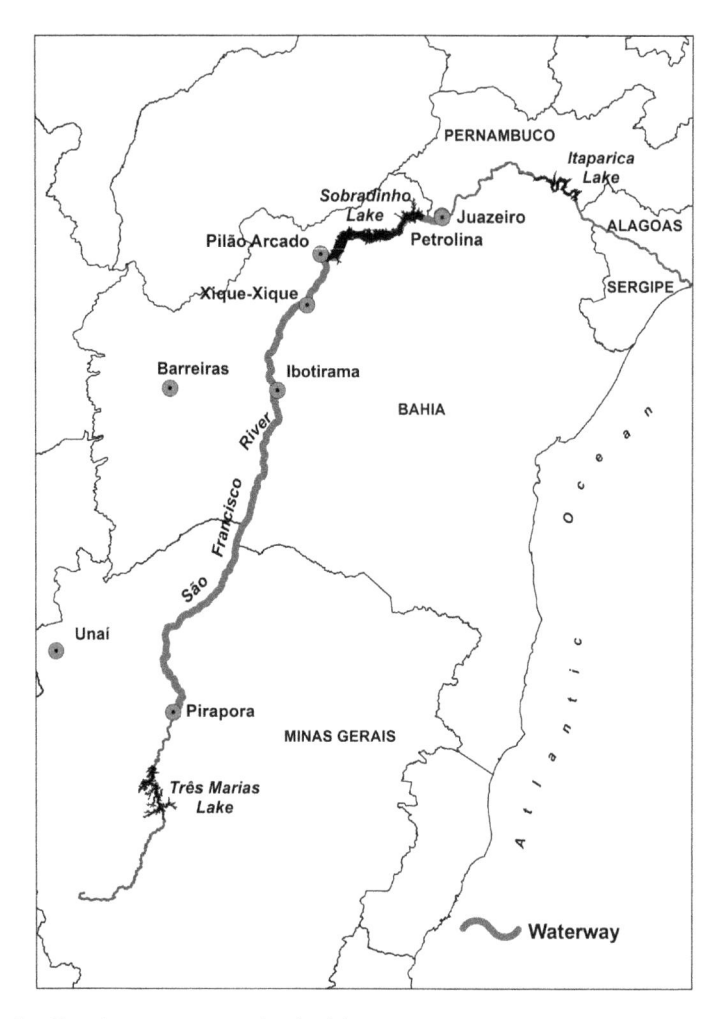

Fig. 7.3 São Francisco waterway and main cities

Table 7.3 Annual cargo demands for 2004—São Francisco waterway

Reach	Cargo demands (tons)	Product	Final destination
Ibotirama-Juazeiro/Petrolina	1,500,000	Soya	NE region and export at Aratu Port
	500,000	Cotton	NE region
Juazeiro/Petrolina-Ibotirama	1,000,000	Gypsum	State of Pernambuco Gypsum Complex

Source: MI et al. (2004)

An intense sedimentation process is occurring in the uppermost region of the Sobradinho Reservoir, generating the so-called delta effect at its entrance. Such an effect interferes with the navigation route, rendering navigation aids (such as signals and buoys) obsolete and causing frequent stranding of vessels near Xique-Xique city (ANA 2004a). As a remedial measure, navigation companies are now utilising global positioning systems (GPS) and bathymetry to dynamically find the best routes for the convoys. On the reservoir, the navigation route lies close the right bank, is well signalled and trouble-free in the portion downstream of Pilão Arcado extending to the dam.

The section in free flow, located between the Sobradinho lock and the cities of Juazeiro/Petrolina, is rocky. The banks are firm and in good condition. In this reach, navigation is controlled by the Sobradinho Reservoir's outflows. Regulated river flows have considerably improved navigation conditions, but there still remain rocks that complicate manoeuvring and hinder the operational safety of the vessels currently used. The minimum discharge flow was set to 1,300 m^3/s until 2003. However, in periods where water storage is critical, a minimum discharge of 1,100 m^3/s flat is compulsory to minimise impacts on power generation for the NE region, and to store extra surplus water flows in Sobradinho Reservoir. In such conditions, navigation will only be feasible after performing civil works in the rocky reaches downstream of Sobradinho.

7.4.2.3 Fishing and Aquaculture

Fishing used to be abundant in the São Francisco River Basin, both in the upper and in the lower reaches, providing a steady supply of food for its inhabitants and attracting fishermen. At present, fishing is declining due to sediment trapping in the reservoir and water pollution from sewage and agriculture.

However, the huge potential for aquaculture in the reservoirs is being explored. A large fish production of 6,321,000 tons/year has been estimated for Sobradinho Reservoir, using only 1% of its surface area.

7.4.2.4 Flood Control

In the São Francisco River Basin, flood control is managed by means of the large HPP reservoirs in combination with longitudinal dikes. The large reservoirs are operated by electric utilities for power generation, but cater to multiple uses of water, including flood control. This is achieved by allocating a particular volume when designing the reservoir, known as 'flood control volume'. The use of such volume mitigates floods and reduces impacts of large run-off inflows to the reservoir, allowing for lower discharges downstream of the dam, as well as guaranteeing the required levels upstream.

Two important cities, Juazeiro and Petrolina, are located just downstream of the Sobradinho Dam. In order to prevent flooding, maximum releases have been established at 8,000 m³/s by the Multi Ministry Flood Control Commission for São Francisco River Valley (Comissão Interministerial de Estudo para o Controle das Enchentes do Rio São Francisco 1980). This flow impacts the operation of the cascade of power-generating reservoirs located downstream of Sobradinho

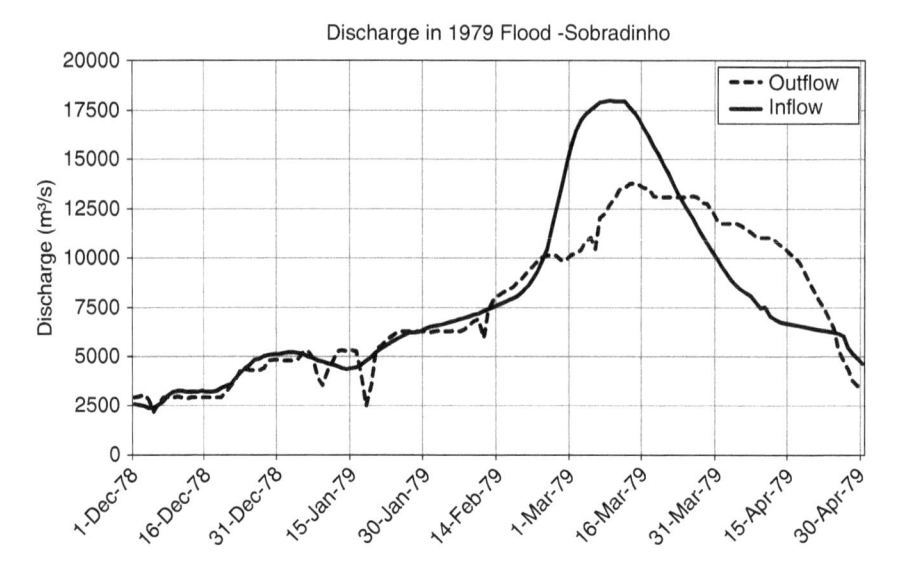

Fig. 7.4 Hydrograph of the 1979 flood event at Sobradinho Reservoir

(Moxotó, Itaparica and Xingó). The importance of the Sobradinho Reservoir in flood control can be clearly seen through an analysis of the major flood in 1979 of the São Francisco River Basin, one of the most severe floods ever recorded. It caused severe social and economical impacts, affecting all kinds of activities along the valley.

Figure 7.4 presents the inflows and outflows at Sobradinho Reservoir. Daily maximum inflows of 17,800 m^3/s were registered in Sobradinho. The flood control operation established in Sobradinho by the CHESF was able to reduce the peak flow to a maximum of 13,700 m^3/s, which was the maximum flow registered at the Juazeiro stream-flow gauge station, located 42 km downstream of the dam.

7.4.2.5 Irrigation

São Francisco currently has 342,712 ha of irrigated land, 30% of which belongs to large companies including the public sector. Irrigated areas include north of Minas Gerais state, particularly the green belts around the cities of Gorutuba, Pirapora, Jaíba and Janaúba, Greater Belo Horizonte in Minas Gerais state; Brasília in Federal District; Formoso/Correntina, Barreiras, Guanambi and Irecê, in Bahia state; and the lower São Francisco River, in the states of Alagoas and Sergipe (Fig. 7.5). Besides these areas, the irrigation complex located in the Juazeiro/Petrolina region is noteworthy for the export of its large fruit production. Construction of the Sobradinho Dam regulated the São Francisco River flows, favouring implementation

7 Impacts of Sobradinho Dam, Brazil

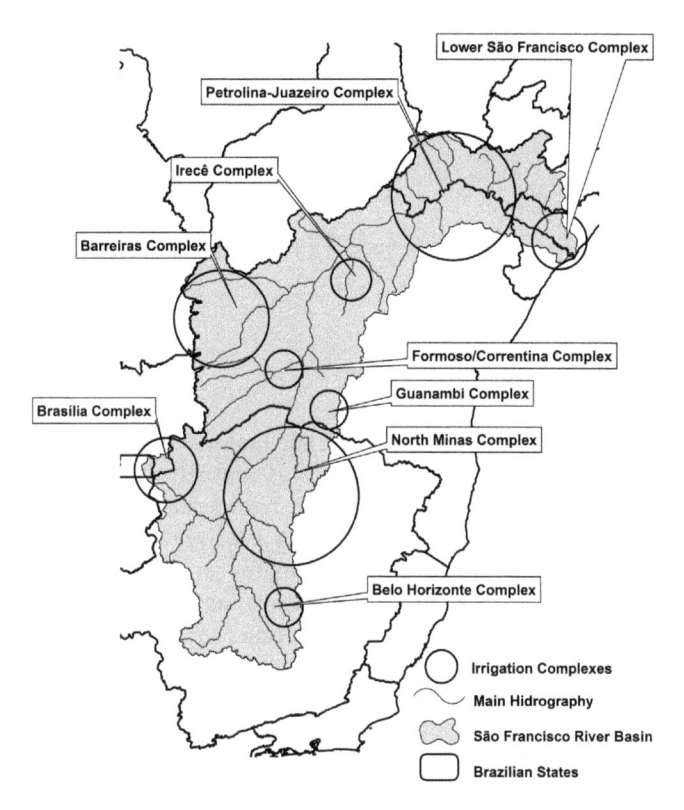

Fig. 7.5 Major irrigation complexes in São Francisco River Basin

Table 7.4 Evolution of total irrigated area in Brazil and in São Francisco River Basin (in 1,000 ha)

Period	Brazil	São Francisco River Basin
Up to 1950	64	0.1
Up to 1960	320	11
Up to 1970	796	60
Up to 1975	1,100	88
Up to 1980	1,600	144
Up to 1985	2,100	206
Up to 1990	2,700	233
Up to 1994	2,800	250
Up to 1995	2,600	300
Up to 1996	2,656	300
Up to 1997	2,756	300
Up to 1998	2,870	330
Up to 1999	–	330

Source: ANA (2004c)

of this important irrigation complex. Table 7.4 presents the evolution of total irrigated area in Brazil and in the São Francisco River Basin, and Table 7.5 shows the quantity of fruits produced annually in the São Francisco River Basin.

Table 7.5 Fruits produced in São Francisco River Basin

Product	Annual production (tons)
Mango	270,000
Grapes	240,000
Banana	160,000
Guava	112,000
Barbados Cherry	22,500

Source: CODEVASF (2004)

7.4.2.6 Inter-basin Water Transfer

In the past, energy was transferred from the São Francisco River Basin to the semi-arid NE region. Today the federal government is similarly considering the transfer of water to the water-scarce NE region. The Inter-basin Water Transfer Project proposes integration of the São Francisco River and river basins located in a semi-arid area of the NE region. The project plans for two water diversion systems located downstream of Sobradinho HPP to basins located in the following northeastern states: Ceará, Rio Grande do Norte, Paraíba and Pernambuco. The project will supply water for several uses (water supply, irrigation and aquaculture) to approximately 12 million people until 2025, in one of the poorest areas of the country.

It will not be a stand-alone work of civil engineering, but rather integrated with the existing infrastructure of the receiving basins. It will be possible to better operate these reservoirs and impoundments, with guaranteed water flows, as a result of the inter-basin transfer project.

Positive impacts of the project include:

- Increase in water quantity and quality (urban and rural settlements)
- Minimise risk of emergencies in situations of prolonged droughts
- Reduce migrations of drought-afflicted people
- Reduce diseases and mortality related to water scarcity
- Social and economic development (MI 2004)

7.5 Environmental Impacts and Mitigation Measures

The Sobradinho Dam was built primarily for hydropower development, a major priority at the time of its design and construction (1970s). In time, the Sobradinho Reservoir has served multiple uses, such as irrigation and flood control. Its construction was completed before legislation requiring environmental licensing was implemented in 1981. Thus, measures to mitigate environmentally negative impacts associated with the dam were not considered at the time.

Current legislation requires that dams obtain an Operating Licence. CHESF (the São Francisco River Hydropower Company) has operated the Sobradinho HPP since its inception, and is presently working towards obtaining an Operating Licence for Sobradinho Dam from IBAMA, the Federal Environmental Management Agency. In anticipation, CHESF has engaged in several initiatives related to environmental concerns in the reservoir's area of influence, such as environmental awareness and education programmes, rehabilitation of the protected areas located within the area of influence, statistical evaluation of fish stocks, survey of local fish species and monitoring of water quality.

Sobradinho Dam has provided several benefits for the NE region, including generation of electricity, reduction of devastating floods, controlled supply of water which can be used to alleviate drought and for a variety of municipal and industrial demands. The economic and social development of the NE region could not occur without electricity or irrigation. However, it is being increasingly recognised that dams constitute an important factor in the overall degradation taking place in estuaries, deltas and various near-shore environments. The Sobradinho Dam is no exception.

Negative environmental effects associated with dams include: alteration of river morphology, deltas and estuaries; possibility of increasing coastal erosion; reduced biodiversity and fisheries productivity, particularly in the estuarine environment; changes in salinity and possible intrusion of saltwater into coastal aquifers. Some positive environmental impacts of damming have also been reported, such as the increase of habitat upstream of dams, in areas of sediment deposition, often creating islands and shallow ponds (Mouvet 2000).

With growing public awareness of such impacts, there is a current trend to focus on mitigation measures so as to restore the natural conditions of the rivers as far as possible.

7.5.1 Impacts Associated with Sobradinho Dam and Reservoir

7.5.1.1 Availability of Fish

The Sobradinho Dam has rendered inaccessible areas of critical spawning habitat for fish that require migration to upstream locations, such as 'piau', 'matrinchão', 'curimatã', 'pacu', 'pira' and two marine species, 'robalo' and 'pilombeta'. Although hydropower development is recognised as one of the main factors responsible for the decline of fish stocks, other actions such as mining, the release of hatchery fish and water pollution also play important roles. However, the reservoir has a huge potential for aquaculture activities. This can offset the lack of fish for local consumption and loss of income for fishermen. A large fish production of 6,321,000 tons/year has been estimated for the Sobradinho Reservoir, using only 1% of its surface area.

7.5.1.2 Erosion in the Lower São Francisco River and Coastal Zone

The São Francisco River now carries only 3% of the sediment load measured in 1968 (ANA 2004d). In 1966–1968, the São Francisco River supplied an estimated 12.5 million tons of suspended sediment to its delta in the state of Sergipe; today only 0.41 million tons of suspended sediment (and regulated freshwater flows) from this river reach the sea. This is causing erosion of the river banks and river bed. In addition, the massive coastal erosion noted in the São Francisco delta with a large advancement of the sea into the land is thought to result partly from the cascade of dams and partly from the natural littoral currents existing on the coast.

7.5.1.3 Potential Reduction of Species Diversity and 'Drying Out' of Wetlands

The Sobradinho Dam and other projects potentially affect sensitive environments in the São Francisco River delta: flood plains, ponds, islands, mangroves and sand dunes. This is due to alteration of sediment and flood regimes, with consequent decline of nutrients and changes in the salinity regime (ANA 2004d). A study of the feasibility of new operational rules for the reservoir to cater for restoration of ecological flows is being considered as part of a project, funded by the Global Environment Facility (GEF), of the National Water Agency of Brazil (ANA).

7.5.1.4 Potential for Increased Salinity of Estuary and Saltwater Intrusion into São Francisco's Coastal Aquifer of Barreiras

Severe coastal erosion in the mouth of the São Francisco River causes loss of sand dunes and beaches, which are natural barriers against saltwater intrusion into estuarine environments and into aquifers. Deterioration of mangroves due to restricted freshwater inflows from Sobradinho Dam can also contribute to saline intrusion as they may be a major source of aquifer recharge. Thus, there is a potential for saltwater intrusion into São Francisco's coastal aquifer of Barreiras and estuarine zones.

7.5.1.5 Sediment Deposition in the Entry of Sobradinho Dam

In the reaches upstream of Sobradinho Dam, an intensive process of sediment deposition has been observed, causing the so-called delta effect, which brings significant problems for navigation. A positive impact of the 'delta effect' is the creation of additional habitats in the areas of sediment deposition.

7.5.2 Mitigation Measures

An important mitigation measure that Brazilian environmental legislation imposes on the implementation of reservoirs is financial compensation. To compensate for the inundation of land the owner of the hydropower plant pays the equivalent of 6.75% of the value of generated power. From the 6.75% thus collected, 6% is distributed to the municipalities and states affected by reservoir construction, in the following manner: states—45%; municipalities—45%; federal government—10%. In turn, the federal portion of 10% is shared between the Ministries of Environment (3%) and Mines and Energy (3%), and finally the National Fund for Scientific Technological Development (4%), managed by the Ministry of Science and Technology.

Of the total revenue generated by this financial compensation, 0.75% is calculated as the actual charge for the use of water resources. This revenue is used by ANA to implement the National System of Water Resources Management.

In 2003, a revenue of around R$ 109 million (US$ 36 million), was generated through financial compensation paid by hydropower plants located in the São Francisco River, with around R$ 8 million (US$ 2.7 million) paid by the Sobradinho HPP (ANA 2004b).

7.6 Conclusions

The hydropower infrastructure of the northeast region, particularly in the São Francisco River Basin, has enabled an economic growth similar to that of the rest of Brazil. It allowed for consolidation of important urban economies, such as the metropolitan regions of Fortaleza, Recife and Salvador and all other capitals of the northeastern states. Presently, hydropower plants located in the São Francisco River Basin account for 80% of the power supply in the northeast subsystem. Besides the benefits generated by the supply of electric energy in the São Francisco River Basin, significant financial resources were captured by the local economy during the construction phase of the Sobradinho hydropower plant. This encouraged the birth of new municipalities and the development of existing ones in the Sobradinho area of influence, such as the municipality of Sobradinho, which today has a population of 38,000 inhabitants.

Another important aspect to consider is flow regulation by the Sobradinho Dam, which favours improvement of the operation of the cascade of hydropower plants located downstream of Sobradinho, along the São Francisco River. This implies an increased financial compensation to the municipalities in the area of influence. Like any major civil work, the Sobradinho Dam and Reservoir has impacted the natural environment of the São Francisco River Basin. One such impact is the reduction of sediment transport and consequent reduction in fishing activities downstream of the dam.

Navigation and water supply to small localities has greatly improved in the Sobradinho/Juazeiro-Petrolina stretch of the São Francisco River. Operation of the Sobradinho Reservoir guaranteed flood control for the cities downstream of Sobradinho, reducing incoming peak flows and supporting irrigation activities in a fruit export centre in the reservoir's area of influence.

References

ANA (Agência Nacional de Águas) (2004a) Projeto de Gerenciamento Integrado das Atividades Desenvolvidas em Terra na Bacia do São Francisco. Subprojeto 4.5.C, Plano Decenal de Recursos Hídricos da Bacia do Rio São Francisco, PBHSF (2004–2013). Estudo Técnico de Apoio ao PBHSF – Nº 8. Navegação. ANA, Brasília, 49 pp

ANA (Agência Nacional de Águas) (2004b) Projeto de Gerenciamento Integrado das Atividades Desenvolvidas em Terra na Bacia do São Francisco. Subprojeto 4.5.C, Plano Decenal de Recursos Hídricos da Bacia do Rio São Francisco, PBHSF (2004–2013). Estudo Técnico de Apoio ao PBHSF – Nº 9. Aproveitamento do Potencial Hidráulico para Geração de Energia Elétrica. ANA, Brasília, 81 pp

ANA (Agência Nacional de Águas) (2004c) Projeto de Gerenciamento Integrado das Atividades Desenvolvidas em Terra na Bacia do São Francisco. Subprojeto 4.5.C – Plano Decenal de Recursos Hídricos da Bacia do Rio São Francisco, PBHSF (2004–2013). Estudo Técnico de Apoio ao PBHSF – Nº 12. Agricultura Irrigada. ANA, Brasília, 108 pp

ANA (Agência Nacional de Águas) (2004d) Projeto de Gerenciamento Integrado das Atividades Desenvolvidas em Terra na Bacia do São Francisco. Relatório Final: Programa de Ações Estratégicas para o Gerenciamento Integrado da Bacia do Rio São Francisco e da sua Zona Costeira – PAE. ANA, Brasília, 333 pp

ANEEL (Agência Nacional de Energia Elétrica) (2002) Atlas de Energia Elétrica do Brasil. ANEEL, Brasília, 153 pp

Centro da Memória da Eletricidade no Brasil (2001) Energia Elétrica no Brasil, Breve Histórico 1880–2001. Centro da Memória da Eletricidade no Brasil, Rio de Janeiro, 224 pp

CODEVASF (Companhia de Desenvolvimento dos Vales do São Francisco e do Parnaíba) (2004) http://www.codevasf.gov.br

Comissão Interministerial de Estudo para o Controle das Enchentes do Rio São Francisco (1980) Relatório da Comissão Interministerial de Estudos para o Controle das Enchentes do Rio São Francisco. DNOS, Brasília, 165pp

IBGE (Instituto Brasileiro de Geografia e Estadística) (1999) Evolução e perspectivas da mortalidade infantil no Brasil. IBGE, Departamento da População e Indicadores Sociais, Rio de Janeiro, 45 pp

MI (Ministério da Integração Nacional) (2004a) Projeto de Integração do Rio São Francisco com Bacias Hidrográficas do Nordeste Setentrional. Relatório de Impacto Ambiental, RIMA, Brasília, 129 pp

MI (Ministério da Integração Nacional) MT (Ministério dos Transportes), MMA (Ministério do Meio Ambiente), MME (Ministério de Minas e Energia) (2004) Revitalização da Navegação de Carga no Médio Curso do Rio São Francisco. MI, Brasília, 31 pp

Mouvet L (2000) Dams in Switzerland: contribution to regional development and some environmental considerations. Dams, development and environment. In: Proceedings workshop on dams, development and the environment – IWRA – International Water Resources Association, São Paulo, pp 51–58

Portal do São Francisco. http://www.portaldosaofrancisco.hpg.ig.com.br/sobradinho/index.html. Accessed 22 Aug 2004

Chapter 8
The Atatürk Dam in the Context of the Southeastern Anatolia (GAP) Project

Dogan Altinbilek and Cecilia Tortajada

8.1 Introduction

The Southeastern Anatolia Region of Turkey has historically been a plateau with low productivity. Although rich in water, land and human resources, the region has lagged behind the rest of the country in terms of development. The development potential of both the Euphrates and Tigris Rivers was recognised in the 1960s, and the idea of harnessing their waters for irrigation and hydropower generation emerged. Towards the end of the 1970s, the General Directorate of State Hydraulic Works[1] (DSI) planned the 'Southeastern Anatolia Project'—a series of land and water resources development projects on the two rivers. Through a Master Plan in 1989, and a significant revision in 2002, the Southeastern Anatolia Project, or Güneydogu Anadolu Projesi (GAP), was transformed from a land and water resources development project into a large-scale, multi-sectoral regional development project to be implemented in nine of Turkey's provinces that came to be known as the Southeastern Anatolia (GAP) Region.

The GAP Project thus came to focus not only on economic growth based on water infrastructural development but also on regional development as a whole, taking into

---

[1] The DSI is responsible for planning, design, construction, operation and maintenance of dams, pumping stations and canals for water supply of large cities, large-scale irrigation systems and hydropower production in Turkey. It is headquartered in Ankara and has several regional directorates in the rest of the country.

D. Altinbilek (✉)
Civil Engineering Department, Middle East Technical University, Ankara, Turkey

C. Tortajada
International Centre for Water and Environment, Zaragoza, Spain

Lee Kuan Yew School of Public Policy, Singapore

Third World Centre for Water Management, Atizapan, Mexico

account industry, transportation, urban and rural infrastructure, environmental protection and social sectors such as employment generation, health, education, capacity building and gender equity. The main objective of the GAP Project was to strengthen social, economic, institutional and technical aspects of human development in an economically disadvantaged region by significantly increasing the living standards of its people (GAP Administration 1999; Altinbilek 1997).

A designated Project Management Unit at the State Planning Organisation (SPO) initiated the Master Plan for the GAP Region. Phase I of the Master Plan Study Completion Report was submitted in July 1988 and it presented a proposal for a suitable project management system to implement activities in the region. The proposal was elaborated further in Phase II of the Completion Report submitted in November 1988. Also in November 1988, an Interim Macroeconomic Plan was submitted that examined alternative development scenarios and frameworks consistent with socio-economic and resource allocation policies at the national level.[2] The Turkish government's increasing emphasis on reducing socio-economic disparity across regions, evident in the GAP Project, was a reflection not only of concern for equitable development but also of the recognition that the realisation of development potentials in less-developed regions would contribute to the national objectives of sustainable economic growth, export promotion and social stability. It was thus acknowledged that, alongside economic growth, provision of social services would have to be improved (Hashimoto 2010,[3] personal communication). The Master Plan was finally completed in April 1989. In November of the same year, Law no. 388 was enacted to establish the Southeastern Anatolia Project Regional Development Administration (also known as the GAP Administration), which largely followed the proposal in the Master Plan[4] (Nippon Koei Co. Ltd., and Yuksel Project A.S., 1990; Hashimoto 2010, personal communication).

The GAP Administration was established in 1989. Its key purpose was to plan and coordinate development efforts in the GAP Region, mostly by integrating multi-sectoral projects.[5] Attached to the Prime Minister's Office, the GAP Administration would approve land use plans and would coordinate the various agencies responsible for implementing development activities in the region. The Administration was initially

---

[2]The Plan incorporated numerous comments that were received during Phase I of the Completion Report.

[3]T. Hashimoto, Director General of Nippon Koei, was a key figure during the preparation of the GAP Master Plan.

[4]Article 1 of Law no. 388 of 6 November 1989 stated that the Southeastern Anatolia Project Regional Development Administration was a juridical entity, affiliated to the Prime Minister's Office and had a duration of 15 years. The Administration would provide, or would organise the provision of, services related to planning, infrastructure, licenses, housing, industry, mining, energy and transport, in order to ensure a rapid development of the region under the Southeastern Anatolia Project. It would take actions, or would organise that actions were taken, to improve the educational level of the local population, and would ensure coordination among the relevant agencies and organisations.

[5]The Southeastern Anatolia Project Regional Development Administration has its headquarters in Ankara and a Regional Directorate in Sanliurfa.

established for a period of 15 years (GAP Administration 1999, 1998). However, this was later modified and the institution is still functional.

The GAP Project became one of the most ambitious regional development projects in the country by encompassing not only hydropower and irrigation infrastructure development as originally planned, but also all related sectors including industry, transportation, rural and urban infrastructure, environmental protection and social sectors in the region. Since GAP was planned as a multi-sectoral integrated regional development project, an integrated approach was essential to achieve its targets and objectives. It was thus understood that construction of energy and irrigation infrastructure, past and future, would contribute to economic growth, social development and environmental change. All these changes would, in turn, have discernible impacts on different parts of the region and would need to be managed with a view to achieving the sustainability of the GAP Project and improving the quality of life for the local population.

Some 20 years on, the progress achieved in the GAP Region is indisputable: per capita income had increased from 47% of average national income in 1985 to 55% by 2001. By the end of 2005, eight hydropower plants had been completed, representing 74% of all energy projects at the national level. In that year alone, the electricity generated in the region was to the order of 253 billion kWh, equivalent to $15.1 billion (1 kWh = 6 cents) and accounting for 47.2% or $1.1 billion of total production at the national level. In addition, seven industrial centres had been completed by the end of 2005 (GAP Administration 2006), and 286,502 ha of land in the Euphrates and Tigris river basins were under irrigation at the end of 2010. In terms of investment, some $19.6 billion (TL 25.6 billion) of public funds had been spent in the region at the close of 2007, representing 62.2% of the total calculated expenditure for the GAP Project as originally projected (Government of Turkey 2008).

In spite of this progress, implementation of the GAP Project as an integrated regional development project is still well behind the targets set out in the Master Plan. Low annual average growth; unemployment; insufficient qualified labour force; inappropriate infrastructure for industry, education, health, drinking water, wastewater and solid waste; low levels of education and low level of regional capital accumulation continue to plague the region. These issues have been further aggravated due to high population growth and environmental problems that have resulted from intensive agricultural practices, excessive and uncontrolled irrigation, insufficient drainage in agricultural areas, and poor water, wastewater and solid waste management in the growing urban centres.

Insufficient funding has been a key issue for the lack of implementation of the project. An example is the investment of DSI in the agricultural sector in the GAP Region, which has always been considered of utmost importance because of its expected positive impacts in the income level of the overall population. In 1998, as requested by the national government, DSI revised its construction programme and annual budget estimates for the 1999–2010 period and submitted them to SPO for their consideration. Nonetheless, the funds allocated by the government to the GAP Project were not enough to implement it by 2010 (DSI 2008).

In order to address the delay in the implementation of the GAP Project, the Government of Turkey has developed the GAP Action Plan for the GAP Region for the period 2008–2012 (Government of Turkey 2008). The objective of this plan is to ensure economic growth, social development and employment creation. Its focus is on meeting basic infrastructure needs, mainly in terms of irrigation, and accelerating economic and social development in the region with the aim to contribute to the national targets of economic growth and development as well as social stability and capacity building. Total expenditures foreseen for the Action Plan 2008–2012 are $20.54 billion of which $20 billion will be spent in infrastructure development mainly for irrigation, energy, transportation and social issues.

This analysis will not attempt an overall evaluation of the socio-economic impacts of the GAP Project. As mentioned by Beleli (2005), the lack of a systematic monitoring and evaluation system of such a complex project makes an assessment of that nature quite impossible. Rather, the objective is to analyse the main impacts of the Atatürk Dam—one of the largest and most relevant water projects in the GAP Region and also one of the most important in Turkey—not only in terms of energy generation and irrigation-related activities, but as part of an overall strategy which aims to achieve energy security for the country. The analysis and discussion are based largely on personal experience as well as on previous studies carried out by the authors in the region.

8.2 The GAP Region

The GAP Region spreads along 75,000 km^2 and includes the Provinces of Adiyaman, Batman, Diyarbakir, Gaziantep, Kilis, Mardin, Siirt, Sanliurfa and Sirnak. The region covers approximately 10% of the country's area and has correspondingly about 10% of the total population.[6] It is estimated that approximately 20% of total irrigable land and 28% of the country's energy potential are in this region.

In 1985, the GAP Region accounted for 4% of the GNP, per capita income of the region was 47% of the national average, the literacy rate was about 55%, and medical facilities and personnel in the region were inadequate with 1,391 doctors and 1,630 nurses per 10,000 population at the regional level compared to 3,631 doctors and 2,758 nurses at the national level for the same number of population. Some 22% of the rural population (living in 3,500 rural settlements) did not have access to clean drinking water, 29% of the villages had telephone access, 66.8% had electricity and 90% were linked to the road networks. Almost 70% of the economically active population was engaged in agriculture, but was generating only 44% of the total value added (Unver 1997).

As originally conceived, the land and water resources programme of the GAP Project included the construction of 22 dams and 19 hydropower plants on the Euphrates and Tigris rivers and their tributaries, as well as very extensive irrigation networks. By the year 2010, the GAP Project was expected to have increased per capita income in the region by 209%, approximately 3.8 million people would have been provided employment opportunities, 1.7 million ha of land would be under

---

[6]According to Aksit and Akcay (1997), the GAP Region does not necessarily denote a uniform and distinct social structure nor does it display a cultural uniformity. It simply refers to a group of provinces included in the area of the Southeastern Anatolia Project.

irrigation and 27 billion kWh of hydroelectric energy would be produced annually (GAP Administration 2006). The situation, however, is quite different. The analysis will explain the reasons thereof.

8.2.1 Development Plans and the GAP Region

The main development instruments used for planning purposes in Turkey are National Five-Year Development plans, Medium Term programmes and annual programmes. All of these plans are prepared by the SPO, Prime Minister's Office.

The National Five-Year Development plans set macroeconomic targets, sectoral and regional objectives and policies, and the overall direction for the country's social development (Say and Yucel 2006; SPO 2008a). The Medium Term programmes are policy documents with a 3-year perspective, which analyse the current status of the country and determine modifications for the macroeconomic, sectoral and regional policies, goals, targets and priorities, and also set the budgets. The SPO is responsible for monitoring their implementation (SPO 2008b).

All the plans and programmes throughout the years have been very ambitious regarding regional development. As is normally the case, the objectives of both development plans and programmes have changed with time, but regional development has always been identified as an important means to ensure the highest economic and social benefits for the country as a whole. Within these varying objectives, land and water resources have been consistently identified as a priority.

The several plans and programmes recognise that imbalances in socio-economic structure and income level across rural and urban settlements, as well as across regions in Turkey, continue to be a major constraint. So far, infrastructure for the provision of services and employment opportunities in most cities remains inadequate for responding to population pressure resulting from both population growth and migration. This requires implementing integrated regional development policies that consider measures which are specific to the problems and potentials of the area, and which focus on strengthening local institutions, developing qualified labour, promoting employment, building social infrastructure and increasing the contribution of the private sector to regional development.

It is also widely acknowledged that institutional arrangements have to be more agreeable for the implementation of national regional policies. For this purpose, in 2002, SPO defined the Nomenclature of Territorial Units for Statistics (NUTS) levels in Turkey and organised the 81 provinces of the country into 26 so-called 'statistical regions' at the NUTS II level[7] (Table 8.1). It was planned that regional development agencies would be established in the NUTS II regions. They were expected to play a

---

[7]NUTS is the name of the statistical region classification used in the European Union. NUTS classification, developed by Eurostat, establishes the framework for regional development policies, collection of regional data, and creation of a comparable statistical database harmonised with the European Union regional statistics system. Regions are classified as NUTS I, II and III depending on their population. Turkey aligned itself with the NUTS classification system in 2002. The country has 12 NUTS I, 26 NUTS II, and 81 NUTS III regions.

Table 8.1 Selected indicators for the first and last five NUTS II regions, ranked according to the Socioeconomic Development Index (SEDI)

Regions	SEDI rank (2003, within 26 regions)	GDP per capita (2001, TR = 100)	Sectoral employment (2005) Share of agricultural sector (%)	Share of industrial sector (%)	Share of services sector (%)	Urbanisation rate (%) (2000)	Net migration rate (per thousand) (2000)
TR10 (Istanbul)	1	143	0.7	37.0	62.4	90.7	46.1
TR51 (Ankara)	2	128	7.3	16.0	76.6	88.3	25.6
TR31 (Izmir)	3	150	18.1	27.7	54.2	81.1	39.9
TR41 (Bilecik, Bursa, Eskisehir)	4	117	18.3	37.8	43.8	76.4	38.7
TR42 (Bolu, Düzce, Kocaeli, Sakarya, Yalova)	5	191	20.4	26.8	52.8	57.2	−9.5
Turkey	–	100	29.5	19.4	51.1	64.9	–
TRA1 (Bayburt, Erzincan Erzurum)	22	50	62.0	3.5	34.5	57.3	−43.5
TRC2 (Diyarbakir Sanliurfa)	23	54	38.1	5.7	56.1	59.1	−39.5
TRC3 (Batman, Mardin, Simak, Siirt)	24	46	29.3	10.0	60.8	59.6	−46.8
TRA2 (Agri, Ardahan, Igdir, Kars)	25	34	61.8	3.1	35.1	44.6	−57.3
TRB2 (Bitlis, Hakkari, Mus, Van)	26	35	48.0	6.3	45.8	49.3	−39.5

TR, TRA, TRB, TRC refer to NUTS codes. For more information see SPO (2006b)

major role in managing regional policies, achieving regional development, and mobilising support and funding for regional development projects, while integrating the public and private sectors with non-governmental organisations (NGOs) working towards regional development (Kayasü 2008).

Six of the nine provinces of the GAP Region are among those considered less developed in the country, in terms of socio-economic development and structure of employment for the agricultural, industrial and services sectors. They are all targeted as priority areas for investment (Table 8.1).

In terms of levels of investment, the 2007–2013 Ninth Development Plan (SPO 2006b) stated that a main constraint in achieving substantial implementation progress in the different regional development plans has been that, except for GAP, plans have been considered only within the scope of sectoral allocation rather than a more integrated point of view, in addition to have been provided with limited funding. Among those plans are the Zonguldak-Bartin-Karabük Regional Development Project (ZBK), the Eastern Black Sea Regional Development Plan (DOKAP), the Eastern Anatolia Project (DAP) and the Yeşilirmak Basin Development Project (YHGP) (SPO 2008a).

However, while the GAP Project has been provided with more funding than other projects, it still needs to be considered as an integrated regional development programme which promotes local initiatives and collaboration among development agencies, rather than a purely infrastructure-based project with investments in energy and irrigation development. Additionally, policies that promote market opportunities and institutional capacity at the local level, and that develop a qualified cadre of human resources, still need to be strengthened (SPO 2006b).

The Long-Term Strategy and Eighth Five-Year Development Plan for 2001–2005 (SPO 2001) consider the reduction of inter-regional imbalances as one of its main objectives and, therefore, list among its priorities the implementation of regional development policies, inter-regional integration and social and economic equity. The plan acknowledges serious disparities between regions in spite of the progress achieved with regard to regional development. Some of the reasons identified are high population growth, low levels of education, insufficient qualified labour and lack of timely fund allocation for development of infrastructure. High inter-regional migration, as well as continuous migration from rural to urban areas, have also been recognised as serious issues since they exacerbate socio-economic problems because of unemployment, lack of urban infrastructure and lack of access to housing, education and health services. In the case of the GAP Region, the adverse impacts of migration have been noted in several provinces: Adiyaman, Gaziantep, Diyarbakir, Batman and Sanliurfa (SPO 2001, 2007a). Nonetheless, these provinces have also been identified as centres for urban growth that could prevent out-migration from the region, and for this reason it is fundamental to promote development of urban infrastructure and services for the increasing population (SPO 2007a, 2008b, 2010).

In 2000, the Long-Term Strategy identified 49 provinces and two administrative regions, including the GAP Region, as First Degree Priority areas (SPO 2001). These areas represent 55% of the country's area and 36% of its population, which indicates the extent of sustained effort that is necessary in order to achieve desired development targets in the country.

Within the frameworks established by Turkey's development and strategy plans, reduction of regional disparities (SPO 2005, 2006a) and fulfilment of regional development (SPO 2007b, 2008a, b, 2009) are also identified as important means to achieve the objectives of the Medium Term Programmes. As before, land and water resources have been acknowledged as priorities, with specific reference to the GAP Project. These plans also emphasise the importance of transforming integrated regional development plans, primarily the GAP Project, into implementable programmes, of ensuring that the allocation of resources is consistent with those programmes, and of setting up monitoring and evaluation mechanisms in order to achieve the targets that have been defined (SPO 2008a, b).

In terms of investment for the GAP Project, the funds needed to finance its implementation during the 2008–2012 period alone are estimated to be approximately $20.54 billion, most of which is expected from the budget of the central government (SPO 2008b). It is also expected that resources from the Unemployment Insurance and Privatisation funds will continue to be transferred to the central government for use within the GAP Action Plan as well as for other investments towards economic and social development (Government of Turkey 2008). It is worth noting that the projected total capital expenditure for the GAP Action Plan for the year 2010 alone would be TL 5.4 billion, or 1.8% of the national GDP (SPO 2010).

The Social Support Programme (SODES) has been developed within the scope of the GAP Action Plan to ensure a focus on social development. Through SODES, 398 projects were financed in 2008 with a total investment of TL 42 million for employment, social integration, and culture, arts and sports purposes. In the following year, TL 92 million was invested in 778 projects (SPO 2010).

The end objectives of the implementation of the development and action plans in the GAP Region are to improve the quality of life of people living in the region by mobilising local resources, eliminating development disparities among the regions, and also by contributing to national targets of economic growth and social stability. Of course, sustained progress in the region cannot be achieved through planning alone—proper implementation is essential.

8.3 GAP Project

The GAP Project is a major attempt to reduce inter- and intra-regional development disparities and accelerate improvements in quality of life for the region's population. In terms of resources allocated primarily for investment in energy and irrigation, it is considered to be the most comprehensive project in Turkey. However, the GAP Project has not been the only regional project of the country. There have been many regional development plans along the years, including the Antalya Project (1959), Eastern Marmara Planning Project (1960–1964), Zonguldak Project (1961–1963), Cukurova Regional Project (1962–1963), Zonguldak-Bartın-Karabük (ZBK) Regional Development Project (1995–1996), Eastern Anatolia Project (DAP) (1999–2000), Eastern Black Sea Regional Development Plan (DOKAP)(1999–2000), Yeşilırmak

Basin Development Project (YHGP) (2005–07) and Konya Plains Project (KOP) (Government of Turkey 2008; SPO 2008a).

The initial aim of the GAP Project was to achieve economic and social development in the region through the implementation of land and water resources projects that involved the construction of 22 dams, 19 hydropower plants with a total installed capacity of 7,500 MW and extensive irrigation and drainage networks. The project was expected to nearly double the area under artificial lakes to 228,136 ha in the country; irrigated land was to increase from 2.9% to 22.8% of the region's total area; and rain-fed agriculture to decrease from 34.3% to 10.7% (Biswas and Tortajada 1999; Unver 1997). So far, its implementation as an integrated regional development project is well behind the targets of the Master Plan. Nevertheless, significant progress has been achieved in terms of the water and land resources programme under which several dams and related projects have been developed and the benefits of which are, for the most part, already tangible in terms of electricity generation and irrigated agriculture. Even though the target of around 1.7 million ha under irrigation is still rather far off, the newly irrigated acreage has transformed the GAP Region into a major producer of cotton contributing nearly 50% of the national output.

The following is an analysis of the land and water resources programme of the GAP Programme, with an emphasis on the main water project for the local area and the region, as well as one of the most important for the country—the Atatürk Dam.

8.4 Land and Water Resources Programme

The GAP Project has multiple programmes for different sectors. The land and water resources programme includes 13 main irrigation and energy projects, seven of which were planned for the Euphrates River and six for the Tigris River. Before discussing the infrastructure development made possible by the project, it is useful to delineate the characteristics of both rivers.

8.4.1 Euphrates and Tigris Rivers

The mean annual flow of the Euphrates (Firat) River is estimated to be 33.6 BCM, and that of the Tigris (Dicle) River is estimated to be 50.9 BCM (excluding the flow of Karun River from Iran). Some 98% of the Euphrates River runoff originates in the highlands of Turkey, while the rest of its catchment in lower arid regions makes little contribution to the river (Altinbilek 2004). The main features that distinguish the hydrologic regime of the Euphrates-Tigris River system are their annual and seasonal fluctuations, with large floods originating from the snow-melt in spring.

The Euphrates River has a catchment area of 127,304 km^2 and a mean catchment elevation of 1,383 m. Its main tributaries, the Karasu and Murat Rivers, flow towards the west along intra-mountain valleys and join the Euphrates River around the

Table 8.2 Characteristics of the Euphrates and Tigris Rivers

	Tigris[a]	Euphrates[a]
Mean catchment elevation (m)	1,451	1,383
Catchment area (km^2)	57,614	127,304
Mean annual discharge (km^3)	21.33	31.61
Mean annual precipitation (cm)	65.8	55.9
Mean air temperature (°C)	12.6	11.0
Number of ecological regions	4	7
Land use (% of catchment)		
Urban	0.0	0.1
Arable	37.7	30.6
Pasture	26.7	31.4
Forest	21.5	17.8
Natural grassland	12.8	18.6
Sparse vegetation	0.0	0.0
Wetland	0.1	0.1
Freshwater bodies	1.2	1.4
Protected area (% of catchment)		
Water stress (1–3)		
1995	3.0	3.0
2070 (estimated)	3.0	3.0
Fragmentation (1–3)	3	3
Number of large dams (>15 m)	11	49
Native fish species	46	42
Non-native fish species	2	1
Large cities (>100,000 people)	2	6
Human population density (people/km^2)	65	57
Annual gross domestic product ($ per person)	1,311	1,535

[a]Only within Turkey
Source: Adapted from Akbulut et al. (2009)

Keban Dam, where they flow first towards the southeast and then the southwest before entering Syria. The Euphrates is 3,000 km long, with 1,230 km in Turkey, 710 km in Syria and 1,060 km in Iraq. Turkey contributes 89% of the river's annual flow and Syria 11%. The Tigris River is 1,850 km long and represents a natural border between Turkey and Syria, later crossing into Iraq. In terms of its annual flow, Turkey contributes 51%, Iraq 39%, and Iran 10% (Altinbilek 2004). It has a catchment area of 57,614 km^2 and a mean catchment elevation of 1,451 m. Its main tributaries within Turkey are the Batman, Garzan, Botan and Hezil Rivers. The Tigris enters Iraq at 300 m mean sea level (msl) and joins the Euphrates to form the Shatt-el-Arab in southern Iraq before discharging into the Persian Gulf (Akbulut et al. 2009) (Table 8.2).

The main difference between the Euphrates and Tigris Rivers is in terms of their discharge: the Tigris receives water from several major tributaries in the middle portion of its course, while all the major tributaries of the Euphrates are in the extreme upper end of the basin. This distinction has a significant effect on the regulation of both rivers. For the Euphrates, a single dam (in this case the Atatürk Dam) in the upper part of the catchment is able to regulate a very large proportion of the flow of the river. Since the Tigris receives water from the Greater Zab, the Lesser Zab, the Adhaim and the Diyala Rivers, its overall water management is more complex than is the case with the Euphrates, requiring the construction of a series of major dams on individual tributaries to provide a control of flow comparable with that of the Euphrates (Altinbilek 2004).

With regard to their river basins, that of the Euphrates is shared by Turkey, Syria, Iraq and Saudi Arabia, while the Tigris Basin is shared by Turkey, Syria, Iraq and Iran. Turkey contributes approximately 98% to the total discharge of the Euphrates River at its mouth, and 53% to the discharge of the Tigris.

8.4.2 Infrastructural Development

Details of infrastructure development under the land and water resources programme as of 2005 are provided in Table 8.3, while the historical development of the GAP Project is traced in Table 8.4.

The first dam to be constructed on the Euphrates River, in what would later become the GAP Region, was the Keban Dam (675 km^2)[8]. Operational since 1974, this dam has great importance for the region because it is able to store 70% of the river's flow within Turkey. Downstream Keban, there are the Karakaya (268 km^2), Atatürk (817 km^2), Karkamis (28.4 km^2) and Birecik (56.25 km^2) dams constructed in 1987, 1992, 1999 and 2000, respectively.

Atatürk Dam, functional since 1992, is widely considered to be not only the largest dam in Turkey, but also one of the largest in the world. The dam generates 8,900 GWh of electric power per year, followed by Karakaya and Keban, with 7,300 GWh and 6,000 GWh respectively. In 1994, water was delivered from the Atatürk Dam to the Harran Plains, an area that has since become the country's major producer of cotton (DSI 2009).

Dams constructed on the Tigris River include the Kralkizi, Dicle and Batman. The main dam, however, will be the Ilisu Dam due to its importance in regulating the river. The construction of this project has been delayed for many years due to strong differences of opinion between the developers and the affected population, as well as credit agencies. The credit agencies set a number of conditions regarding resettlement of affected people, possible pollution of the future reservoir due to discharge of untreated wastewater by various cities upstream, as well as protection of the cultural heritage of Hasankeyf, an ancient town with numerous archaeological sites that was declared a natural conservation area in 1981. When adequate measures were not implemented within the given timeline, the credit agencies decided to cancel their loans. At present, the Turkish government has decided to continue

---

[8] All figures in brackets refer to the reservoir area.

Table 8.3 Status of the GAP Project as of 2005

Project status	The Euphrates Projects	The Tigris Projects	Total
Total			
Installed capacity (MW)	5,318	2,172	7,490
Energy production (GWh)	20,140	7,247	27,387
Irrigated land (ha)	1,188,135	632,913	1,821,048
Number of dams	14	8	22
Number of HEPP	11	8	19
In operation			
Installed capacity (MW)	5,066	402	5,468
Energy production (GWh)	19,464	927	20,391
Irrigated land (ha)	175,571	38,353	213,924
Number of dams	6	3	9
Numbers of HEPP	4	3	7
Under construction			
Installed capacity (MW)	50	0	50
Energy production (GWh)	124	0	124
Irrigated land (ha)	103,246	57,014	160,260
Number of dams	1	0	1
Numbers of HEPP	1	0	1
Completed design			
Installed capacity (MW)	202	1,770	1,972
Energy production (GWh)	552	6,320	6,872
Irrigated land (ha)	909,318	537,546	1,446,864
Number of dams	7	5	12
Numbers of HEPP	6	5	11

Source: Akyürek (2005)

the project with funds from international commercial loans secured by local contractors without guarantee from the treasury. The government will pay only the value-added-tax portion of the expenditures. Some 48 houses that were built within the construction site by the government for the affected people of Ilisu village were distributed to owners during October 2010. A new town will be built to resettle Hasankeyf village in the upstream area. Sites of cultural heritage that will be flooded due to the dam will be relocated above the level of the future lake (535 m). A new museum will be built to exhibit the findings of salvage excavations which took place over a decade. Construction is now progressing with a deadline to withhold water in 2014 and to produce energy in 2015.

The hydropower plants in the GAP Region are of great importance at the national level from the viewpoint of energy, since they produce half of Turkey's hydroelectrical energy as well as very significant revenues for the country (DSI 2009). In fact, the main impact of the GAP Project so far is considered to be the production of energy, the market value of which is calculated to be more than the total investment for energy purposes in the region. Table 8.5 shows the total energy generation in the GAP Region, and its monetary equivalent from the time the different dams became functional and until the end of 2004.

8 The Atatürk Dam in the Context of the Southeastern Anatolia (GAP) Project

Table 8.4 Historical development of the GAP Project

Year	Event
1936	Research on the Firat River was initiated under directives from President Atatürk
1938	Geological and topographical studies were begun and flow stations were established in Keban
1954	The General Directorate of State Hydraulic Works (DSI) was established
1961–1971	The Firat Planning Authority, established in 1961, published the Firat Basin Reconnaissance Report in 1964. The Lower Firat Feasibility Report was prepared in 1970 and the Dicle Basin Reconnaissance Report was published in 1971
1966	Foundation work for the Keban Dam was started
1974	The Keban Dam started operations with the great benefit that a regular flow of water would be supplied to any dam downstream
1976	Construction of the Karakaya Dam was begun
1980	The Lower Firat and Dicle Projects were combined, and named the GAP Project
1981	Construction of the Atatürk Dam's diversion tunnels as well as the Sanliurfa tunnels was initiated
1987	The Karakaya Dam began to produce electricity
1990	Water was stored in the Atatürk Dam
1992	The Atatürk Dam started to produce electricity
1994	Water reached the Harran Plains via the Sanliurfa tunnels
1997	Water was stored in the Kralkizi and Dicle dams
1998	Water was stored in the Batman Dam
1999	The Karkamis Dam and HEPP were completed
2000	The Birecik Dam and HEPP were completed

Source: DSI (2009)

Table 8.5 Hydropower energy production and revenues in the GAP Region until the end of 2004

Dam and HEPP	Year of operation	Installed capacity (MW)	Total energy production (billion kWh)	Monetary equivalent (million $)
The Euphrates River				
Karakaya	1987	1,800	127.83	7669.80
Atatürk	1992	2,450	94.10	5,646
Karkamis	2000	189	1.67	100.20
Birecik	2001	672	7.96	477.60
Total		5,111	231.56	13893.60
The Tigris River				
Kralkizi	1999	94	0.58	34.80
Dicle	2000	110	0.96	57.60
Batman	2003	198	0.59	35.40
Total		402	2.13	127.8

1 kWh = 6 US cent
Source: Akyürek (2005)

Investments in the energy sector of the region have more than paid off. The total value of energy generated in the region has been higher than the total investments made in the energy sector within the framework of the GAP Project which, up to 2002, were $4.17 billion (Ercin 2006).

8.5 Atatürk Dam

The Lower Euphrates Project, one of the 13 main irrigation and energy projects in the GAP Region, is the largest and the most comprehensive one. It includes the Atatürk Dam and Hydroelectric Power Plant (HEPP), Birecik Dam and HEPP, Sanliurfa tunnels, Sanliurfa-Harran irrigation, Mardin-Ceylanpinar irrigation, Siverek-Hilvan pumped irrigation and Bozova pumped irrigation (DSI 2009, 2000; Unver 1997).

The Atatürk Dam is considered to be the main undertaking within the GAP Project. The dam is a multi-purpose project for both hydropower generation and irrigation. It is a rock-fill dam with clay core. It has a crest length of 1,664 m, a crest width of 15 m and a crest elevation of 549 m. Its height from the foundation is 169 m, and the maximum water elevation is 542 msl. The volume of the reservoir is 48,700 hm^3 and it has a reservoir area of 817 km^2. The volume of embankment is 84.5 million m^3 (MCM), the spillway capacity is 16,800 m^3/s and it generates 8,900 GWh per year.[9]

The dam consists of eight Francis-type turbine and generator groups of 300 MW each, supplied by Sulzer Escher Wyss and ABB (Asea Brown Boveri), respectively. The turbines have a rated discharge of 8 × 300 m^3/s and a head of 151 m (Table 8.6).

With the completion of the Atatürk Dam, some 81,700 ha of land were inundated (GAP Administration 1999).

Water reaches the Sanliurfa-Harran Plains, the largest irrigation area in the GAP Region, through the Sanliurfa twin tunnels system. The Sanliurfa tunnels are concrete-lined, each 7.62 m in diameter and 26.4 km long (DSI 2000). There are 52 connection tunnels and 23 ventilation shafts, with one shaft every 1,500 m and depths which vary between 65.24 and 207.95 m, which facilitate excavation and concreting activities. The T1 (left-side tunnel, in terms of direction of water flow) has been used since May 1997, while T2 (the right-side tunnel) was finalised later. (Yesilnacar 2003).

8.5.1 Irrigated Agriculture

One of the main goals of the GAP Project has been to transform the region into a base for agricultural exports. Irrigation is expected to increase crop yields and promote diversification, thus contributing to increased economic activities and development of agro-industries and other agricultural services.

---

[9]Pöyry Energy Ltd, www.poyry.com, accessed on 30 March 2010, and General Directorate of State Hydraulic Works, DSI, www.dsi.gov.tr/baraj/, accessed on 30 March 2010.

8 The Atatürk Dam in the Context of the Southeastern Anatolia (GAP) Project

Table 8.6 Atatürk Dam and hydropower scheme

Key data	
Hydrology	
Catchment area	92,240 km^2
Annual inflow	26,585 MCM
Reservoir	
Retention water level	EL 542.0 m
Minimum operating level	EL 526.0 m
Maximum water level	EL 544.15 m
Active storage	12,700 MCM
Reservoir capacity	48,700 MCM
Reservoir area	817 km^2
Diversion structures	
Number of tunnels, concrete-lined	3
Length	4,100 m
Diameter	8.0 m
Discharge capacity	3,900 m^3/s
Dam	
Type	Rock-fill with central core
Height above lowest foundation	169 m
Crest length	1,664 m
Elevation of top of dam	EL 549.0 m
Volume of dam	84.5 MCM
Spillway	
Type	Controlled overflow spillway
Type of gates	6 radial gates
Size of gates	16 m × 18 m
Discharge capacity	16,800 m^3/s
Power intake	
Type	Concrete gravity dam, 8 blocks
Type of gates	8 roller gates
Size of gates	7.5 m × 7.5 m
Penstocks	
Number of penstocks	8
Length	Between 515 and 640 m
Diameter	7.25 m
Powerhouse	
Location	Adjacent to the dam toe
Size (l × w × h)	257 m × 53 m × 49 m
Net head	151 m
Valves	8 Butterfly
Turbines	8 Francis
Rated space	150 rpm
Turbine discharge	1,748 m^3/s
Installed capacity	8 × 300 = 2,400 MW
Annual energy total	8,900 GWh

Source: Adapted from Pöyry Energy Ltd (no date)

Most of the irrigated agriculture in the GAP Region is being developed in Sanliurfa Province, which has 11 districts and 1,080 villages and settlements. The districts include Akcakale, Birecik, Bozova, Ceylanpinar, Halfeti, Harran, Hilvan, Sanliurfa City, Siverek, Suruc and Viransehir. The Province of Sanliurfa ranks ninth in terms of population at the national level. It is the most populated area in the region, with 21.8% of the region's total; within the Province Sanliurfa City has the highest population, of approximately 37%.

The GDP of Sanliurfa is the second highest in the region. In the period 1990–2001, it increased from TL 597 billion to TL 1,236 billion (at 1987 prices) (Ercin, 2006). This rise is directly attributed to economic activities resulting from irrigation.

8.5.1.1 Sanliurfa-Harran Plains

Irrigation of the Sanliurfa-Harran Plains was the first project to become operational among several such projects in the GAP Region in 1995. Irrigation of the plains has resulted in numerous social, economic and environmental impacts throughout the years, both positive and negative. While the impacts are manifold, the main ones are mentioned below:

- Overall improvement in the welfare of the population, although health and education services still need to be significantly enhanced.
- Change in crop patterns to an emphasis on cotton, following the introduction of irrigation. This change has reversed the patterns of out-migration from the region since farmers do not have to go to other provinces for seasonal work.
- Land consolidation, which has improved the systems of land tenure and increased the value of land.
- Negative environmental impacts in some areas of the Harran Plains, mainly in terms of salinity.

Sanliurfa is the province with the highest agricultural output in the region (mostly in Harran), with the Atatürk Dam and irrigation projects playing a major role. Agricultural yield increased from 2 million tons in 1994 to 3 million tons in 2003. This improvement in the quantity of agricultural production is also evident in its market value, which rose from $459 million in 1994 to $1 billion in 2003. Although cereals have a low market value, the total cereal production in the GAP Region increased by nearly one million ton between 1994 and 2003, from 3.5 to 4.5 million tons. The market value of cereal output also increased from $404 million before irrigation in 1994 to $902 million in 2003 (Ercin 2006).

Cotton has become the dominant crop in the area, accounting for over 85% of all summer-irrigated crops. Its production has increased from 160,000 tons before irrigation to 1,135,886 tons in 2003. The net result has been that the GAP Region has become the main producer of cotton in Turkey with almost 50% of the country's cotton production. In 2003 alone, the region produced 1.13 million of tons with a total value of $582.6 million. Within the region, the Province of Sanliurfa (Harran Plains)

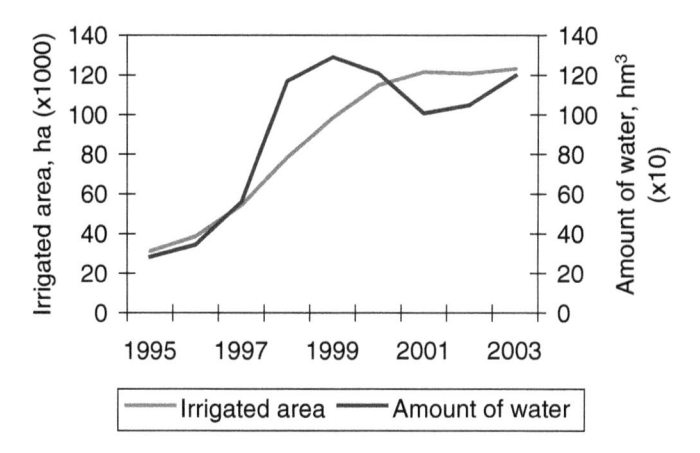

Fig. 8.1 Distribution of irrigated area and water carried to the Harran Plains, 1995–2003. *Source*: Yesilnacar and Uyanik (2005)

has become the largest producer of cotton (51.1% of the region's total), followed by Diyarbakir (18.6%) and the rest of the provinces (Ercin 2006).

In 2000, the gross agricultural output value (GAOV) in the GAP Region was approximately $262 million, representing $2,347 per ha and $2,547 per capita.

The change from rain-fed to irrigated agriculture has meant that water requirements for cotton in the Harran Plains have also increased agricultural water use from about 370 MCM in 1993 to over 1 BCM in 2002. The expansion of irrigated land also seems to have resulted in a steady decrease in potential evaporation due to increased roughness and decreased humidity deficit in the Harran Plains. If changes in future evaporation conditions are of a similar nature, it is possible that demand for irrigation water will decrease by more than 40% in future (Ozdogan et al. 2006). The decrease in the use of water for irrigation in the Harran Plains despite the increase in irrigated area is also mentioned by Yesilnacar and Uyanik (2005) (Fig. 8.1). These are important findings when planning for water use in future irrigated areas.

The socio-economic benefits of irrigation in the Harran Plains have also been studied over the years (Harris 2008; Miyata and Fujii 2007; Akyürek 2005; Kundat and Bayram 2000; Oklahoma State University et al. 1999). While assessment of the impact of irrigation, and of related opportunities, has varied by economic activity, age, gender and marital status of persons interviewed, there seems to be general agreement that the overall quality of life in the villages has improved significantly. However, the interviewees also appear to agree that irrigation schemes should be organised more effectively and efficiently, that the performance of water users associations should be improved, and that the problems related to salinity in the area should be addressed and solved in a targeted way.

Transformations in the area are said to have had positive impacts not only for farmers working on their own land, but also for the landless workers whose conditions and returns were reported as having improved over what they were before. The assessments also indicate the importance of supporting small farmers' organisations and of enhancing the access of small producers to technical information, credit and technology. There was much room for improvement in the provision of overall services to the population as well as in the cooperation and coordination of organisations and agencies in the region (Oklahoma State University et al. 1999).

Studies indicate that the Adiyaman Province has lost a great portion of its agricultural land because of the Atatürk Dam's construction. Nonetheless, the share of agricultural product has increased, mainly due to irrigation. For instance, in 1994 the output and value of crop production were 568,821 tons and $166 million respectively; this increased to 614,350 tons and $201 million in 2003 (Ercin 2006).

On the whole, a major constraint for agricultural development in the overall GAP Region has been the lack of investment. So far, most of it has been channelled to the Harran Plains without much consideration of the agricultural potential of other provinces in the region.

In terms of environmental consequences, there are serious problems in the area because of increasing water logging and salinity, largely due to unsustainable irrigation methods, insufficient drainage and poor land management. Salinity distribution studies for 1987, 1997 and 2000 show that salinity has increased significantly in some parts of the plains, mainly in Akcakale which is located in the lower level in the plain. The total area in the plains affected by salinity was 5,500 ha in 1987, 7,498 ha in 1997 and 11,403 in 2000 (Çullu et al. 2002) (Fig. 8.2). Salinity problems in the Harran Plains are also confirmed by other studies (see Kapur et al. 2009; Yesilnacar and Gulluoglu 2007; Kendirli et al. 2005; Çullu 2003).

8.5.1.2 Assessment of Economic and Social Impacts

In order to understand and appreciate the changes that the construction and operation of the Atatürk Dam have brought about through economic and social development, for the people living in the project area as well as in the region, a study was carried out to determine the extent and magnitude of the actual social, economic and environmental impacts of the dam and the reservoir some 8 years after their construction. The emphasis of the study was on economic, social and environmental issues, both direct and indirect, over the short- to medium-terms, which could be objectively estimated and evaluated with reasonable accuracy. It included an evaluation of the direct impacts (positive and negative) on people living in the two provinces most directly affected, Adiyaman and Sanliurfa, as well as on the region as a whole. The analysis did not include an evaluation of impacts at the national level. The detailed findings of the study have been published elsewhere (see Biswas and Tortajada 1999; Tortajada 2000, 2004).

The assessment of the economic and social impacts included fieldwork and discussions, both in Ankara and the project area. There were interviews with senior members of national and international institutions within the region and beyond (GAP Administration and other planning and implementing institutions, especially

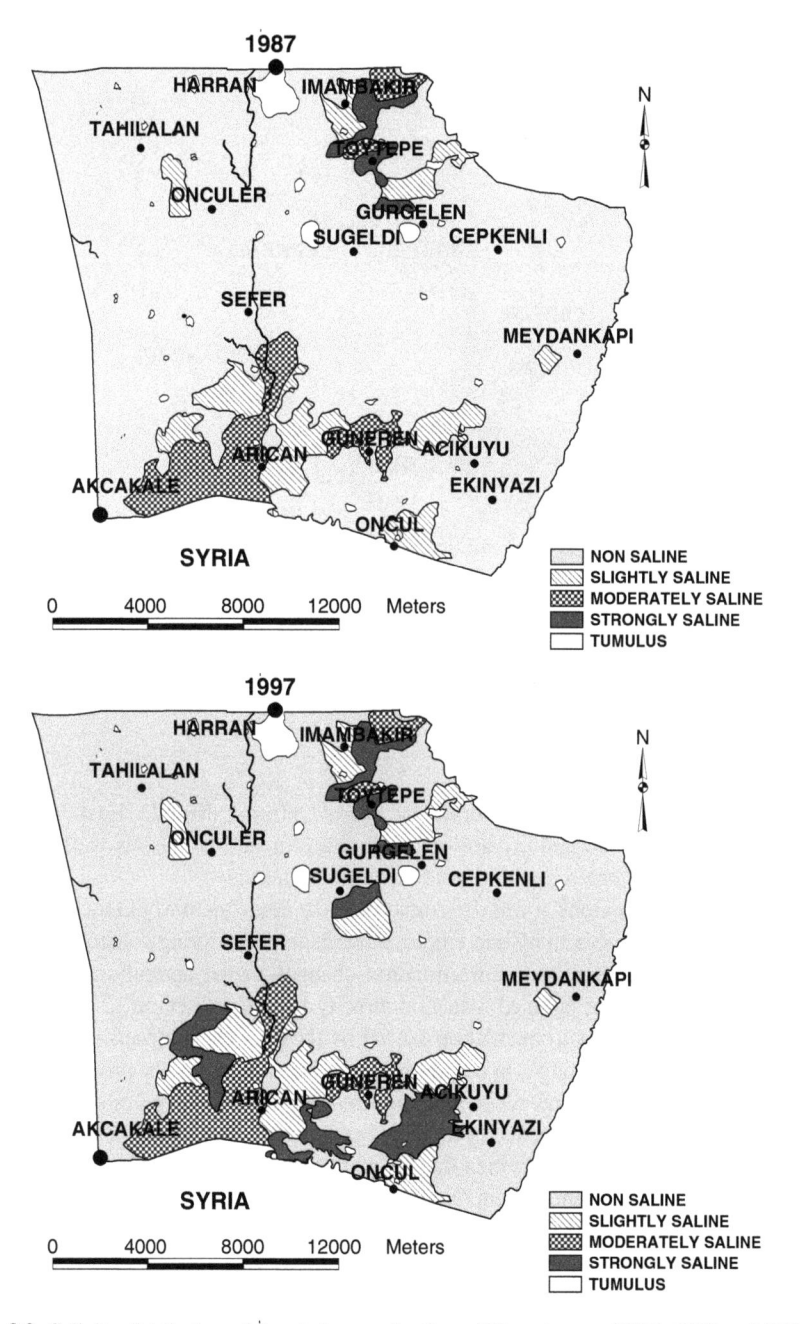

8 The Atatürk Dam in the Context of the Southeastern Anatolia (GAP) Project 189

Fig. 8.2 Salinity distribution of the study area for three different years (1987, 1997 and 2000). *Source*: Çullu et al. (2002)

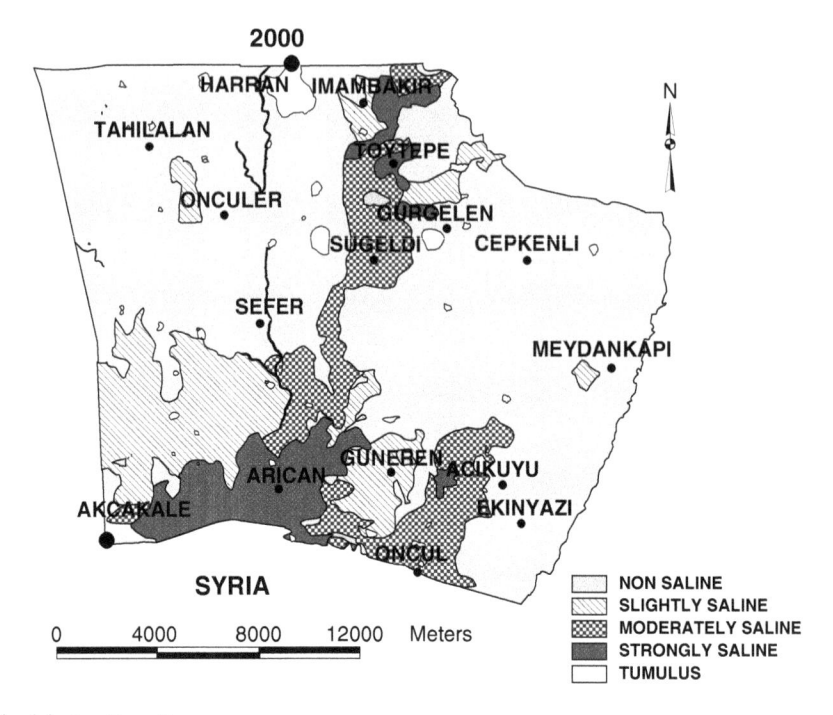

Fig. 8.2 (continued)

the DSI, SPO, General Directorate of Rural Affairs, Middle East Technical University etc.), concerned private sector institutions and NGOs, as well as representatives of the affected population at different locations.

After initial discussions, it was determined that the assessment would focus on issues such as new economic activity and employment generation during construction of the dam, reservoir and associated infrastructure; changes in the agricultural yields and incomes of farms using pumped irrigation directly from the reservoir; a review of the resettlement process due to inundation caused by the reservoir; impacts on health and education; and overall changes in quality of life for those living in the project area.

During the study, it was clear that the lifestyle and working conditions of the local population had improved, employment opportunities had significantly increased, and expanded economic activities had in many cases encouraged migration from rural to urban areas of the region. New urban growth poles were developing, as in Sanliurfa, which helped somewhat in evening out the urbanisation process. The areas around the dam were primarily rural, with limited infrastructural facilities. Before the dam was constructed, transportation and communication networks between the various population centres in the vicinity of the site were inadequate. When construction began a good road network was built, which considerably improved communications in the area. The movement of people and goods from one place to another became much easier and less time consuming. Commercial activities also increased.

8 The Atatürk Dam in the Context of the Southeastern Anatolia (GAP) Project   191

In general, the magnitude and extent of the social and economic impacts generated by the Atatürk Dam and its reservoir have been positive not only for the project area but for the country as a whole. The benefits accruing to the country through the increase in electricity generation alone are substantial. Equally, there has been a marked improvement in lifestyle for those living in the project area, and especially for the majority of people living near and around the reservoir. In retrospect, on the basis of currently available data, it is clear that construction of the Atatürk Dam, which is the main infrastructural project in the GAP area, has acted as an engine for economic growth and development in a historically underdeveloped area. The dam has enhanced the quality of life and working conditions of local people, and significantly increased employment opportunities for communities. Expanded economic activities have encouraged migration from rural to urban areas. The region's semi-urban and urban areas are now facing a surge in population, with the attendant need for more and more housing, water, education, health services, employment opportunities, and efficient and reliable transportation and communication systems.

The area has flourished and employment has been generated during both the construction and the subsequent operation of the dam. Many of the labourers, who were initially unskilled, were trained during dam construction and acquired skill through the process. Many of them have since been employed in the construction of dams all over the country resulting in an increase in income for the region, since the labourers remit their incomes to their families.

The benefits of irrigation are visible mainly in Sanliurfa Province, where both formal and informal jobs have increased exponentially. However, the benefits of dam construction were not limited to employment. The daily exposure of the villagers to the different traditions and practices of 'outsiders' over more than a decade of construction resulted in noticeable social changes, evident in new ambitions for better and more housing and transportation, altered food and health habits, a desire for higher education, the decision of some of the local population to send their children (including girls) to school, a demand for information and communication, etc. The lifestyles of the area's population have thus begun to change, which will likely lead to a better quality of life for many people.

Construction of this dam and its associated hydraulic structures has also had adverse direct and indirect social and environmental impacts, for example, resettlement of a large number of people from the inundated area, impoverishment of those who did not manage their expropriation funds properly, loss of productive agricultural land, and increase in environmental contamination due to higher levels of economic activity. The affected population has, however, generally been fairly compensated for their losses.

Regarding resettlement and rehabilitation, information was collected from DSI (central office and regional office in Sanliurfa) and from the General Directorate of Rural Affairs. The Regional Directorate of DSI in Sanliurfa Province provided information on the status of urban and rural resettlement as of 1993, and also the status of expropriation as of 1996. DSI data did not include information on resettlement after 1993. On the basis of the information collected from DSI and the General Directorate of Rural Affairs, the expropriation of properties up to the height of 542 m had been completed by 15 September 1997. The total cost was TL 13,057 trillion (at 1995 prices). By the end of 1995, TL 2,979 trillion (at 1995 prices) had been paid to settle disputes with the resettled population through the decisions of the courts. The General Directorate of Rural Affairs estimated that 1,129 families had to be displaced due to the Atatürk Dam from 1988 to 1997. Of these displaced families, 44% were to be resettled in rural areas and the balance 56% in urban areas. By 1998, only 30% of the population had been resettled (344 families), and 70% were yet to be resettled (369 families in rural areas, and 416 families in urban areas) (Tortajada 2004). By 1999, 375 families affected by the dam were still living in rented houses waiting to be resettled in rural areas (Altinbilek et al. 1999a, b).

According to Bayram (2000, personal communication), by July 2000 an additional 36 families had been resettled in Ayrancilar village and six other families had decided not to wait any longer to receive the support of the government. For the most part, the main problem in relocating the population was scarcity of land rather than a lack of funds for compensation.

In order to obtain a clearer picture of the efficacy of the resettlement process from the perspective of the affected population, extensive discussions were conducted with project-affected people in several villages in Adiyaman and Sanliurfa Provinces. Collective meetings with resettled communities were organised in a few villages. All the meetings included the Mukhtar (village head) and every head of household (men) in that town. At least 50% of the heads of the households were interviewed in detail.

The main issues discussed during the interviews and meetings were the effectiveness of the resettlement process, status and level of paid compensation, quality of housing and services provided, as well as the impacts of the construction of the Atatürk Dam on the lives of the men, their families and their villages. The three villages studied in detail were New Samsat, Akpinar and Kizilcapinar. (A detailed analysis of the resettlement process from the perspective of the resettlers is available in Tortajada 2004.)

The overall benefits and costs of the dam were viewed differently at the local level depending upon whether the people were from Sanliurfa or Adiyaman and whether or not they had access to irrigation. In all cases, however, people confirmed that their quality of life in the new settlements was better than what they had before. However, the vast majority of them were not aware of the resettlement process as a whole, nor did they have much knowledge of the relevant resettlement or expropriation laws and their entitlements under these laws.

Properly planned and implemented, resettlement programmes can become part of an overarching national strategy for poverty reduction. Well-planned investments in new infrastructure and services (water, electricity, schools, hospitals, roads, etc.) represent an opportunity to improve the standard of life of populations directly affected by development projects. Since it is unrealistic to avoid or reject involuntary resettlement altogether, it is essential to improve the knowledge base for the planning and implementation of projects, with a view to protecting the entitlements and livelihoods of those affected. Resettlement should be approached as part of a development process and not as the neglected stage of an infrastructure construction project.

8.6 The Impact of the GAP Project on the GAP Region

As mentioned above, the GAP Project was planned as a large-scale, multi-sector, integrated regional development project aimed at economic growth and regional development of the GAP Region, through a focus on industry, transportation, urban and rural infrastructure, environmental protection and social sectors such as employment generation, health, education, capacity building and gender equity. While important progress has been achieved, it has mostly been limited to electricity generation and irrigated agriculture. The project still lags behind the targets of the Master Plan in terms of its objective of integrated regional development. Issues such as low annual average growth; unemployment; insufficient qualified labour force; inadequate infrastructure for industry, education, health, drinking water, wastewater and solid waste services; and low level of regional capital accumulation continue to be major concerns in the region. These problems have been further aggravated by high demographic growth. Environmental problems have also resulted from intensive agricultural practices, excessive and uncontrolled irrigation, insufficient drainage in agricultural areas, and poor water and solid waste management in growing urban centres (Government of Turkey 2008). Thus, as an integrated regional development project, the GAP Project still has a long way to go.

According to the socio-economic development ranking of Turkey's provinces carried out in 1996 and 2003 (Dincer 1996; Dincer et al. 2003), the GAP Region is still the least developed area in the country in spite of many local efforts and achievements.

In an extensive analysis of the social and economic impacts on the GAP Region, Ercin (2006) has shown the progress, or lack thereof, in terms of socio-economic, agricultural, industrial and energy development in the region's nine provinces. The overall findings reveal that the GAP Project has fallen behind the targets of the Master Plan mainly in terms of public and private investments and in the quality of the projects implemented, mostly in the social sector, which are considered to be very poor. The main factors are considered below, with the exception of agricultural development which was discussed earlier in the chapter.

8.6.1 High Population Growth

High population growth has made it more difficult to achieve the targets of the GAP Project, and the general infrastructure in place has proved to be inadequate for the provision of services to people in the region. The annual growth rate in the region's urban areas is the highest in the country, with the population having increased 36.8% from 1990 to 2003. This high rate of urbanisation is caused by employment opportunities, industrial development and better health and education facilities in urban areas, which are insufficient in themselves but superior when compared to conditions in the region's rural areas. This has encouraged people from both within and outside the region to move to urban areas within the GAP. Security concerns in the rural areas and resettlement in urban areas after the construction of dams have also contributed to the high urbanisation rate.

8.6.2 Economic Development

Economic development in the GAP Region is based mostly on agriculture, industry and service-related activities. Agriculture, which is the main economic activity in the region, accounted for 26.6% of the region's economy in 1987 but fell to 24.6% in 2000. Similarly, the share of industrial activities declined to 17.9% in 2000 from 21.1% in 1987, with the electricity, gas and water sub-sectors increasing during this period. Finally, only the services sector in the region grew between 1987 and 2000, from 52.4% to 57.5%, mainly in terms of transportation, communication and government services (Table 8.7).

The GDP resulting from services between 1987 and 2000 was TL 2 billion, agriculture was TL 1,773 billion, and industry was TL 1,189 billion (all at 1987 prices) in the region. The values of both imports and exports have increased in the region from the start of the GAP Project. Exports have increased by $1 billion between 1989 and 2005, and imports have increased from $100 million in 1989 to $600 million in 2004.

Public and private investments, or lack of them, have been a major problem in achieving the objectives of the GAP Project. In the case of public investments, there has been a decrease from 6% of the total investment in the country in 1990, to 2% in 2001. Within the region, Sanliurfa Province has benefited the most, since 50% of the investments have been directed to this province for infrastructural development.

Despite new economic activities in the region, unemployment has increased: it was at 12.13% in 2000, four times the national average, from 4.32% in 1980, which was close to the national average. Only the agricultural sector in Sanliurfa and Adiyaman Provinces and the industrial sector in Gaziantep Province have seen some improvement. Stagnation in the Turkish economy is one of the main reasons for this, but population growth is also a factor.

8.6.3 Industrial Development

Industrial development, which was intended to develop the region into an agriculture-based export centre, has been slow and has depended mostly on public investment with almost no participation from the private sector. According to SPO, the GAP Region has not shown a significant improvement from 1996 to 2003 (Dincer et al. 2003; Dincer 1996) even though the number of factories and plants (mostly small to medium-size) has almost doubled. The manufacturing sector, which was already strong in Gaziantep Province before the project began, does not appear to have benefited from the GAP Project.

Table 8.7 Economic structure and change in GAP and Turkey[a]

	GAP region 1987	GAP region 2000	Turkey 1987	Turkey 2000	Change (%) GAP	Change (%) Turkey
Agriculture	26.5	24.5	18.2	14.0	−2.0	−4.2
Agriculture and livestock production	25.9	24.4	16.7	13.3	−1.5	−3.4
Forestry	0.6	0.1	1.1	0.3	−0.5	−0.8
Fishing	0.0	0.0	0.4	0.4	0.0	0.0
Industry	21.0	17.9	26.3	24.0	−3.1	−2.3
Mining and quarrying	8.7	4.4	2.0	1.1	−4.3	−0.9
Manufacturing	9.5	9.1	22.3	19.9	−0.4	−2.4
Electricity, gas and water	2.8	4.4	2.0	3.0	1.6	1.0
Services	52.5	57.6	55.5	62.0	5.1	6.5
Construction	7.7	5.2	7.4	5.2	−2.5	−2.2
Trade, wholesale and retail trade	16.1	13.9	17.5	16.5	−2.2	−1.0
Hotels, restaurants services	1.1	1.2	2.7	3.5	0.1	0.8
Transportation and communication	8.8	13.5	11.7	14.2	4.7	2.5
Financial institutions	1.3	1.5	3.1	3.8	0.2	0.7
Business and personal services	1.3	1.7	2.4	3.9	0.4	1.5
Government services	6.9	16.8	5.1	10.1	9.9	5.0
Ownership of dwelling	9.3	3.8	5.6	4.8	−5.5	−0.8
GDP	100.0	100.0	100.0	100.0		

[a]The figures represent the percentage share of each sector in the economies of GAP Region and Turkey
Source: Ercin (2006)

8.6.4 Energy Development

The development of the energy sector has been most relevant. Investments for energy development in the region have more than paid off, since the total value of the energy generated from investments in the GAP Region is higher than the total investments into the energy sector in the GAP Project.

To take just one example of the scale of the project's potential in terms of energy production, in 2002 alone a total of 7.06 billion kWh of energy were produced by three of the dams in the region (Atatürk, Birecik and Karkamis), with a monetary value of almost $424 million. Also, by 2002, the Atatürk Dam alone had produced a total of approximately 80 billion kWh worth $4.7 billion. This figure is higher than the total investments made in the energy sector in the GAP Region up to 2002, which is $4.17 billion (Ercin 2006). This achievement is very important not only in terms of electricity generation, but also in terms of a broader strategy of energy security for the country.

8.6.5 Health Services

In terms of health, the SPO considered the GAP Region to be the least developed in the country in 1996 (Dincer 1996). In 2003, even though all relevant indicators had improved, all provinces of the region were still below the national averages. In

2003, the region was ranked sixth (out of seven) in overall health at the national level; sixth in infant mortality and number of pharmacies (1.85) per 10,000 people; and last in terms of number of doctors (5.49), dentists (0.52) and hospital beds (13.26) per 10,000 people (Dincer et al. 2003).

8.6.6 Education and Training Activities

The literacy ratio in the GAP Region in 2000 was the lowest in Turkey at 72.2%, compared to the national average of 87.32%. However, literacy of women in the region has increased from 29% in 1985, to 38% in 1990 and 52% in 2000, representing a significant improvement.

Training programmes have been implemented for years for disadvantaged population living in poor urban neighbourhoods and in rural communities. The most successful example in the region so far has been the Multi-purpose Community Centres, known as CATOMs. These centres have focused mainly on women development with modular programmes on literacy, hygiene, nutrition, access to public services, income and employment generation activities (handicrafts, rug weaving, knitting, embroidery, silver works, stone working, computer skills, and so on), micro-credits, management skills for them to start their own businesses, etc. (Tigret and Altinbilek 2003). According to Unver and Gupta (2004), in 2001 alone, 4,512 people in the region benefitted directly from the training programmes implemented at CATOMs.

8.7 Conclusions

The GAP Project was established as an initiative to improve one of the less developed regions in Turkey. Initially conceived as a series of land and water resources development projects on the Euphrates and Tigris Rivers, it was later transformed into a water- and land-based large-scale, multi-sectoral regional development project. The objectives of the project were manifold, complex and very ambitious, and attempted to influence the social, economic, institutional and technical aspects of human development in a large area of the country. However, the implementation of the project as an integrated regional development project is well behind targets established in the Master Plan of the Region in 1989, and in 2002 when it was reviewed.

The project faces many challenges if it is to be implemented by 2012 as laid down in the GAP Project Action Plan 2008–2012. As mentioned by Kayasü (2008) and Beleli (2005), some of the major constraints in the implementation of the project have been a highly centralised system which is reflected in regional policies and projects as well as in the institutional arrangements for their execution; a sectoral focus on public investment planning; limited public investments in less-developed areas; vaguely defined division of tasks and responsibilities between planning, implementing and coordinating institutions; non-institutionalised coordination efforts; and inadequate administrative capacity of local and regional planning and implementing institutions.

These issues are only mentioned here and not analysed in detail because they are not within the scope of this chapter. They are included, however, because it is important to be aware of the reasons why the GAP Project may not have reached its planned goals both in the initial and the revised Master Plans, and in order to understand the limitations that surround its implementation.

In spite of these shortcomings, there has been significant progress in some sectors of the GAP Region as a direct result of the project. These are the energy and the agricultural sector. The development of the energy sector has transformed the region into a fundamental component in an overall strategy for the country's energy security. In terms of agricultural development, the region has become the main producer of cotton at the national level, with the related positive and negative impacts of extensive irrigation practices.

Within the water projects, the Atatürk Dam has resulted in multiple benefits for people living in the project area as well as for the country as a whole. Its direct benefits have been in terms of energy generation and irrigated agriculture, with indirect benefits through the promotion of urban, industrial, agricultural and commercial activities, mostly in Sanliurfa Province.

The social and economic impacts of the construction of the Atatürk Dam and its reservoir have been substantial through a variety of pathways. As mentioned earlier, both the dam and the reservoir have acted as an engine for economic growth and development in a historically underdeveloped area that has flourished since the construction of the dam. In addition, the daily exposure of the villagers to the traditions of 'outsiders' over more than a decade of construction, resulted in social and cultural changes evident in new ambitions for better and more housing and transportation, altered food and health habits, a desire for higher education, the decision of some of the local population to send their children (including girls) to school, a demand for information and communication, etc. The lifestyles of the area's population have thus begun to change, which will likely lead to a better quality of life for many people.

References

Akbulut N, Bayar S, Akbulut A (2009) Rivers of Turkey. In: Klement T, Urs U, Christopher TR (eds) Rivers of Europe. Academic Press, London, pp 643–672

Aksit B, Akcay AA (1997) Sociocultural aspects of irrigation practices in Southeastern Turkey. Int J Water Resour Dev 13(4):523–540

Akyürek G (2005) Impact of Atatürk Dam on social and environmental aspects of the Southeastern Anatolia Project. Master dissertation, Graduate School of Natural and Applied Sciences, Middle East Technical University, Ankara

Altinbilek D (1997) Water and land resources development in Southeastern Turkey. Int J Water Resour Dev 13(3):311–332

Altinbilek D (2004) Development and management of the Euphrates and Tigris Basin. Int J Water Resour Dev 20(1):15–33

Altinbilek D, Bayram M, Hazar T (1999a) The new approach to development project-induced resettlement in Turkey. Int J Water Resour Dev 15(3):291–300

Altinbilek D, Bayram M, Hazar T (1999b) The new approach to reservoir-induced resettlement and expropriation in Turkey. In: Turfan M (ed) Benefits of and concerns about dams. Case

Studies, 67th Annual Meeting of the International Commission of Large Dams (ICOLD), Antalya

Beleli O (2005) Regional policy and EU accession: learning from the GAP experience. Turkish Policy Q 4(3):87–96

Biswas A, Tortajada C (1999) Rapid appraisal of social, economic and environmental impacts of the Ataturk Dam. GAP Regional Administration and UNEP, Ankara

Çullu MA (2003) Estimation of the effect of soil salinity on crop yield using remote sensing and geographic information system. Turkish J Agric Forestry 27:23–28

Çullu MA, Almaca A, Sahin Y, Aydemir S (2002) Application of GIS for monitoring soil salinisation in the Harran Plain, Turkey. Proceeding of the international conference on sustainable land use and management, Çanakkale, Turkey, pp 326–332

Dincer B (1996) Socio-economic development index ranking of the provinces-1996. State Planning Organization, Ankara

Dincer B, Ozaslan M, Kavasoğlu T (2003) Socio-economic development index ranking of the provinces-2003. State Planning Organization, Ankara

DSI (General Directorate of State Hydraulic Works) (2000) DSI in brief. General Directorate of State Hydraulic Works, Ministry of Energy and Natural Resources, Republic of Turkey, Ankara

DSI (General Directorate of State Hydraulic Works) (2008) DSI in brief, 1954–2007. Ministry of Energy and Natural Resources, Republic of Turkey, Ankara

DSI (General Directorate of State Hydraulic Works) (2009) GAP 2008, Dicle-Firat Basins and the Southeastern Anatolia Project. General Directorate of State Hydraulic Works, Ministry of Environment and Forestry, Republic of Turkey, Ankara

DSI (General Directorate of State Hydraulic Works). Available at www.dsi.gov.tr/baraj/. Accessed 30 Mar 2010

Ercin AE (2006) Social and economic impacts of the Southeastern Anatolia Project. Master dissertation, Graduate School of Natural and Applied Sciences, Middle East Technical University, Ankara

GAP Administration (1998) Law no. 388 concerning the establishment and duties of the Southeastern Anatolia Project Regional Development Administration, 6 November 1989, and Amending Law 4314 on the Decree in force of Law on the Institutions and Duties of the Southeastern Anatolia Project Regional Development Administration, 14 December 1997. Prime Minister's Office, Ankara

GAP Administration (1999) Latest status in Southeastern Anatolia Project. GAP Regional Development Administration, Prime Minister's Office, Ankara

GAP Administration (2006) Latest situation on Southeastern Anatolia Project: activities of the GAP Administration. Prime Minister's Office, Ankara

Government of Turkey (2008) Southeastern Anatolia Project Action Plan 2008–2012, Ankara

Harris LM (2008) Water rich, resource poor: intersections of gender, poverty, and vulnerability in newly irrigated areas of Southeastern Turkey. World Dev 36(12):2643–2662

Kapur S, Kapur B, Akca E, Eswaran H, Aydin M (2009) A research strategy to secure energy, water, and food via developing sustainable land and water management in Turkey. In: Brauch HG, Oswald SU, Mesjasz C, Grin J, Behera NC, Chourou B, Kameri-Mbote P, Liotta PH (eds) Facing global environmental change: environmental, human, energy, food, health and water security concepts. Springer, Berlin, pp 509–518

Kayasü S (2008) Institutional implications of regional development agencies in Turkey: an evaluation of the integrative forces of legal and institutional frameworks. 42nd ISoCARP (International Society of City and Regional Planners) Congress, 14–18 September, Istanbul

Kendirli B, Cakmak B, Ucar Y (2005) Salinity in the Southeastern Anatolia Project (GAP). Issues Options Irrigation Drainage 54(1):115–122

Kundat A, Bayram M (2000) Sanliurfa-Harran Plains On-farm and Village Development Project. In: Kudat A, Peabody S, Keyder C (eds) Social assessment and agricultural reform in Central Asia and Turkey. World Bank Technical Paper No. 461, The World Bank, Washington, pp 255–302

Miyata S, Fujii T (2007) Examining the socioeconomic impacts of irrigation in the Southeast Anatolia Region of Turkey. Agric Water Manage 88:247–252

Nippon Koei Co Ltd and Yuksel Project AS (1990) The Southeastern Anatolia Project Master Plan Study. Final Master Plan Report, Volume 1. GAP Administration, Prime Minister's Office, Ankara

Oklahoma State University, SU-YAPI Engineering, Consulting Company and KKGV Foundation for Rural and Urban Development (1999) Sanliurfa-Harran plains On-farm and Village Development Project: a social evaluation. GAP Regional Development Administration, Ankara

Ozdogan M, Woodcock CE, Salvucci GD, Demir H (2006) Changes in summer irrigated crop area and water use in Southeastern Turkey from 1993 to 2002: implications for current and future water resources. Water Resour Manage 20:467–488

Pöyry Energy Ltd (no date). Available at: www.pyry.com. Accessed 30 Mar 2010

Say NP, Yucel M (2006) Strategic environmental assessment and national development plans in Turkey: towards legal framework and operational procedure. Environ Impact Assess Rev 26:301–316

SPO (2001) Long-term strategy and Eight Five-Year Development Plan 2001–2005. Prime Minister's Office, Government of Turkey, Ankara

SPO (2005) Medium Term Programme (2006–2008), Accepted by the decision of Council of Ministers No. 2005/8873 on 23 May 2005. Official Gazette (2nd repeated) on 31 May 2005, No. 25831, Ankara

SPO (2006a) Medium Term Programme (2007–2009), Cabinet Decree No. 2006/10508, 30.05.2006. Official Gazette No. 26197, 13.06.2006, Ankara

SPO (2006b) Ninth Development Plan (2007–2013). Prime Minister's Office, Government of Turkey, Ankara

SPO (2007a) 2007 Annual Programme, Ninth Development Plan (2007–2013). T.R. Prime Minister's Office, Republic of Turkey, Ankara

SPO (2007b) Medium Term Programme (2008–2010), Cabinet Decree No. 2007/12300, 28.05.2007. Official Gazette No. 26559, 21.06.2007, Ankara

SPO (2008a) 2009 Annual Programme, Ninth Development Plan (2007–2013). T. R. Prime Ministry, Republic of Turkey, Ankara

SPO (2008b) Medium Term Programme (2009–2011), Cabinet Decree No. 2008/13834, 28.06.2008. Official Gazette No. 26920, 28.06.2008, Ankara

SPO (2009) Medium Term Programme (2010–2012), Cabinet Decree No. 2009/15430, 14.07.2009. Official Gazette No. 27351, 16.09.2009, Ankara

SPO (2010) 2010 Annual Programme, Ninth Development Plan (2007–2013). T. R. Prime Ministry, Republic of Turkey, Ankara

Tigret S, Altinbilek D (2003) Sustainable human development in the Southeastern Anatolia Project. Middle East Technical University, Ankara

Tortajada C (2000) Evaluation of actual impacts of the Atatürk Dam. Int J Water Resour Dev 16(4):453–464

Tortajada C (2004) South-Eastern Anatolia Project: impacts of the Atatürk Dam. In: Biswas AK, Ünver O, Tortajada C (eds) Water as a focus for regional development. Oxford University Press, Delhi, pp 190–250

Unver O (1997) South-Eastern Anatolia Integrated Development Project (GAP), Turkey; an overview of issues of sustainability. Int J Water Resour Dev 13(2):187–207

Unver O, Gupta R (2004) Participative water-based regional development in the South-Eastern Anatolia Project (GAP): a pioneering model. In: Biswas AK, Unver O, Tortajada C (eds) Water as a focus for regional development. Oxford University Press, New Delhi, pp 154–189

Yesilnacar MI (2003) Grouting applications in the Sanliurfa Tunnels of GAP. Turkey, Tunnelling Underground Space Technol 18:321–330

Yesilnacar MI, Gulluoglu SM (2007) The effects of the largest irrigation of GAP Project on groundwater quality, Şanliurfa – Harran Plain, Turkey. Fresenius Environ Bull 16(2):206–211

Yesilnacar MI, Uyanik S (2005) Investigation of water quality of the world's largest irrigation tunnel system, The Sanliurfa Tunnels in Turkey. Fresenius Environ Bull 14(4):300–306

Chapter 9
Impacts of King River Power Development, Australia

Roger Gill and Morag Anderson

9.1 Introduction

The King River Power Development, on the rugged west coast of Tasmania, is a good example of excellence in hydroelectric power development, and in many ways provides a guide for the overall development and sustainable operation of other hydroelectric power schemes. The scheme demonstrates sustainable environmental and economic development, and has noteworthy technical innovations. It was developed and is owned, operated and maintained by Hydro Tasmania.

The King River Power Development, with an installed capacity of 143 MW, contains the following classic elements of a hydro scheme: two dams, one 83 m high; a large diameter 7 km long headrace tunnel; a single machine, remotely controlled power station; 50 km of 220 kV transmission line and some 36 km of road works. The lake created by the scheme (Lake Burbury) has been developed to support both recreation and fishing and is of benefit to the whole west coast community of Tasmania.

This hydropower development was born in controversial circumstances. A more ambitious development, that would have produced competitively priced energy, raised community concerns regarding the environmentally sensitive nature of the area to be inundated. In 1982, a decision was taken to proceed instead with the substitute King River Power Development, incorporating environmental lessons learned from this experience.

Notable environmental measures were implemented during construction, for example, those relating to water quality, timber salvage and land rehabilitation. These were the catalysts for development of a formalised environmental manage-

R. Gill (✉)
Hydro Focus Pty Ltd., Taroona, TAS, Australia

M. Anderson
RDS Partners, Hobart, TAS, Australia

ment system within Hydro Tasmania that underpins the maintenance of a sustainable environment for both hydro schemes and their associated catchments. A major benefit of this scheme was the recognition of the need to develop an expert environmental management team to support Hydro Tasmania's operations. The activities of the environmental team have grown to the point where the operation of hydro schemes within Tasmania can be considered as an exemplar of environmental best practice within the industry.

The scheme is significant for the future business direction of Hydro Tasmania. It plays a very important role in Tasmania's participation in the Australian National Electricity Market, contributing to Hydro Tasmania's export of peak power across the 630 MW, 270 km long Basslink undersea cable that electrically connects Tasmania to the eastern Australian grid. In addition, because it has a large water storage with a single large turbine, it enhances Hydro Tasmania's ability to store wind energy from the recently commissioned wind farm developments.

A notable feature is the Crotty concrete face rock-fill dam that incorporates an articulated chute spillway on the downstream face. This world-leading innovation continues Hydro Tasmania's pioneering developments in large concrete face rock-fill dams.

The hydro scheme was designed and built by Hydro Tasmania's own workforce and constructed between 1983 and 1993. It features benefits accruing not only to Hydro Tasmania, adding to its renewable energy portfolio which stands at approximately 60% of Australia's total, but also to the wider community.

This chapter presents first the actual scheme and the socio-political context surrounding the King River Power Development. It then describes the technical features of the scheme, notable aspects of the scheme's investigation stage and development of an Environmental Management Plan as well as social benefits, business contribution, operational considerations and aspects of the scheme's ongoing environmental management.

9.2 Socio-Political Context

The King River Power Development was constructed in a climate of intense public disquiet about the environmental effects of Tasmania's future power schemes. This disquiet arose from events relating to the Gordon River Power Development Stage 2 in Tasmania's southwest that was proposed to be constructed downstream from the Gordon River Power Development Stage 1 (Figs. 9.1 and 9.2).

Opposition to hydro schemes in Tasmania began in 1967 with the proposed flooding of Lake Pedder, as part of Stage 1 of the Gordon River Power Development. The emerging conservation movement ran a strong 'Save Lake Pedder' campaign with a high national profile. This anti-dam campaign sparked major public unrest at both state and national levels, and was the catalyst for the formation of the Tasmanian Wilderness Society and The Greens. These groups grew to become major community and political influences and remain so today. Following an official inquiry into the flooding of Lake Pedder, the Australian Commonwealth government ratified the

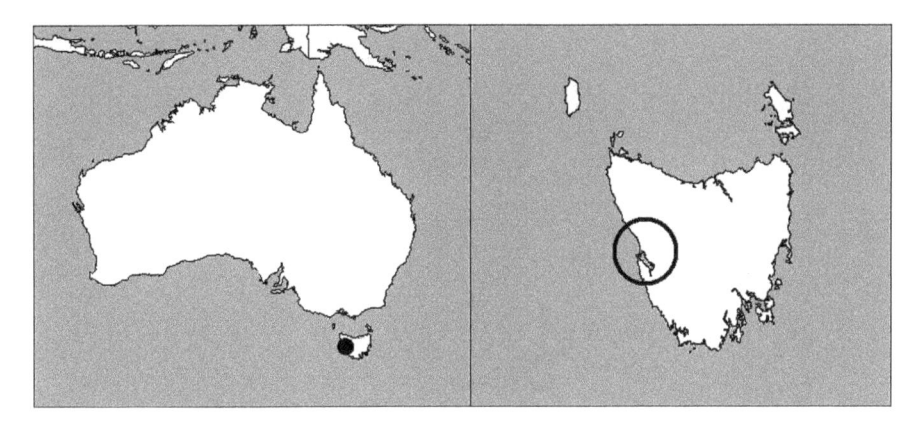

9 Impacts of King River Power Development, Australia

Fig. 9.1 Location of the state of Tasmania, and the King River Power Development in Australia

World Heritage Convention in 1974, thus giving the Commonwealth government extra powers over State governments.

Stage 2 of the Gordon River Power Development (the 'Lower Gordon Scheme') was proposed in 1981 and would have involved flooding of the Lower Gordon and Franklin Rivers. This proposal was strongly opposed by the Tasmanian Wilderness Society and Green groups from around Australia. Significantly, in 1982, the then South West National Park, Franklin–Lower Gordon Wild Rivers National Park and the Cradle Mountain–Lake St. Clair National Park were designated as the 'Tasmanian Wilderness World Heritage Area' (see Fig. 9.2, 1981 World Heritage Area Nomination). The Lower Gordon power development proposal would have led to inundation of parts of this area, including significant Huon River pine habitat and caves containing aboriginal artefacts.

Demonstrations continued against Stage 2 of the Gordon River Power Development despite it receiving state parliamentary approval. When the national Labor Party came to power in 1983, it used the external powers under the World Heritage Convention to override the State government, and construction of the Lower Gordon power scheme was permanently halted.

Hydro Tasmania had planned the development of hydropower in Tasmania to meet the forecast rise in demand for electrical energy. The future ability of the state of Tasmania to meet the demand for electricity was thrown into doubt by the blocking of the Lower Gordon development. Up to 1,000 jobs were jeopardised in a region which offered little alternative employment for the skilled workforce.

After swift but careful consideration, two smaller hydro schemes were proposed that were both outside the Tasmanian Wilderness World Heritage Area—the 'King' and the 'Anthony'. Together these schemes would have an average annual output of 120 MW average, somewhat less than the 180 MW average planned for with the Lower Gordon scheme. The King and Anthony schemes were also less economic than the Lower Gordon scheme. The Commonwealth government agreed to fund part of the capital cost so that the unit price of power from the two alternative schemes would be the same as that forecast for the Lower Gordon scheme.

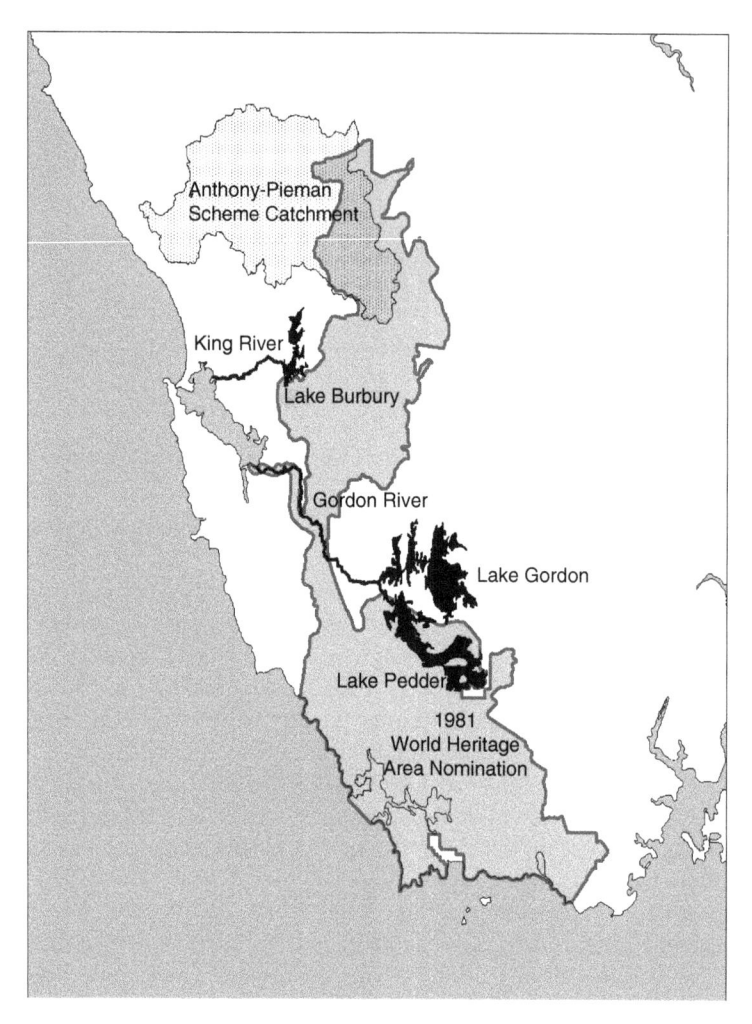

Fig. 9.2 King River and Lake Burbury (King River Power Development) in relation to Lake Gordon and Lake Pedder (Gordon River Power Development, Stage 1); the Anthony-Pieman Scheme catchment and the Tasmanian Wilderness World Heritage Area (1981 WHA nomination)

These events set the stage for the King River Power Development. Both Tasmanian houses of parliament gave their approval promptly, as did the wider Tasmanian community, and Hydro Tasmania embarked on the engineering investigation, design, and construction activities concurrently instead of in the normal planned sequence. While there was no legal requirement for Hydro Tasmania to produce an Environmental Management Plan, the decision to do so reflected a business recognition of the need to demonstrate more socially and environmentally responsible development.

9 Impacts of King River Power Development, Australia 205

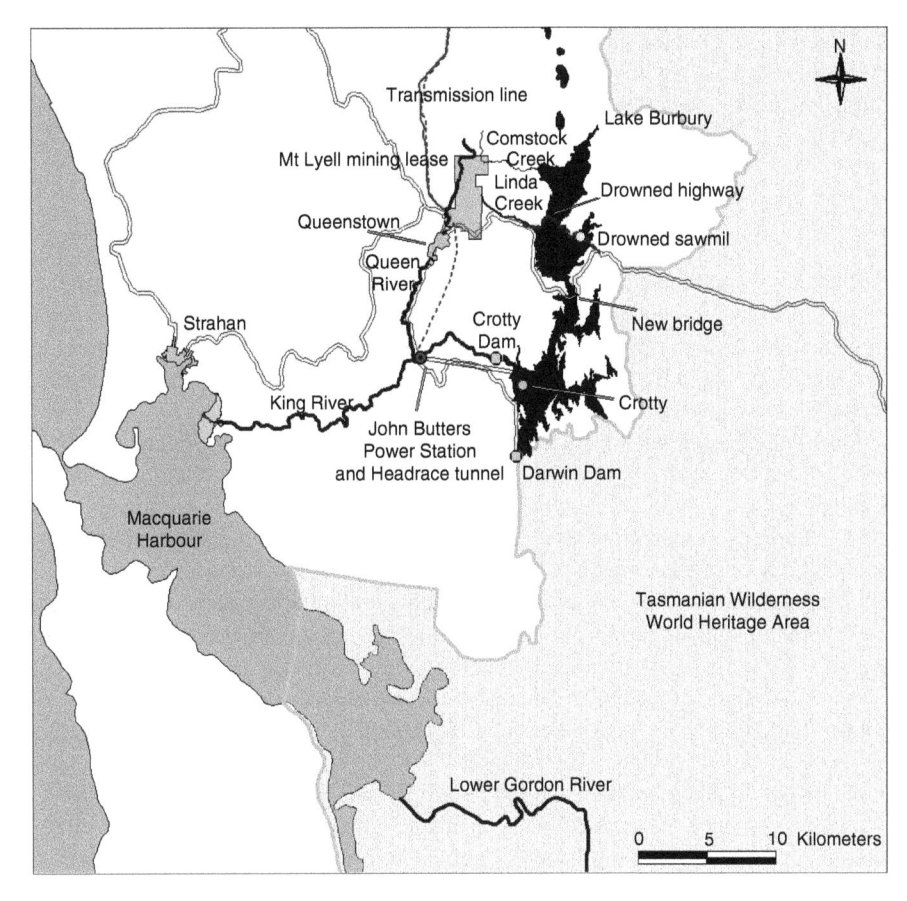

Fig. 9.3 The King River Power Development and associated environs

9.3 Technical Description

The King River Power Development (Fig. 9.3) has the following components:

- Two dams—the 83 m high Crotty Dam, and the 20 m high Darwin Dam (a saddle dam), which together create the 53 km² Lake Burbury
- A 7 km long headrace and 7.2 m diameter power tunnel
- The John Butters Power Station, containing a single 143 MW machine which is controlled remotely from Hobart
- A 50 km long 220 kV transmission line, connecting the John Butters Power Station to the state grid
- A 12 km deviation of the Lyell Highway which includes a major bridge over Lake Burbury
- The Mt Jukes Road, a 24 km main access road crossing the West Coast Range from Queenstown

Hydro Tasmania has won a number of Engineering Excellence Awards from the Australian Institute of Engineers for technical excellence in relation to the King River Power Development.

9.3.1 Dams

Located at the head of the King River Gorge, the main dam, Crotty Dam, is 83 m high. It is a concrete-faced gravel and rock-fill embankment. The fill is local river gravels from the King River floodplain upstream of the dam. Their use avoided the need for quarrying and thus reduced costs and visual impacts, but required special techniques to avoid saturation in a high rainfall area.

The spillway consists of a 3.6 m diameter bottom outlet in the diversion tunnel and, contrary to traditional practice for rock-fill dam design, a chute spillway is located on the downstream face of the embankment. This choice significantly lowered the cost of construction and eliminated any scarring to the adjacent environment. The reinforced concrete chute is designed to accommodate settlement of the embankment and ensure safe operation. Hydro Tasmania was emboldened to proceed with this innovative step, a world first, because of its 20-year history of building and monitoring very successful concrete face rock-fill dams.

Darwin Dam is a gravel dam located on a saddle between the King and Andrew Rivers. The foundations were found to be very complex, with several geological faults and some karstic limestone, all hidden below a thick mantle of gravel. Defensive measures have been taken against the possible development of sinkholes in the storage.

Hydro Tasmania has a world-class dam safety programme focused on ensuring that the dams perform in a manner that does not place the community or assets at risk. Ongoing programmes of surveillance, performance analysis and emergency preparedness assure safety.

9.3.2 Tunnels

Normally, long headrace tunnels are excavated from the downstream end on a gradually rising grade so that water inflows can drain away. In contrast, the King headrace tunnel was excavated from upstream for both technical and environmental reasons. The result is that the tunnel intake is permanently submerged and the highest point in the tunnel is near its downstream end.

Where the headrace tunnel becomes the power tunnel and descends to the power station, a vertical surge shaft would normally be provided. The cost of a surge shaft was avoided by constructing a tunnel rising to the surface at road grade from this point. Surges take place in the access tunnel, which also meets the need for road access when inspection or maintenance is required. The geology of the region was complex, with numerous faults and folds in the predominantly metamorphic rocks. The tunnel was permanently supported using dowels, avoiding concrete lining, which resulted in substantial savings.

9.3.3 Power Station

The power station is of circular slip-form construction. It is located at the downstream end of the narrow King River Gorge and is tucked into a left bank excavation that is clear of any flows which might deposit gravel in the tail race. With an effective head of 184 m on the turbine, the rated output is 143 MW for a flow of about 85 m^3/s. The station is designed for unmanned operation and is controlled remotely from Hobart. Oil bunding was custom designed and built into the transformer yard, which ensures that any spills will be captured and contained.

9.3.4 Transmission Line

A 220 kV line connects the power station to the state grid at the Farrell Switchyard 50 km away. Community consultation and careful planning ensured that throughout its length the line was sited to make it as visually unobtrusive as possible. The use of naturally brown (Austen) steel towers, non-reflective conductors and brown insulators helps to make the line generally unobtrusive across various terrains.

9.3.5 Roads

Well-researched route selection and careful revegetation maximised the scenic vistas of the area to take full advantage of its tourism potential and virtually nullified visual impacts of construction on the environment. The Mt Jukes Road provides access from Queenstown to all parts of the scheme. In crossing the rugged West Coast Range, it presents visitors with views of the mountains, Lake Burbury and the World Heritage Area.

- On the Mt Jukes Road, parking areas were constructed at vantage points so that the vistas could be enjoyed and photographed clear of passing traffic. Both this road and the Lyell Highway command splendid views of the lake and the surrounding mountains.
- An interpretation site for cool temperate rainforest on the Mt Jukes Road explains the forest ecology and names the main species.
- Timber was salvaged from Mt Jukes Road in conjunction with the Tasmanian Forestry Commission.

9.3.6 Highway Deviation

As Lake Burbury would inundate part of the original Lyell Highway linking Hobart with the West Coast, 12 km of new highway was constructed, together with a seven-span bridge across a narrow point of the lake.

- The Lyell Highway Deviation was built to state highway standards. This new section contrasted with the adjoining section and the Tasmanian government Department for Main Roads decided to upgrade several kilometres on the eastern side. Thus, there were additional benefits for the motoring public and local workforce.
- After consultation with the Queenstown Council, a new picnic area was built near Lake Burbury to replace one inundated by the lake.

9.4 Environmental Aspects

9.4.1 Prior to Construction

The King River Power Development presented some particularly unique site investigation and design challenges, primarily arising from a suite of environmental issues associated with a large copper mine which had been operating in the catchment over the past 100 years.

At the time of site investigations, the Mount Lyell Copper Mine in Queenstown was still continuing its 78-year practice of directly discharging tailings (fine-grained waste sediments) into the Queen River. This river drains into the King River approximately 15 km from Macquarie Harbour (Fig. 9.3). Substantial accumulations of tailings and smelter slag could be found stored in the bed, banks and delta of the lower King River. Additionally, considerable sulphidic rock is found exposed to air and rainfall on the mining lease. Heavy metals associated with this sulphidic rock, notably copper, aluminium, and zinc, are liberated due to the creation of acid drainage and are present in high concentrations in the run-off from the lease site.

At the time the Lower Gordon scheme was halted, significant studies had already been carried out on the King River catchment as part of Hydro Tasmania's explorations of alternative power development options. These studies were released for public scrutiny in 1980, and included:

- Identification of the main heavy metal pollution sources arising from the historical and ongoing mining activities in the catchment
- Estimates of the amounts of heavy metals that would enter a proposed storage
- Modelling of the heavy metals within a proposed storage
- The status of the biota in the river systems and assessment of the effect of heavy metals on the biota
- Proposed methods of reducing pollution off the lease site to acceptable levels

Based on these studies, the dam and power station were located upstream from the Queen River tributary that delivers tailings to the lower King River, but acid drainage from the lease site into the proposed storage would still occur. Diversion works were undertaken on the lease site to address this occurrence, and are described below (see Water Quality Diversion Works).

Although in 1983 Hydro Tasmania was not formally required to prepare environmental impact statements, the business voluntarily prepared an Environmental Management Plan for the construction of the King scheme. It was prepared in recognition of the need to address environmental issues in a rigorous manner. This was the first hydropower scheme built in Australia to have a formal Environmental Management Plan, and this plan influenced the design and construction activities of the development. An Environmental Committee was formed for the construction period to ensure that unnecessary impacts were avoided and unavoidable ones were minimised and treated.

Parts of the King River Valley, now inundated by Lake Burbury, have a rich history associated with the North Mt Lyell Mining Company, which built a railway from Macquarie Harbour to the mine via a village and copper smelter at Crotty (Fig. 9.3). The village had a population of 900 in the year 1900, but virtually disappeared overnight when the smelter failed to produce any metal, and was uninhabited at the time of dam construction and subsequent inundation. Relics of that era were recorded in an archaeological study commissioned by Hydro Tasmania.

Considerable effort was devoted to the sensitive issue of studying the potential impact of the inundation of the King River valley on Aboriginal heritage. Archaeological experts of national repute were engaged to investigate and report on this potential impact, with findings carefully and exhaustively chronicled. The investigations proceeded with the involvement of representatives of the aboriginal community. The findings of these comprehensive studies were provided to the aboriginal community and the regulatory authorities and the permit to inundate the valley was granted.

9.4.2 During Construction

The power scheme is located very close to the Tasmanian Wilderness World Heritage Area. In recognition of this, great care was taken to minimise visual damage to the landscape. Measures in the Environmental Management Plan to minimise visual impact included:

- Locating temporary access tracks, camps, works areas and quarries for rock-fill below minimum water levels so they would not be visible after commissioning
- Placing excess road spoil and material from other engineering works below minimum water levels
- Locating permanent roads and structures as unobtrusively as possible, including:
  - Reducing the size of roadside excavations (batters)
  - Spacing horizontal steps more closely in the batters to allow greater coverage by vegetation
  - Having a single lane road (rather than two lanes)
  - Lowering the design speed of the road

- Stockpiling of peat and topsoil removed in the construction process for reuse during restoration and revegetation of road cuttings, spoil dumps and disturbed areas
- Using native species for revegetation
- Developing hydro-mulching strategies effective in the local environment for the revegetation of rocky or steep areas
- Salvaging timber from Mt Jukes road
- Salvaging timber from the storage area before commissioning to avoid unsightly dead trees and hazards for boats using the lake
- Using appropriate materials for and locating the transmission line to make it as visually unobtrusive as possible

Further environmental issues addressed in the Environmental Management Plan included:

- Designation of special 'no-go' areas
- Strategies to limit waterway siltation and erosion
- Considerations for drainage and the disposal of effluent
- Guidelines for drilling and blasting
- The removal of litter and rubbish
- Fire management

Of particular note are the environmental measures that were implemented relating to water quality diversion works, timber salvage and land rehabilitation.

9.4.2.1 Water Quality Diversion Works

Water quality in the new lake was a concern due to acid drainage and heavy metal run-off from old mine workings and spoil dumps in the catchment. In particular, high copper concentrations were identified as having potential to interfere with the development of a recreational trout fishery in the lake.

Following intensive studies into the relative contributions of the various sources of the pollutant in the catchment, remediation works were designed and implemented to reduce a significant amount of the copper flux entering the new storage. The work involved passive diversion of some flows to the already heavily polluted Queen River, and the sealing and revegetation of old tailings dumps to reduce leaching by rainwater. The Queen River joins the King River downstream of the power station (Fig. 9.3).

9.4.2.2 Timber Salvage

48,900 m^3 of timber was salvaged from Mt Jukes Road and the storage area in conjunction with the Tasmanian Forestry Commission.

The storage area was cleared from full supply level to 2 m below the minimum operating level, to avoid unsightly dead trees and remove water hazards for boats using the lake. By maximising timber salvage, Hydro Tasmania generated favourable publicity, created jobs, provided the State government with revenue and minimised the areas that remained for storage clearing.

9.4.2.3 Land Rehabilitation

Immense effort was directed towards ensuring that the scheme should be as invisible as possible. Many environmental management plans were drawn up and adhered to from design through to construction and commissioning. Measures included stockpiling of peat and topsoil generated during construction for later reuse, use of native species for revegetation and developing and utilising hydro mulching for rocky and steep sites.

9.4.3 Commissioning

The most important considerations in commissioning of the King River Power Development were environmental. These were, in particular, concerns with:

- Heavy metals and water quality associated with the filling of the storage, Lake Burbury
- Rescuing of stranded animals during lake filling
- Oxygen levels in the power station discharges

9.4.3.1 Water Quality in Lake Burbury

Hydro Tasmania, working with the Inland Fisheries Commission Biological Consultancy (IFCBC), began a Lake Burbury water quality monitoring programme with the commencement of filling of the lake in August 1991. The programme was implemented in response to concerns about heavy metal contamination via polluted Linda and Comstock Creeks, which drained into the new storage, and as a means to monitor the effectiveness of ameliorative measures such as diversion works which had been put in place.

The main objectives of the programme included the following:

- Monitoring the development of the physico-chemical environment of the new storage, including changes associated with thermal stratification and decay of flooded vegetation
- Assessing the spatial and temporal distribution of metal contamination in the water and sediments of Lake Burbury
- Assessing the effect of any contamination on the productivity of the lake
- Monitoring the fish populations and development of the trout fishery in the lake
- Assessing the effect of releases from the lake on water quality in the King River and Macquarie Harbour

The monitoring programme included ongoing water, soil and sediment sampling, lake productivity measurements, and investigations into aspects of the fishery such as heavy metal levels in trout, distribution of fish in the lake, growth rates, physiological condition and reproductive success of the fish.

By 1996, ongoing monitoring had established that there were no major problems with either copper toxicity or the development of the Lake Burbury fishery. Testing by the Tasmanian IFCBC found that even though mean soluble copper concentrations were 2–3 times higher than recommended Australian and New Zealand Environment Conservation Council (ANZECC) levels, the toxicological response in fish was much lower than predicted. This was attributed to the presence of high levels of naturally occurring dissolved organic matter in the water (creating the 'tea' colour associated with river systems on Tasmania's west coast). Dissolved organic matter binds with copper, rendering it biologically unavailable, and so protects organisms from any toxic effect of this heavy metal. Based on these results, the remediation works described earlier were found to be adequate for the protection of a recreational fishery in Lake Burbury. Lake Burbury presently supports a successful recreational trout fishery, and Hydro Tasmania now conducts its own copper surveys on a routine basis to ensure these conditions do not deteriorate.

9.4.3.2 Animal Rescue Programme

As the lake filled, Hydro Tasmania implemented an animal rescue programme. For this programme, the business created a fauna recovery team that removed stranded animals from islands. The rescued animals were relocated to suitable habitats in the surrounding area, thereby ensuring minimum casualties amongst the native species.

9.4.3.3 Oxygen Levels in the Power Station Discharges

Soon after commissioning of the power station in 1992, water containing very low levels of oxygen accompanied by hydrogen sulphide was noted in the tail race, a common problem in the early life of a reservoir. The low level of oxygen was caused by thermal stratification in the reservoir. This resulted in the release of cold, deoxygenated bottom water into the King River. After investigation and consideration of ameliorative measures, water discharged downstream from the power station was aerated by operation of a jet pump installed on the turbine. This is now utilised to increase dissolved oxygen concentrations at appropriate times. Ongoing and continuous monitoring of the water quality leaving the tail race ensures adequate notice of low dissolved oxygen levels, and timely utilisation of the air injection facilities in the turbines.

9.4.4 Post-commissioning

Since commissioning, Hydro Tasmania has put considerable effort into understanding the interactions of power station operations with downstream water quality and tailings transport, particularly in light of the growing aquaculture industry in the receiving waters of Macquarie Harbour.

Discharge of tailings into the river ceased in December 1994 when a tailings dam was built. However, power station operations still affect the transport and deposition of tailings and the metal-laden acid drainage arising from lease site run-off.

Hydro Tasmania was a major contributor to a large-scale environmental investigation programme in the lower King River and Macquarie Harbour between 1992 and 1995. The King River–Macquarie Harbour Environmental Study was a collaborative study of the Tasmanian Department of Environment and Land Management, the Mount Lyell Copper Mine and Hydro Tasmania. It involved extensive water quality surveys, sponsorship of doctoral research into storage and transport of the mine tailings and hydrodynamic and water quality monitoring in Macquarie Harbour.

These investigations led to the essential understanding of how power station operating patterns influence patterns of pollutant transport to the downstream environment. An important consideration is the highly successful aquaculture industry in Macquarie Harbour, producing Ocean (rainbow) Trout and Atlantic Salmon. 'Worst-case' conditions for the fish farms are now well understood; a monitoring system allows early detection of the conditions and they can be mitigated by appropriate operation of the station.

9.5 Economic Aspects

The King River Power Development cost A$ 463 million (January 1993 dollars) and the assessed long-term average energy output is 67 MW average representing a capacity factor of 46%.

Since commissioning in 1993, the scheme has operated satisfactorily and has incurred no abnormal costs. Annual power production has averaged 90% of the original project estimate due to lower than average rainfall (about 4% below the long-term average) and because 40% of the time the power station is run in power system support mode rather than at its most energy efficient setting. This output represents about 6% of the annual power generated in Tasmania.

The scheme plays an important role in the integrated generating system operated by Hydro Tasmania. The power station can be operated to vary its generation within a particular range to meet the fluctuations of the daily load curve not met by other stations, and so discharge out of the power station tends to be variable over very short periods. This scheme, therefore, provides considerable flexibility in the operation of the system and contributes to system stability, hence making the overall Hydro Tasmania generating system more economic.

9.5.1 Role in National Electricity Market

Hydro Tasmania began selling peak power into the National Electricity Market when the Basslink undersea cable was completed. The 270 km, high-voltage direct current (HVDC) Basslink cable enables the export of up to 630 MW into the well-established Australian wholesale electricity market. John Butters Power Station, on the King River Power Development, is a key contributor in this peak power export. There are significant financial benefits for Hydro Tasmania being in the National Electricity Market. This includes maximising the value for water in the Hydro Tasmania generating system, and providing a market for expansion of renewable energy development in Tasmania.

9.5.2 Wind Energy Development

Significant potential for additional renewable energy development in Tasmania is brought by the prevailing westerly winds that circle the earth's southern latitudes. The wind resource is amongst the best in the world. In the northwestern corner of Tasmania, Roaring Forties, a joint venture between Hydro Tasmania and China Light & Power has commissioned one of Australia's largest wind farms. It comprises 62 turbines with a total generating capacity of 140 MW. Plans are well advanced for the development of additional wind farms.

The synergy between hydropower and wind power offers improved returns compared with stand-alone wind developments. When the wind blows, energy can be stored in Hydro Tasmania's substantial water storages by curtailing hydro turbine operation, and when the wind is calm the fast-response hydro plant can supply the market. This mechanism enables energy production to be shifted to the most valuable time of the day, thereby targeting premium-priced peak demands. The King River Power Development is an ideal example of a scheme that contributes strongly to an integrated hydro and wind operation.

9.6 Social Aspects

During design and construction, significant community consultation took place with interested stakeholders including the people of Queenstown, those seeking access to the proposed lake, the local councils, the Mount Lyell mine, recreational fishermen interested in the potential for a trout fishery in Lake Burbury and aquaculture interests in Macquarie Harbour.

The power scheme provided a very welcome boost to local employment opportunities during construction, as the copper mining industry was winding down. With Queenstown so close to the scheme, Hydro Tasmania made increasing use of local labour and contractors.

The building of a new transmission line also provided the opportunity to upgrade the power supply to Queenstown and to prepare for the future line from Anthony

Power Station. When construction of the King scheme was completed, the local workforce was utilised to build the nearby Anthony Power scheme.

A number of public amenities were built into the scheme, including picnic areas, boat ramps, and public viewing points. The power scheme has provided some additional tourism opportunities for existing and potential businesses in Queenstown. Comprehensive and early revegetation of disturbed areas enabled many areas to recover by the time tourist visits began.

The intended future of the lake as a major recreational trout fishery was a consideration that guided numerous design and construction activities affecting the lake. The local community has greatly appreciated Hydro Tasmania's efforts to establish a recreational fishery in the lake in particular through construction of the diversion works designed to reduce the inflow of heavy metals into Lake Burbury. Several construction roads have been converted into convenient boat ramps for public use, and the clearing of timber from the lake has made it safer for boating.

9.7 Corporate Role in Sustainability of Scheme

Hydro Tasmania has always had good standing in the international community as a developer of innovative and technically excellent engineering works. Due to the highly political nature of the events surrounding the development of the King River scheme, Hydro Tasmania recognised that a holistic approach to best practice was needed—engineering excellence alone was not enough. The organisation recognised that community consultation and a complete understanding and effective management of the environmental impacts of its developments were necessary core competencies for a hydropower developer. New approaches were introduced including the development of an environmental management system and a comprehensive aquatic environment programme with substantial stakeholder engagement. The King River Power Development and the events surrounding it can be regarded as the catalyst for this important evolution in corporate culture.

9.7.1 International Hydropower Association Sustainability Guidelines and Compliance Protocols

Through its membership of the International Hydropower Association, Hydro Tasmania has been heavily involved in drafting the Sustainability Guidelines and Sustainability Assessment Protocol for new and existing hydropower developments. These documents encourage the incorporation of sustainability principles and good practice in the development and operation of hydropower schemes across the globe.

9.7.2 Sustainable Environmental Management

Hydro Tasmania takes its environmental responsibilities very seriously. It was the first government business enterprise in Tasmania to introduce a statement of environmental policy in 1992, and has continued its commitment to promoting best practice environmental management of its operations. For example, in 1995, a formalised Environmental Management System (EMS) was put in place, and the Aquatic Environment Programme was developed in 1998 to enable sustainable management of water resources. This programme has already delivered a number of changes to Hydro Tasmania's water management practices, and anticipated the requirements of Tasmania's new Water Management Act 2000 and associated water licence by 2 years.

9.7.2.1 Environmental Management System ISO 14001

Since its statement of environmental responsibility was introduced in 1992, Hydro Tasmania has continued its commitment to promoting best practice environmental management as part of its operations.

The EMS was independently certified by Bureau Veritas Quality International (BVQI) to the international standard ISO 14001 during 1998, and re-certified in 2001, 2004 and 2007. Continued satisfactory performance has been monitored by six-monthly surveillance audits by BVQI. Regular internal reviews and audits ensure that conformance to the standard is maintained and environmental performance is subject to continual improvement.

9.7.2.2 Aquatic Environment Programme

Hydro Tasmania manages an extensive network of modified lakes, rivers, streams and canals, flowing through a diverse range of landforms and land-use zones, each of which have unique aquatic issues. Hydro Tasmania recognises that water is central to its business and that for its business to be sustainable, the aquatic environment must contain healthy ecosystems.

Hydro Tasmania has developed a comprehensive Aquatic Environment Programme. This programme utilises skills in environmental assessment and management in conjunction with other areas of expertise within the organisation, including survey/geographic information systems (GIS), hydrology and engineering, to provide balanced, well-considered solutions that incorporate community needs and expectations. This comprehensive programme, involving over AUD $1 million annually, incorporates a major commitment to research and consultation.

The Aquatic Environment Programme includes:

- An aquatic environment policy
- Reviews of a range of aquatic environment issues in catchments that affect Hydro Tasmania's water storages
- Strategies, priorities and long-term programmes for issues such as threatened species and fish migration
- A broad-based multidisciplinary water monitoring programme of lakes and rivers impacted by Hydro Tasmania projects, involving physico-chemical, biological and physical habitat monitoring
- Setting up of and active involvement in catchment management studies and water management plans
- Increased liaison with the community, university research and other government and research agencies

Hydro Tasmania has demonstrated its commitment to sustainable waterway management by such means as:

- Minimum environmental flow releases
- Constraints on management of water levels in lakes
- Construction of barriers to exotic fish migration and ladders for native fish passage
- Modifications to water transfer infrastructure to improve water quality and consequently threatened species habitat
- Commitments to ongoing environmental investigations and monitoring

One of the most proactive aspects of the Hydro Tasmania Aquatic Environment Programme is the Water Management Reviews. Hydro Tasmania is progressively undertaking reviews of its current water management practices in all its seven catchments with the aim of developing sustainable water management strategies. The process involves gathering background information, consulting the community and stakeholders to identify important issues, researching solutions to these important issues, and developing Hydro Tasmania water management plans.

9.8 Conclusion

The King River Power Development was born in controversial political circumstances. From this experience, Hydro Tasmania has emerged as an organisation willing to engage in developing more sustainable hydropower practices. Significant cultural change has taken place within the organisation, mirroring that of a greater awareness within the community at large, of the issues surrounding triple bottom-line accountability and sustainability (people, planet, profit). Hydro Tasmania now proudly champions the concept of sustainability and is actively promoting this both at home and on the wider international stage.

Acknowledgement The authors wish to acknowledge the support of Hydro Tasmania in the preparation of this chapter.

Bibliography

Supporting information can be sourced from the following web sites:

www.basslink.com.au – for development of the Basslink Cable
www.hydro.com.au (for Hydro Tasmania information)
www.hydropower.org/iha_blue_planet_prize/information.html (for International Hydropower Association (IHA), creator of the Blue Planet Prize and promoter of hydropower Sustainability Guidelines and Sustainability Assessment Protocol)
www.entura.com.au for Hydro Tasmania's Consulting business Entura

Chapter 10
Resettlement in China

Guoqing Shi, Jian Zhou, and Qingnian Yu

10.1 Introduction

In the past 56 years, China has seen more than 70 million involuntary resettlers due to its large-scale economic growth, including the construction of reservoirs, railways, roads, power plants and power transmission projects, airports, watercourses, urban water supply and sewage treatment projects, urbanisation and reconstruction and land acquisition for development and real estate. After more than 50 years of research, study and summing up of experiences, it can be said that the development-induced involuntary resettlement projects in new China have achieved great success: relevant policies and laws have been made from scratch and gradually enriched and perfected; institutional frameworks and responsibilities have been more clearly defined; resettlement projects are moving towards standardisation and the working methodologies are increasingly based on scientific knowledge. The number of people affected by various kinds of projects is shown in Table 10.1.

10.2 Policies, Laws and Regulations

10.2.1 Legal System and Legislation in China

The legal framework of China includes the constitution, laws, state and local rules and regulations, international practice and international pacts, etc. China pays great attention to legislation on land and real estate, and more than 30 laws and rules

G. Shi (✉) • J. Zhou
National Research Center for Resettlement (NRCR), Hohai University, Nanjing, China

Q. Yu
Land Management Institute, Hohai University, Nanjing, China

Table 10.1 Estimates of resettlement of various kinds of projects (millions of people)

Category	1950s	1960s	1970s	1980s	1990s	2000s	Subtotal
Reservoir	4.6	3.2	1.4	1.0	2.5	2.3	15
Transportation	2.5	0.9	2.7	1.3	3.2	4.4	15
City construction	1.5	1.3	2.6	8.5	12	14.1	40
Total	8.6	5.4	6.7	10.8	17.7	20.8	70

Source: Shi et al. (2001, 2007)

pertaining to land use have been formulated and promulgated in the past 50 years. The history of this legislation can be divided into five phases (Shi et al. 2001):

- Phase 1. Land legislation in 1950s. The land use legislation in this phase solved mainly three problems: land ownership, land acquisition for construction of national projects and regulation of urban real estate.
- Phase 2. Land legislation from 1960s to late 1970s. In this phase, legislation focused on consolidating and perfecting state land use ownership, rural collective land ownership and legal frameworks for urbanised land for housing and industrial development.
- Phase 3. Land legislation of the late 1980s. The main purpose during this period was to regularise various legal matters related to land used for construction, improving the legal system, bringing into effect the nationalisation of urban land, and in 1986, promulgating the Land Administration Law of the PRC (People's Republic of China).
- Phase 4. Land legislation in 1990s. The key of land legislation lies in reform of the system land use, i.e. establishing and fostering the land market, and strengthening laws on the management of land property. For this purpose, the Law of the PRC on Administration of Urban Real Estate was adopted in 1994.
- Phase 5. Since 2004, the central government has implemented the strictest policies to minimise the impact of land acquisition in order to protect farmers' interests (Tang and Jia 1999). A number of policies and regulations were revised, particularly the Regulations on Compensation for Land Acquisition and Resettlement on Large and Medium-sized Water Conservancy and Hydroelectric Projects (first issued in 1991) which was updated in August 2006 and the New Resettlement Policy on Post-relocation Support for Reservoir Projects, which was issued by the State Council in July 2006.

10.2.2 Development of Policies on Land Acquisition, House Demolition and Resettlement

The main laws on which the country's land acquisition, demolition, removal and resettlement rely began in November 1953 with the Government Administration Council's promulgation of the 'Methods of National Construction Land Acquisition'. Following more than 40 years of practice and development, the state promulgated five administration laws related to land acquisition. The basic principles of these five laws were to secure land for the construction of public projects, to economise on the use of land, and at the same time, to properly take care of resettlers' livelihood and productive activities. Accordingly, there are certain procedures for land acquisition.

In the past 56 years, China has revised the essential land acquisition approval limit and compensation standard numerous times. For instance, the limit on land acquisition approval at the level of county governments, which was 1,000 mu (66.7 ha) or below in 1953, was adjusted to 300 mu (20 ha) in 1958, further adjusted in 1986 to 3 mu (0.2 ha) and finally nil in 1998 which meant that county governments had no authority to approve land acquisition. The transfer of basic farmland into construction land needs the approval of the State Council.

Changes in Rights and Limits regarding Approval of Acquisition of Land for Construction in China are as follows:

Methods of National Construction Land Acquisition, first version, 1953

- County people's government: The land acquisitioned is less than 1,000 mu (15 mu = 1 ha), or the number of resettlers is smaller than 50 households.
- Provincial people's government: The land used is more than 1,000 mu, and the resettlers constitute more than 50 households.

Methods of National Construction Land Acquisition, second version, 1958

- County people's government: The land used is less than 300 mu, and there are fewer than 30 households of resettlers.
- Provincial people's government: The land used is more than 300 mu, and there are more than 30 households of resettlers.

Regulations on Land Condemnation for National Construction, 1982, Land Administration Law of the PRC, 1986

- County people's government: The cultivated land to be acquired is less than 3 mu, or other kinds of land are less than 10 mu.
- Provincial people's government: The cultivated land and garden plot to be used are more than 3 mu, cultivated land and grassland are more than 10 mu, or other type of land is above 20 mu.
- The State Council: Cultivated land to be used is above 1,000 mu, or other type of land is above 2,000 mu.

Land Administration Law of the PRC, 1998. This law has cancelled the right of the people's government below provincial level (at city, county as well as the cities at county level) to analyse and approve land acquisition.

- The State Council shall analyse and approve: (i) capital farmland; (ii) cultivated land is excluded from the category of capital farmland when exceeds 35 ha and (iii) other type of land that exceeds 70 ha.
- The provincial people's government shall analyse and approve: acquisition of land other than the one analysed and approved by the State Council.

The approval limit for county governments regarding the acquisition of construction land in China is shown in Fig. 10.1 as an example.

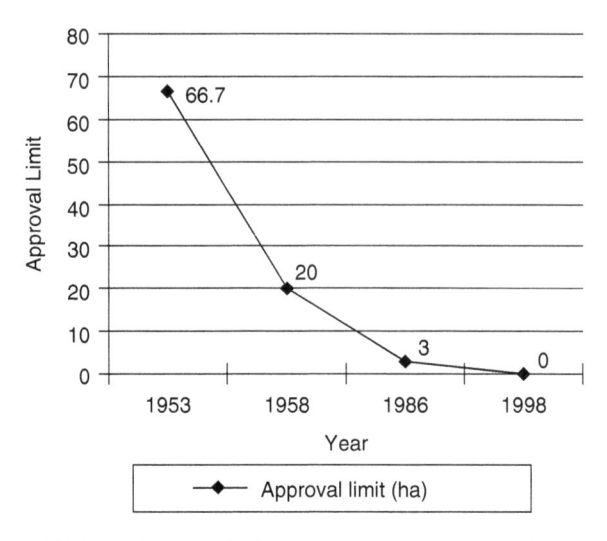

Fig. 10.1 Approval limits for land acquisition by county governments (ha)

The compensation standards for land acquisition for construction by the state were 3–5 times the annual average output value (AAOV) of the land acquired in 1953, 2–4 times in 1958, 5–20 times in 1986 and currently stand at 10–30 times from 1999. Since 2004, according to the State Council Decision to Deepen Reform and Strictly Enforce Land Administration (Document 28, November 2004) compensation can be more than 30 times the AAOV, with the amount above 30 times the value taken from the government's income from land sale. As a matter of fact, in China 30 times the AAOV is more than 60 years' net income from land-related activities.

Following are the changes in compensation standards for the acquisition of land for construction in China at different years.

Methods of National Construction Land Acquisition, first version, 1953

- Compensation for the acquisition of unclassified land shall be for the total yield value of the last 3–5 years as the standard. The compensation standard of special land shall be handled separately.

Methods of National Construction Land Acquisition, second version, 1958

- Compensation was based on the total yield value of 2–4 years as the standard.

Regulations on Land Condemnation for National Construction, 1982 and Land Administration Law of the PRC, 1986

- Land compensation fee for cultivated land: 3–6 times the average annual output or AAOV (last 3-year average) of acquired land.
- Resettlement subsidy for cultivated land: 2–3 times the AAOV (last 3-year average) of acquired cultivated land. The highest resettlement subsidies shall not exceed 10 times the AAOV (last 3-year average).
- Total land compensation and resettlement subsidies shall not exceed 20 times the AAOV of the acquired land.

10 Resettlement in China

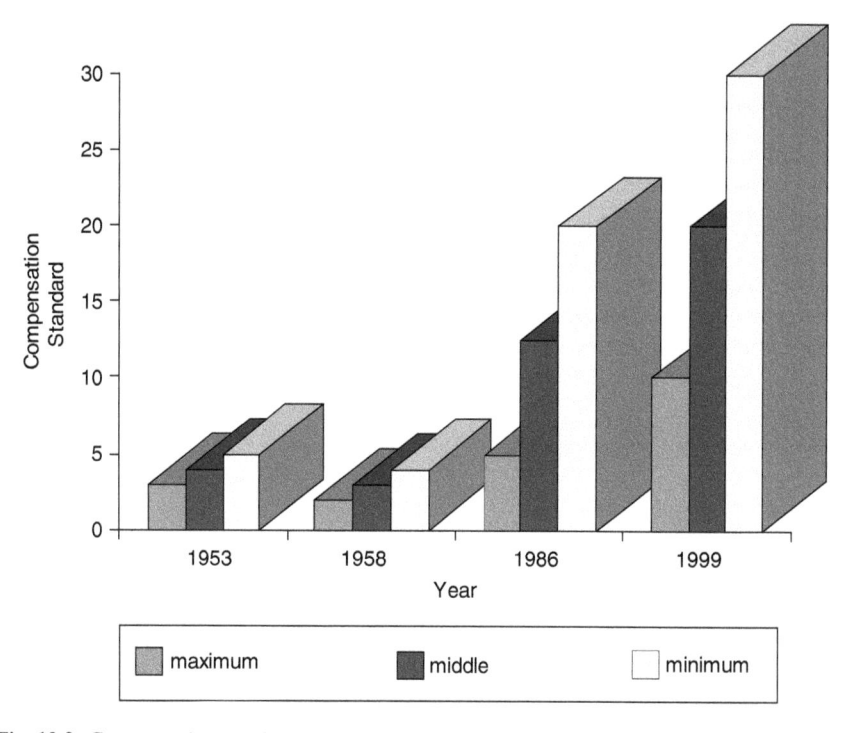

Fig. 10.2 Compensation standards for land acquisition (Unit: AAOV)

- The compensation for young crops, houses and wells shall be prescribed by the provincial people's government.

Land Administration Law of the PRC, 1998

- Land compensation fee for cultivated land: 6–10 times the AAOV (last 3-year average) of acquired land.
- Resettlement subsidy for cultivated land: 4–6 times the AAOV (last 3-year average) of acquired cultivated land for each farmer to be resettled. It should not exceed 15 times per hectare of cultivated land to be acquired.
- Total land compensation and resettlement subsidies shall not exceed 30 times the AAOV of the acquired land.
- Compensation for young crops shall be prescribed by the provincial people's government.

State Council Decision to Deepen Reform and Strictly Enforce Land Administration (Document 28, November 2004)

- Compensation can be set at more than 30 times the AAOV.
- The amount above 30 times the AAOV will be taken from the government's income from land sale.

Figure 10.2 also compares compensation standards at different years.

10.2.3 Main Policies on Land Acquisition, Demolition, Relocation and Resettlement

The aim of China's resettlement policy is to minimise the use of occupied land for projects and the number of displaced persons. But if this cannot be avoided, no effort shall be spared in resettling displaced persons properly, under the spirit of 'complete responsibility' and appropriate reconciliation between the state, the collective and individuals, with due consideration to all sides as far as possible. The general principles of the policies on project resettlement are the following: social mobilisation, economic compensation, preferential government policy for resettlers, support from the various partners, self-reliance, development of resources, follow-up support, etc. The aim of the resettlement policy is to gradually restore or surpass previous living standards. For details, see Box 10.1.

10.2.3.1 General Policies

Limit the Use of Land for Construction and Protect Cultivated Land

The state protects cultivated land and strictly controls the conversion of cultivated land into non-cultivated land. Care will be taken to minimise the use of land for

---

Box 10.1: Main Points of Land Acquisition, Demolition, Relocation and Resettlement Policies in China

- Land should not be used for non-agricultural activities. Agricultural land should not be converted into land for construction of housing.
- The fundamental object of resettlement is to restore the original living standards of the affected people.
- Provide compensation and jobs to collective landowners.
- All compensation and resettlement costs should be included in the budget of the main development project.
- Land acquisition and resettlement plans should receive government approval prior to execution.
- Affected people in rural areas shall be given first option of resettlement land for agriculture.
- Affected people in urban areas may choose to have houses in appropriate locations, exchange houses or receive cash compensation.
- Parties interested in land acquisition or house demolition should sign contracts on compensation and resettlement.
- Disputes over land acquisition and resettlement can be settled through an administrative and lawful grievance channel.
- The above principles are applicable to all development projects that require land acquisition.

non-agricultural purposes. If it is possible, the use of wasteland and land of inferior quality should be used instead of cultivated land or land of superior quality.

Legal Acquisition of Land for Construction

The state may, in the public interest, lawfully acquire land owned by the communities. All units and individuals that require land for construction purposes shall, in accordance with the law, apply for the use of state-owned land, including land owned by the state and land originally owned by communities but acquired by the state. Where land for agriculture is to be used for construction purposes, examination and approval must be undertaken for the conversion of use.

Compensation for Land Acquisition

Compensation for acquired cultivated land shall include compensation for land, for houses, stalls, toilets and greenhouses and young crops on the land as well as resettlement subsidies. Compensation for the cultivated land shall be 6–10 times the value of the average AAOV of the last 3 years of the land. Resettlement subsidies shall be 4–6 times the same value for each farmer to be resettled, but cannot exceed 15 times this value per hectare of cultivated land to be acquired. The total land compensation and resettlement subsides shall not exceed 30 times the same value. The State Council may, in light of the level of social and economic development and under special circumstances, raise the standards of land compensation and resettlement subsides for acquisition of cultivated land. Compensation standards for houses, stalls, toilets and greenhouses, as well as young crops on the land to be acquired, shall be prescribed directly by provinces, autonomous regions and municipalities. For acquisition of vegetable plots in urban suburbs, the land user shall pay to establish a development and construction fund for new agricultural plots in accordance with the relevant regulations of the state. For compensation of farmland acquired for water resources and hydropower projects, the minimum compensation rates should be 16 times the AAOV of farmlands, as of 1 September 2006. The compensation rates could be higher than this rate, on the basis of the income restoration needs of the resettlers.

Compensation and Allocation

Any construction unit which wants to use state-owned land should pay for all types of compensation as required. However, land to be used by public institutions or military purposes, urban infrastructure projects, public welfare activities, and major energy, transportation, water conservation and other infrastructure projects supported by the state, may be allocated with the approval of a people's government at or above the county level.

Participation and Consultation

The administrative departments in charge of demolition and removal need to proclaim the demolition and removal of properties through appropriate announcements, and explain the concerned policies and negotiate compensation and resettlement agreements with the local population. Local government uses public boards to disseminate information regarding the compensation resettlement scheme and consult with resettlers and landowners on their views.

Government Responsibility

The government is responsible for resettlement implementation. For example, the resettlement of the Three Gorges Project is controlled by both central and local governments. The demolition and relocation of urban houses can be organised by local government according to a 'universal plan'.

Advocating and Supporting an Ongoing Policy on Resettlement with Development

The policy aims to support rural communities which land is acquired, and also to support the population in their efforts to engage in business development or start new enterprises.

10.2.3.2 Sector-Specific Policies

Post-Relocation Support

The state adopts methods to provide compensation and subsidies during the preparatory stage to development, as well as support for economic activities after the relocation of resettlers. The period for such support has been set at 20 years from the date of relocation, and funds are obtained through the establishment of a 'reservoir area construction fund' which is covered from project profits. However, this is a special support provided by the Water Department and especially for reservoir projects.

10.2.3.3 Policies in Practice

Agricultural Resettlement

Most resettlers displaced due to project construction are farmers. For them, China has adopted a resettlement strategy based on land which includes developing new and existing land for agricultural purposes and conducting large-scale agriculture so as to make it possible for resettlers to live of agricultural activities restoring and improving their original living standards as much as possible. Especially in regions that are economically and culturally underdeveloped, such agricultural resettlement that suits local situations is the best way to achieve a successful resettlement. At the same time, the government shall encourage the development of secondary and tertiary sectors to develop the economy for resettlement.

Preferential Policies

To properly arrange the livelihood of resettlers, the state and local governments at various levels shall provide preferential treatment to displaced persons within their powers, such as reduction or exemption of relevant taxes (e.g. housing and land taxes). In addition, preferential policies should be formulated for improved agricultural production, provision of grains, credit application, employment opportunities and appropriate land allotment. For example, the resettlement of the Three Gorges Project is specially prescribed by the State Council. For details, see key resettlement policy issues below.

Mobilising Social Support for Resettled Communities

The policy aims to mobilise all kinds of social support to aid resettlers, e.g. social groups (such as industrial or mining enterprises, troops, etc.) who help resettlers to move and transport their belongings; the departments of civil affairs and women's organisations who help the elder, weak, sick, disabled, women and children and so on. For the resettlement of the Three Gorges Project, the central government also formulated and implemented a policy encouraging support from more developed areas (provinces and cities) near the Three Gorges Reservoir area.

10.2.3.4 Special Policies for Projects

According to national and local laws and regulations, special land acquisition and resettlement policies are formulated for particular projects. The policies adopted for the Xiaolangdi multi-purpose dam project were as follows:

- Compensation for houses should be paid at replacement cost.
- Compensation should be paid at replacement cost for equipment, structures and other facilities in industries and mines.
- Infrastructure should be reconstructed to restore its original function.
- Resettlement of farmers should be land based wherever possible.
- Sharing land with host villages should be based on the principle of mutual acceptance. It should also be planned so as to provide higher incomes (from all sources) for resettlers and hosts.
- Houses and community facilities in new town and village sites should be constructed with higher standards than previously existed.

- The financial resources to carry out relocation and development proposals should be available when and where required. Development plans should be prepared according to relocation plans.
- All daily-wage labourers should receive compensation equal to 3–6 months' wages.
- If shared farmlands do not improve the income of the population, non-agricultural employment opportunities should be offered to some of the labour force.
- The affected population should not only maintain their present living standards but also be entitled to benefit from the project.
- Every possible assistance should be rendered, socially and economically, to minimise and shorten the transition period of resettlement.
- Resettlement should be realised through employment in agricultural activities, in newly built or expanded industrial enterprises and in the tertiary sector. Re-establishment of the social and economic life of the affected population should be based on reliable and practical employment solutions.
- Changes in occupation will have to take into account the original background of the affected population in terms of education and employment. Changes of livelihood should be on a voluntary basis.
- With the objective to increase the income level of both the affected population as well as the local population in the host areas, land shared with hosts should be organised in such a way that is acceptable to both sides and according to plans.
- Rehabilitation proposals should be accepted by the affected population. They should also present their comments.
- The relocation site should be as close as possible, and an attempt should be made to relocate the same community together.
- Resettlement and rehabilitation plans should try to reduce existing land loss.
- Resettlement and rehabilitation plans should include detailed institutional arrangements so as to ensure the implementation of the resettlement plan in a timely and effective way.
- Funds should be provided in a timely fashion for resettlement, rehabilitation and development plans, and development plans should be prepared based on the resettlement plans.
- The impact of resettlement on the natural, social, economic and environmental conditions of relocation sites should be acceptable.
- Only economically viable enterprises should be relocated. Compensation provided to unviable enterprises should be used in the creation of employment opportunities.
- The affected infrastructure such as roads and bridges should be restored or even upgraded.
- Dependents of staff or workers of state companies may be transferred to non-agricultural areas on a voluntary basis. Compensation to these staff or workers will be paid to the agencies or government organisations for which they work. These agencies should provide employment to resettlers.

Table 10.2 New policies for reservoir resettlers

Key points	New policy	Old policy
Minimum compensation standard	• 16 times the AAOV • The compensation standard can be increased if the compensation cannot restore the affected person's livelihood	• 10 times the AAOV
Compensation scope	• Compensation for individual assets in the inundated area as well as in areas above line of inundation for resettled households	• Compensation for individual assets (house, trees, etc.) within the inundated area
Public participation	• The quantity of all individuals' affected assets should be confirmed by affected households and county people's government • During the resettlement planning, affected persons' opinions need to be sought and incorporated into the final resettlement plan • The standard and scope of compensation, funds for compensation and the rehabilitation plan should be disclosed to the public • The houses need to be rebuilt by affected persons themselves	• While the norms listed under the new policies were implemented in many projects, they had not been issued in formal policies or regulations
Post-relocation support	• Post-relocation fund of CNY 600 per year per capital, for 20 years	• CNY 250–400 per year, for 5 years

Note: Exchange rate as of November 2007, c. USD: CNY = 1:7.5
Source: Shi et al. (2007)

10.2.4 New Policies on Reservoir Resettlers

As mentioned earlier, the Regulations on Compensation for Land Acquisition and Resettlement on Large and Medium-sized Water Conservancy and Hydroelectric Projects (1991) was updated in August 2006. The New Resettlement Policy on Post-relocation Support for Reservoir Projects was issued by the State Council in July 2006. Table 10.2 compares the differences between the old and new policies.

10.3 Institutional Arrangements

There are now three levels in China's resettlement and rehabilitation organisations. The first is the administrative level of the central government, i.e. the State Council and its ministries and institutions. The second consists of the management sector and concerned government agencies of provincial, municipal and county governments, including the land management sector, house removal management sector, supervisory committee and judicial institutions. The third level is made of the institutions for implementing and servicing the project, such as the project owner (including institutions that are employed to work for the project owner, for design, monitoring and evaluation) and the resettlement and rehabilitation institution trusted by the owner. In China, the differences in resettlement and rehabilitation between projects of different sectors of the economy lie in different implementing agencies and sector management department.

The major institutions involved in resettlement operations are the project owner and implementing agencies, which play key roles in resettlement activities and are responsible for policy formulation, plan approval and monitoring. At present in the water resources and hydropower sectors, the Resettlement Bureau of the Ministry of Water Resources, Resettlement Development Bureau of the Three Gorges Project and the Resettlement Department of South to North Water Diversion project are responsible for resettlement and rehabilitation.

Resettlement Planning and Design Unit: In China, the technical works for resettlement planning related to reservoirs are mainly prepared by institutions affiliated with the Ministry of Water Resources and National Power Corporation. They are prepared also at province/ministry levels or at experienced research institutions since they have staff knowledgeable on socio-economic surveys and resettlement planning. These institutions have significant practical experience and are the main technical consultants for project owners and local governments.

Resettlement Monitoring and Evaluation Institutions: The resettlement monitoring and evaluation institutions in China are usually research institutions, consultant services, some colleges or universities and professional divisions of the Chinese Academy of Social Sciences. In addition to projects financed by the World Bank and the Asian Development Bank, the Three Gorges Project and the South to North Water Diversion project also conducted monitoring and evaluation activities. The system for supervising resettlement processes in water and hydropower is being explored and practised in development projects such as the Three Gorges, Xiaolangdi, Wanjiazhai, Shanxi and Mianhuatan. The audit department is one of the supervising units for resettlement; audit departments at different levels are responsible for auditing the utilisation of funds as well as the economic responsibilities of the staff.

In China, the main non-governmental organisations (NGOs) engaged in resettlement work are the National and Provincial Reservoir Economic Committee and the Urban House Removal Committee, yet their main functions are academic exchange and consultation.

10.4 Resettlement Implementation

Generally speaking, the implementation of resettlement in China can be divided into four phases.

10.4.1 Preparation

This involves project selection, feasibility study, preliminary design and design approval. The focus in this phase is to carry out a socio-economic survey and prepare the resettlement plan.

10.4.2 Implementation

The second phase begins when the preliminary design has been approved. Generally, the contract is signed first between the project owner and the local provincial government. Then the concerned province-city-county-township-villages sign agreements with each other. In some cases, affected families can directly sign agreements with the county government. In this phase, owners and the local government prepare an implementation programme according to the approved preliminary design. Many complex tasks are completed during this phase such as house demolition, land clearing, relocation, labour arrangements, infrastructure and house reconstruction as well as reconstruction of special facilities, fund management, dispute management and coordination.

10.4.3 Checks, Acceptance and Post-support

After relocation has been completed, the project will be checked and accepted by the owner, design unit, implementation unit and government at each level according to the design standards established for the project. A formal document must be prepared after checks and acceptance. Once this has been prepared, the project enters a 20-year post-support period. If necessary, a post-evaluation can be made after the post-support period.

10.4.4 Income Restoration

As of now, the livelihood recovery of rural resettlers in China is handled in the following ways: the readjustment of the cultivated land which remains after acquisition for the project within the villages and groups with the objective to ensure that each villager has a plot of necessary land; exploiting wasteland and protecting the reservoir area to increase cultivated land; improving low-yield land with irrigation; establishing towns, villages enterprises and joint ventures; offering employment in state-operated, collective-owned enterprises; providing endowment insurance and living allowances; self-employment and non-agricultural family businesses.

10.5 Comparison Between World Bank's and China's Involuntary Resettlement Policies

As a whole, the resettlement policies of the World Bank are basically the same as those of China (see Table 10.3). However, the resettlement policy of the World Bank pays more attention to resettlement planning, overall consultation and participation, agreement by the population with the processes followed, resettlement of vulnerable groups, impacts on residents in host areas and monitoring and evaluation (Chen 2001).

10.6 Case Study: Three Gorges Dam Project

10.6.1 Basic Information

The Three Gorges Dam Project is the largest hydropower project in terms of both power capacity and the number of involuntary resettlers. The project affected about 1.3 million people, of which 44% were farmers and 56% were urban citizens in two provinces and 21 districts or counties. In addition, 1,599 enterprises, 11 towns and two cities were inundated, and 175,000 rural affected people were relocated further away in 11 coastal or downstream provinces, while others were relocated within the counties. Most of the affected rural farmers were resettled with land for land compensation. The resettlement budget was CNY 40 billion based on May 1993 prices, which was 44.6% of the project budget (CNY 97 billion). About 1.1 million affected people were relocated before July 2006.

10.6.2 Key Resettlement Policy Points

The main points of the resettlement policy for the Three Gorges Dam Project were outlined as follows:
- The central government will carry out a policy of resettlement with the development for the Three Gorges Project. Concerned local governments will directly organise and manage the resettlement work through comprehensive plans, which will ensure that the standard of living of displaced persons is at least restored or exceeds former levels. This will create conditions for the long-term economic development and improvements in the living standards of project-affected people in the Three Gorges Reservoir area.
- The Three Gorges Project upholds the resettlement principles of national support, preferential policies, support from all institutions and self-reliance. It encourages good relationship among the state, the communities and the individuals.

Table 10.3 Comparison of World Bank's and China's resettlement policies

World Bank resettlement policies	Current stipulations of laws and policies adopted in China
Involuntary resettlement should be avoided where feasible	The Land Administration Law (1998) aims to 'strictly control transferring farmland to construction land', and states that 'special protection should be offered to cultivated land', 'strictly control transferring farmland to non-farmland' (article 2)
Where population displacement is unavoidable, it should be minimised by exploring all viable project options	Policies aim to minimise land use for non-agriculture construction, use waste or inferior land instead of cultivated and good quality land and make full use of existing land for construction in urban development. There should be no or minimal occupation of farmland
People displaced should be compensated and assisted so that their economic and social future will be generally as favourable as it would have been in the absence of the project	Land Administration Law (1998), Urban House Demolition Administration Regulations and other relevant laws require that compensation and aid should be supplied when land is acquired and houses are demolished
People affected should be fully informed and consulted on resettlement and compensation options	The Land Administration Law (1998) regulates that after the determination of compensation and resettlement plans for land acquisition, the decisions should be announced to the public by the local government authorities. The opinions of the people and the organisations should be taken into consideration. It is stipulated in the Urban House Demolition Administration Regulations that there should be written agreements on compensation, resettlement, management and other issues between the agency undertaking demolition and the owners of the units being demolished
Existing social and cultural institutions of resettlers and the host community should be supported and used to the greatest extent, and resettlers should be integrated economically and socially into host communities	There are no similar stipulations in existing laws. Resettlers should be resettled in local towns and local counties as much as possible
The absence of a formal legal title of land among some affected groups should not be a limitation for compensation; particular attention should be paid to households headed by women and other vulnerable groups, such as indigenous people and ethnic minorities, and appropriate assistance should be provided to help them improve their livelihood	Compensation policies for women are the same as those for men. At the same time, the women's rights insurance laws are also applicable during resettlement. The same policies are applied to ethnic minorities as to other groups, and the Law of Regional Autonomy by Ethnic Minorities (2001) is also applied to ethnic minorities. Specific laws have been made for women and the ethnic minorities in China. The Regulations Governing Urban House Demolition and Relocation (2001) stipulate that illegal buildings and the temporary buildings for which permits have expired should not be compensated

(continued)

Table 10.3 (continued)

World Bank resettlement policies	Current stipulations of laws and policies adopted in China
As much as possible, involuntary resettlement should be conceived and executed as a part of the project	The Land Administration Law (1998) and the Regulations Governing Urban House Demolition and Relocation require that land acquisition and house demolition should be approved. It is also stipulated that the construction of reservoir projects can commence only after approval of the resettlement plan
The full costs of resettlement and compensation should be included in the presentation of project costs and benefits	There are no specific stipulations in national laws. It is stipulated in various regulations and norms that the budget for land acquisition and house demolition should be included in project budgets. However, special subsidies from local governments and social districts are not to be included in project budgets
Costs of resettlement and compensation may be considered for inclusion in bank financing for the project through loans	There are no specific stipulations in national laws
Stakeholder participation in resettlement planning and implementation	The Land Administration Law (1998), Regulations Governing Urban House Demolition and Relocation and relevant laws require that compensation and aid should be granted when land is acquired and houses are demolished. Any adjustments to agreements on contracted lands should be agreed by two-thirds of villagers. The Regulations Governing Urban House Demolition and Removal stipulate that the agency demolishing a house and the owner of the house should sign a written agreement on compensation and resettlement
Clear mechanisms for grievance redressal	The Land Administration Law (1998) has explicit stipulations regarding dispute adjudication on land ownership and land access, and the Regulations Governing Urban House Demolition and Relocation has similar terms
Full disclosure of resettlement plans and implementation information to affected persons	The Land Administration Law (1998) requires local governments to openly publicise land acquisition, compensation and resettlement plans. The Regulations Governing Urban House Demolition and Relocation requires the public announcement of the agency conducting the demolition as well as the scope and deadline of demolition. The resettlement scheme should include consulting affected people

Source: Zou (2002); World Bank (1990, 2001)

- Resettlement operations shall take into account progress in construction of the project and the Three Gorges Reservoir area, in addition to undertaking efforts to open up the project area to the outside world, soil and water conservation and environmental protection. These actions will help to promote the regional economy, improve the environment and create an attractive investment climate in the Three Gorges Reservoir area.
- It is necessary to prepare a resettlement plan for the Three Gorges Project. The resettlement plan of each county (or city) shall be analysed and approved by the Three Gorges Project Resettlement Development Bureau of the State Council. The people's governments of Hubei Province and Chongqing Municipality and lower levels of people's government in the Three Gorges Reservoir area are responsible for implementing the approved resettlement plan. They are under the supervision of the Three Gorges Project Resettlement Development Bureau. Some 85% of the resettlers due to Three Gorges Dam are in the 15 counties of Chongqing Municipality.
- People's governments of Hubei Province, Chongqing Municipality and other cities or counties located in Three Gorges Project area will, according to the plans, jointly arrange the compensation to be paid for land acquisition. The compensation is to be used for land development, resettlement of displaced people and re-establishment of their livelihood.
- The resettlement of rural displaced persons depends mainly upon the development of large-scale agriculture, through improvements of cultivated land, enhancement of medium- and low-yield fields, cultivation of high-yield grain fields and commercial crops, orchards and forests, and development of forestry, animal husbandry, fisheries and ancillary occupations. Town and village enterprises are encouraged, and if conditions permit, secondary and tertiary sectors shall be actively developed. These are sound approaches for resettlement.
- Project-affected people should be relocated in or near their original villages, towns, counties or cities. Where local resettlement is impossible, displaced persons may be resettled within the same province. If even this is impossible, they should be resettled in a suitable location that is both rational and economic.
- When displaced persons of rural areas have to be resettled with populations other than those of their own village, the concerned resettlement communities of cities (or districts) must sign agreements in advance with the host villages. The rural communities shall arrange the livelihood and housing for resettlers according to the agreements.
- When such communities receive a resettlement fee from the government to reclaim wasteland with a slope below 25°, and to improve medium- and low-yield fields owned by them, they can accept resettlers and make egalitarian arrangements with their original members. The people's governments of concerned cities (or counties) may allocate newly reclaimed land in a certain proportion to communities as compensation, while the remainder will be used to resettle displaced persons and then establish new communities. The ownership of land can be transferred in accordance with the law. With regard to land contracted to individuals and other kinds of land, the rural communities can modify the land

use and the management rights; the rights and obligations of both parties to a land contract can also be adjusted.
- When resettlers have to be resettled outside their original county or city, the local governments and host government shall negotiate and sign agreements with the displaced persons and complete the necessary procedures. Local governments shall transfer commensurate resettlement funds to the host governments to arrange resettlers' livelihood and residence.
- When rural housing has to be moved, proximity to and convenience for economic activities and daily life must be considered during planning. Construction plans for new housing must be prepared according to the law and carried out in stages. Compensation for removal of dwellings should be contracted with individual households in accordance with rural housing compensation standards, and affected people should be allowed to build new houses themselves. Where conditions permit, the state government encourages centralised house construction.
- Infrastructure for newly constructed rural dwelling sites such as roads, water and electricity shall be developed within integrated plans which shall be included in the master resettlement plan of the project. The infrastructure may be constructed by villagers.
- When construction of the project requires the removal of cities and towns, it is important to select new sites and prepare city reconstruction plans according to the legal requirements. Compensation for removal and construction, as per the resettlement plan, will be listed in the resettlement budget of the project. Those items of construction and any other items excluded from or higher than resettlement compensation standards shall be financed by the concerned local people's governments.
- Compensation standards for urban houses demolished and removed due to the project will be made according to the Regulations Governing Urban House Demolition and Relocation.
- Compensation for enterprises and institutions that require removal due to the project shall be according to their original scale and standards at replacement price, considering adjustments for overall technical innovation and industrial structure. Any actual reconstruction cost that exceeds the approved resettlement budget due to expansion in scale or improvement of standards shall be borne by the concerned local government.
- Special facilities such as highways, bridges, ports, wharfs, water conservation projects, electrical power facilities, transmissions lines and broadcast lines or cultural relics that are flooded due to project-related impoundment shall be rebuilt above the line of inundation at a reasonable cost. Investment needed to rebuild such facilities on their original scale and at original standards, or to restore the original functions (including the actual length of new highways, transportations lines or broadcast lines), shall be included after verification in the resettlement budget. Any additional investment due to an expansion of scale or the raising of standards shall be managed by the concerned parties.
- Project resettlement funds, separate from the general project estimate, shall be listed in the budget plan of the central government. Resettlement funds must be used for specific purposes and should not be diverted to other uses. Local

10 Resettlement in China

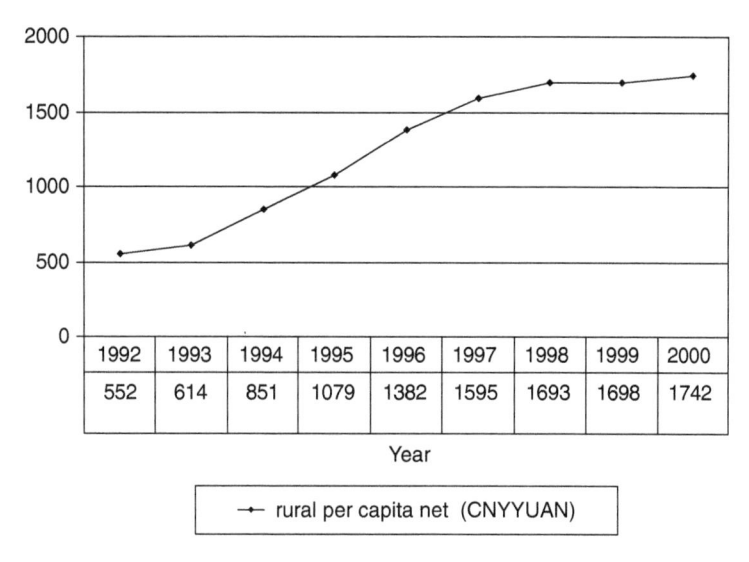

Fig. 10.3 Rural per capita net income of resettlers in Chongqing Reservoir area in CNY. (*Source*: Chongqing Yearbook 2000, Chongqing Municipality Statistic Bureau)

Year	1992	1993	1994	1995	1996	1997	1998	1999	2000
	552	614	851	1079	1382	1595	1693	1698	1742

— rural per capita net (CNYYUAN)

governments responsible for resettlement activities and personnel in charge of resettlement institutions will be audited regularly by the central government.

10.6.3 Progress and Effectiveness

By the end of 2008, 1.128 million people have been relocated, accounting for 99.1% of the planned task, 34.11 million m² of houses have been constructed, which is 98.8% of the plan and 1,397 industrial and mining enterprises have been relocated, accounting for 100% of the plan.

10.6.3.1 Reservoir Areas

Great social and economic changes have taken place in the reservoir area, and the quality of life of the resettlers' has been improving steadily. The average annual increase in GDP of the 15 counties in the reservoir area is 10.3%, which is 0.2 percentage points higher than the average level of Chongqing Municipality. The per capita domestic space for rural and urban resettlers is 41 m² and 25 m² respectively, both of which have increased by about 10 m². Rural per capita net income of resettlers in Chongqing reservoir area in 2000 was nearly three times that the level in 1992 (see Fig. 10.3). The population in absolute poverty has reduced from 1.022 million to 327,000 last year. Generally speaking, in the concentrated resettlement sites water supply, power, roads, education and other municipal services are better than before.

Table 10.4 Sample size and distribution of nation-wide survey in 2003

	Province										
	Guangdong	Shandong	Anhui	Hunan	Fujian	Sichuan	Jiangsu	Zhejiang	Shanghai	Hubei	Jiangxi	Total
Sample size	31	40	42	30	40	65	40	39	51	37	25	440
Proportion (%)	7.0	9.1	9.5	6.8	9.1	14.8	9.1	8.9	11.6	8.4	5.7	100.00

Table 10.5 Distribution of living conditions [convenience (in percentage)]

	Extremely convenient	Relatively more convenient	Common convenience	Inconvenient	Very inconvenient	Total
Purchasing daily necessities	40.5	42.0	13.4	2.5	1.6	100
Purchasing non-staple food	38.4	43.4	14.8	2.0	1.4	100
Purchasing means of production	34.2	44.2	13.9	5.9	1.8	100
Medical services	35.3	40.5	14.6	6.4	3.2	100
Transportation	45.0	33.6	16.8	3.2	1.4	100

10.6.3.2 Resettlement Outside Chongqing

Great changes have also taken place for resettlers who were relocated in other provinces. A nation-wide survey was conducted in 2003 to determine whether these resettlers' living standards and livelihood had been restored. The target population consisted of 1,756 rural households with 7,321 persons, who migrated from Chongqing Municipality to provinces such as Shanghai, Zhejiang, Anhui, Fujian, Jiangxi, Shandong, Hubei, Hunan, Guangdong, Jiangsu and Sichuan. The sample size and distribution is shown in Table 10.4.

Generally speaking, the findings of this survey indicated that resettlement for the Three Gorges Project has achieved its targets. Most of the resettlers surveyed were satisfied with the resettlement policies, with their livelihood and economic activities having improved significantly.

Economic Situation

- Per capita land holding of resettlers was 1.6 mu (0.11 ha), while that before resettlement was only 1.3 mu (0.09 ha).
- Some 63% of interviewees thought that the level of farming mechanisation had improved.
- While most resettlers were still involved in agricultural production, more and more earned an income from non-agricultural jobs.

Living Conditions

- Resettlers' houses had improved a great deal. Some 60.3% were living in multi-storied houses, compared to 44.1% before resettlement.
- With regard to the convenience in various living conditions, more than 75% of interviewees responded positively. For details, see Table 10.5.
- More than 90% of interviewees had become adapted to the new living environment. See Fig. 10.4.

10.7 Experiences and Lessons

10.7.1 *Experience of Success*

According to the resettlement policies and practices of the past 40 years, particularly the last 20 years, we can conclude that the successes of resettlement are mainly attributed to a higher degree of attention paid to resettlement, forceful leadership and efficient government institutions at each level; well-developed and implemented

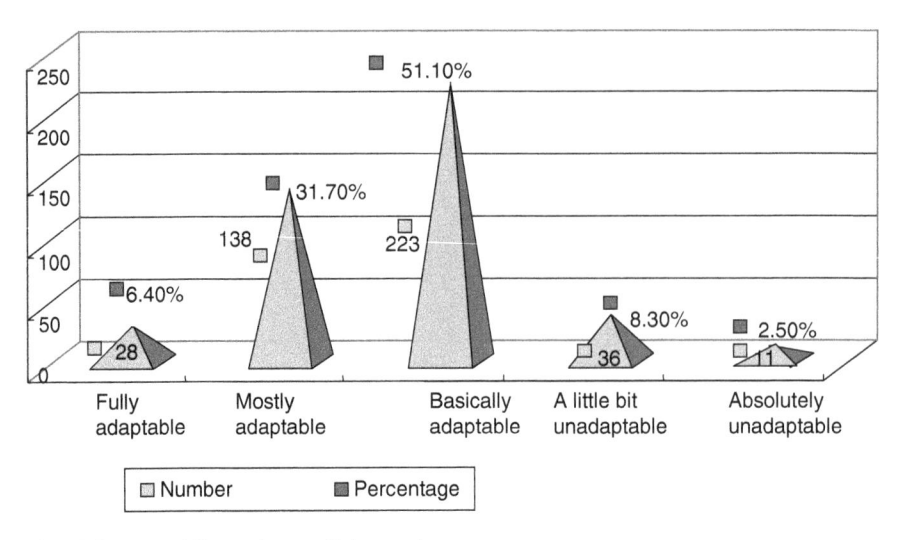

Fig. 10.4 Adaptability to the new living environment

resettlement policies and legal frameworks which ensure that resettlement is carried out in an orderly and efficient way; higher importance attached to resettlement planning, socio-economic research and improvement of lifestyle, income included, for displaced persons; use of various methods and development-oriented resettlement strategies to restore or improve the quality of life of displaced persons; active involvement of displaced persons and external monitoring and evaluation systems.

10.7.2 Lessons from Failures

Resettlement policies and practices prior to the 1980s showed that the failures of resettlement are mainly attributed to the ideology of valuing 'civil works of projects above resettlement' (Tang 2002); lack of efficient agencies on the part of project owners and governments; lack of laws and regulations for housing demolition from the 1950s to the early 1980s which led to inconsistent and unreliable compensation and rehabilitation practices; neglect of resettlement planning; incorrect administrative practices; failure in planning for resettlement funds which in a few cases led to difficulties in income recovery; failure to disseminate information on compensation standards for land acquisition and resettlement in some projects; inadequate attention to the opinion of displaced persons on the resettlement programmes; insufficient involvement of displaced persons and the failure of research and training on resettlement in meeting actual needs.

References

Chongqing Yearbook (2000) Chongqing Municipality Statistic Bureau. China Statistics Press, Beijing
Chen X (2001) Involuntary resettlement: involuntary resettlement policy of World Bank and its inspiration. Dev Strat Countermeas Res 12:42–44
Land Administration Law of the PRC (1986) People's Republic of China
Law of the PRC on Administration of Urban Real Estate (1994) People's Republic of China
Law of Regional Autonomy by Ethnic Minorities (2001) People's Republic of China
Methods of National Construction Land Acquisition (1953, 1958) People's Republic of China
New Resettlement Policy on Post-relocation Support for Reservoir Projects (2006) People's Republic of China
Regulations on Compensation for Land Acquisition and Resettlement on Large and Medium-sized Water Conservancy and Hydroelectric Projects (1991) People's Republic of China
Regulations on Land Condemnation for National Construction(1982) People's Republic of China
Regulations Governing Urban House Demolition and Relocation (2001) People's Republic of China
Shi G (2007) New policies and mechanism on reservoir resettlement in China. Proceeding of international workshop in involuntary resettlement
Shi G, Chen S, Xun H, Xiang H (2001) China resettlement policy and practice. Ningxia Renmin Press, Yinchuan
Shi G, Ruiqiang Z, Chunmei M (2007) Evaluation on "large and medium sized hydropower engineering land requisition compensation and resettlement regulations". Water Econ 4:75–77
State Council Decision to Deepen Reform and Strictly Enforce Land Administration (2004), Document No. 28, September, Beijing
Tang C (2002) Resettlement policy and practice for dam, resettlement and social development. Hohai University Press, Nanjing
Tang J, Jia Y (1999) The comparative research on reservoir resettlement between China and foreign countries. Guangxi Education Press, Nanning
World Bank (1990) Operational Directive 4.30. Involuntary resettlement. World Bank, Washington DC
World Bank (2001) Operational Policy 4.12. Involuntary resettlement. World Bank, Washington DC
Zou Y (2002) China's practice of World Bank Resettlement Policy. Resettlement and social development. Hohai University Press, Nanjing

Chapter 11
Officials' Office and Dense Clouds: The Large Dams that Command Beijing's Heights

James E. Nickum

11.1 Introduction

In China one speaks not of dams and their heights but of reservoirs (*shuiku*) and their capacities. Beijing claims to have over 80 large reservoirs (storage capacity over 0.1 km^3), but the municipality's surface water systems are dominated by two: the Guanting (capable of many literal translations into English, but perhaps most closely, 'Officials' Office'), controlling over 90% of the Yongding River catchment in the west, and the Miyun ('Dense Clouds') controlling 88% of the Chaobai River catchment in the northeast.

This chapter takes a long view of the co-evolution of these two large dams/reservoirs, which control over three-quarters of the surface water supply of Beijing, with that municipal area and upstream administrations. In particular, even in the short span of 50 years, there have been continuous shifts in the purposes and uses of these impoundments, driven by urban development, watershed changes and the nature of the water supply mix available to Beijing's various sectors.

11.2 A Short Water History of a Water-Short Capital

Neither the Yongding nor the Chaobai River system flows by nature to the urban core of Beijing. This is by design. Historically, the principal means of flood control was to locate the capital as much as possible out of harm's way, and to focus on diverting waters to the city for supplementary water supply and navigation, especially grain transport from the south. In the past three centuries or so, as the city settled in one place and river management became more feasible, attention focused on river control.

J.E. Nickum
Asian Water and Resources Institute at Promar Consulting, Tokyo, Japan

When the capital of China returned to Beijing in 1949 after a brief hiatus, it was much a metropolis as before whose primary reason for existence was not economic but rather that it was the capital of China. The surrounding farmlands were largely rain fed, producing drought-resistant crops such as maize and Chinese sorghum. Urban residents relied on 'virtual water' from the wetter south, embodied in foodstuffs, some still brought in at great cost along the Grand Canal that had been dug in previous centuries for just this purpose (Sternfeld 1999:148). For most of its uses, even in times of drought, the urban part of the city could turn to its groundwater, which was the most abundant on the North China Plain, and close enough to the surface that it came out in springs in the western foothills.

Hence, as with previous governments, the primary water control concern of the new government in 1949 was flood control. What was new was the capability to build large dams. The focus on flood control was well timed. The 1950s were extraordinarily wet. Then came ever-drier periods and water-intensive development. While floods remained a concern in China's concentrated monsoon climate, other purposes grew in relative importance.

11.3 The Guanting and the Permanently Fixed River

The most popular stretch of the Great Wall outside Beijing, and for a long time the only one open to sightseers, is at Badaling in the driest part of Beijing Municipality, where the average annual precipitation is 500 mm or below. Looking from the side of the wall away from the city, one can still imagine it when it was a wild open land, a broad valley from where nomadic warriors sized up the gate and dust storms overwhelmed the ramparts. These days the scene is mitigated by an increasing number of buildings and a long blue line—the Guanting Reservoir.

The dam and almost the entire reservoir lie outside Beijing's municipal boundaries, in Hebei Province, at the lower end of an interior valley draining the upper Yongding River. The catchment area, with an average precipitation of about 400 mm/ann and an evapotranspiration rate many times that, extends westwards into Shanxi Province and in the northeast into Yanqing County of Beijing. It is separated from Beijing proper by the Western Hills (Xishan).

The removal of the original forest cover in the early centuries of the second millennium significantly increased the silt load and instability of the river that was optimistically renamed from Wuding ('Unfixed') to Yongding ('Permanently Fixed') in an act of wishful thinking in 1698. It stayed stable for 40 years but then changed course six times in just over 30 years (Li 2005; Beijing Shi Yongding He Guanlichu 2002:11). Perched above the city behind the mountains to its northwest, and spilling out in high water times through a narrow and crooked traversal of those mountains into a flickering flow that continually threatened the area to the southeast of the capital, Yongding River's flood waters reached the city proper in 1801 (flood peak 10,400 m^3/s) and 1890 (Beijing Shi Shuiliju 1999:37).

One month after the People's Republic of China was declared in October 1949, a plan for the control of the Yongding River was drafted that focused on soil erosion control in the headlands and the construction of a dam primarily for flood storage and power generation. Construction of the clay core Guanting Reservoir began in October 1951. The dam was commissioned in May 1954 with an initial dam height of 45 m and designed storage capacity of 2.27 km^3. Its catchment area of 43,400 km^2 is 92.3% of that of the Yongding. To address concerns that its flood standards were too low, the dam was raised by 7 m and storage capacity increased to 4.16 km^3 in the late 1980s to reach a design capacity to retain the waters of a flood statistically expected an average of once every 1000 years (11,640 m^3/s) (Beijing Shi Yongding He Guanlichu 2002: 33, 110).

11.4 The Miyun

In October 1958, Beijing Municipality was expanded to 16,800 km^2, only about 1,000 km^2 of which was the urban core (Peisert and Sternfeld 2005:33). Interestingly, the catchment areas of the Guanting Reservoir and the reservoir itself remained split between Beijing and neighbouring Hebei Province to the west, but Miyun County of the same province to the northeast, the site of Beijing's other large dam, was incorporated into the capital.

The Miyun Reservoir was begun the previous month in the heat of the Great Leap Forward and basically completed two years later in the midst of the subsequent famine.[1]

The principal focus in the first half of this period was on stabilisation of project components (notably the Zuomazhuang auxiliary dam and the No. 1 discharge tunnel) and on blocking seepage. A major flood in Henan Province in August 1975 led to a reassessment of Miyun's flood resistance and the construction of a third spillway. Less than a year later, the devastating July 1976 earthquake in Tangshan, only 150 km away, caused a slippage of the upstream face of the main Bai River dam, inducing a number of reinforcement projects in the Miyun reservoir complex, as well as the addition of a diversion tunnel with a capacity of 924 m^3/s to protect downstream populations. Since the inlet of the original spillway tunnel was over 10 m higher than the bottom of the reservoir, two new emergency discharge tunnels, each 3 m in diameter, were dug in 1977–1980 in case extreme events require the emptying of the reservoir (Beijing Shi Shuiliju 1999:111).

The Miyun Reservoir has two main dams, with heights of 56 m and 66.4 m and a total storage capacity of 4.375 km^3, only slightly more than the enlarged Guanting,

---

[1]When I visited the reservoir in 1974, we were told that it had been built on the principle of the 'three simultaneities' (synchronised design, construction and operation) in order to speed up its construction. With a sense of relief, the hydraulic engineers who were our hosts noted that it had held up quite well, except for termite damage. Nonetheless, Sternfeld (1999:130), citing Chinese sources, lists 11 major repair projects that were carried out between 1964 and 1985, which she claims were mostly due to deficiencies in planning and construction. In some cases, however, it is difficult to separate design flaws from upgrading of standards in light of new information from nature.

but for much of its history filled with more and cleaner water than its counterpart. Development in the upper catchment has been slower than on the Yongding, and the average precipitation in the mountains immediately surrounding the reservoir is 639 mm, considerably higher than for the Guanting (Miyun Xian 2003:4).

11.5 Water Displacements

Information on displaced people is limited. A total of 65,000 people were resettled from the area occupied by the Guanting Reservoir, beginning in 1952. This was organised by a committee set up for the task by Hebei Province. There is some indication that villages were relocated in their entirety, but that many people, especially the elderly, had to be persuaded by local officials that they should yield their place for the greater good endowed by the project (Sternfeld 1999:122). Even though Miyun is larger, the area was more lightly populated, so a slightly smaller figure of 56,900 displaced people is given for it (Sternfeld 1999:127). Like most statistics of the Great Leap Forward period, this number must be viewed with a particularly critical eye, but it seems to be of a reasonable magnitude.

Some of the able-bodied among the displaced persons may have worked on the construction of the reservoir facilities. Actually, a surprisingly modest total of 5,000 civil workers from Miyun County are reported as having participated in the construction of the Miyun Reservoir (Miyun Xian 2003:17), out of a project workforce that peaked at 200,000 (Sternfeld 1999:127). During this period in particular, China favoured labour-intensive construction methods for water projects, using a modified corvee levy that remunerated rural dwellers with work points rather than wage goods (Nickum 1978).

Most likely, relocation was done in the near vicinity of the reservoir, given the 'subsidiarity' policies of the times, the low population density in the area, and the subsequent concentration of Miyun's lower watershed population in the near-shore areas. This caused some problems in the 1980s and beyond, when watershed degradation became a major concern and new resettlement programmes were inaugurated in 1999 (Peisert and Sternfeld 2005:39).

11.6 Watershed Degradation

Much of the recent history of Guanting, and later Miyun, is concerned with addressing the thorny issues of watershed development. Watershed effects were poorly anticipated in the planning and construction of the two reservoirs, perhaps due to their intractability and the limitations of perspective and time of the development-obsessed decision making process. A watershed component centred on erosion control was part of the original design of the Guanting Reservoir, but even in this area, achievements were limited. The economic development of relatively poor upstream areas in other provinces came to haunt both reservoirs.

11.6.1 Guanting

In order to handle monsoonal floods, Guanting was designed with considerable excess storage capacity, even before expansion. After its expansion, out of its 4.16 km³ total storage capacity, 2.99 km³ were reserved for flood storage, an order of magnitude more than the 0.25 km³ allocated to beneficial storage (Beijing Shi Yongding He Guanlichu 2002:111). Up to 1990, it had successfully retained seven moderate flood crests in excess of 1,000 m³/s (Beijing Shi Yongding He Guanlichu 2002:46).

Much of this capacity has been lost to siltation, however. By 1960, 353 million m³ (MCM) of silt had accumulated in the reservoir. While the rate of accumulation slowed significantly after the wet 1950s, a total of 650 MCM had gathered by the year 2000, about one-third of which occupied the effective storage capacity. The pattern of siltation has been of greater concern than the quantity, as it threatens to block the inflow of the Gui He, the short, north-western tributary originating in Yanqing County, Beijing (Beijing Shi Yongding He Guanlichu 2002:89, 111).

Initially, in the construction of both the Guanting and Miyun, concerns regarding the watershed appear to have been limited to the problem of sedimentation, and even there building the dam came before erosion control, in a reversal of priorities that was all too common at the time (Shapiro 2001:62–63). The possibility of increased degradation from upstream irrigation, urbanisation, chemical agriculture and mining does not appear to have been factored in at all.

There was a continuous decline in the inflow of water to be stored in the Guanting Reservoir, due in small part, presumably, to climate change but largely to upstream withdrawals.

By the 1970s, the water of the Guanting had already been seriously degraded by sediment load, untreated urban sewage, industrial effluents and agricultural chemicals entering from upstream. Administration of the reservoir by the central Ministry of Water Resources from 1954 to 1970 does not appear to have been successful in controlling adverse developments in the watershed. It may even have been complicit in the construction of two large upstream reservoirs that contributed to the reduction in inflow into the Guanting—the Cetian Reservoir in Shanxi (580 MCM storage capacity) and the Youyi ('Friendship') Reservoir in Hebei (116 MCM capacity) (Beijing Shi Yongding He Guanlichu 2002:111).

Transfer of control to Beijing Municipality in 1971 probably did little to ameliorate inter-jurisdictional conflicts with upstream Hebei and Shanxi Provinces (Sternfeld 1999:179). The Leading Group for the Protection of the Guanting Reservoir was established a year later. It is said to have successfully improved the quality of the water in the reservoir after three years of water resource protection work (Beijing Shi Shuiliju 1999:209). The success was only temporary, however. A water resource protection agreement was ratified by the three administrations in 1985, but possibilities of consensus ran aground of disagreement over who should clean up the water (Peisert and Sternfeld 2005:35). Presumably, this was a classical Coasean problem of 'imperfect' property rights, with Beijing arguing that it had a

Table 11.1 Main pollutants entering Guanting Reservoir (mg/l)

COD					Ammonia nitrogen (NH$_3$-N)				
1980	1985	1988	1990	1995	1980	1985	1988	1990	1995
3.26	13.26	38.6	14.6	15.7	2.35	2.35	5.36	8.09	7.09

Note: Measured at Bahaoqiao Station, a regulating station at the intake of the reservoir after the confluence of the Yang and Sanggan Rivers (Yi 2007)
Source: Beijing Shi Yongding He Guanlichu (2002:167)

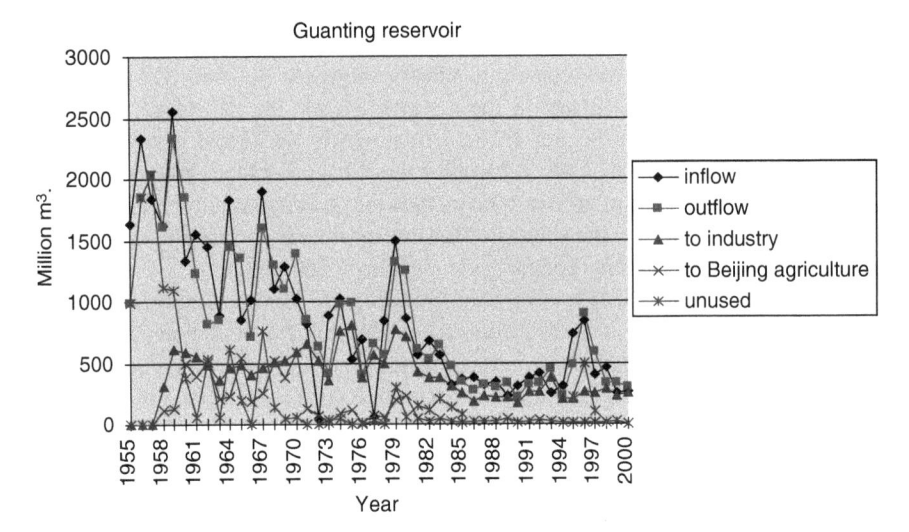

Fig. 11.1 The decline and industrialisation of Guanting's water (*Source*: Beijing Shi Yongding He Guanlichu 2002:149–150)

historical right to clean water and the upstream provinces pleading relative poverty and an absolute right to use water within their borders.

In the 1970s, the main pollutants detected in the reservoir water were phenol, cyanide, arsenic, mercury, chromium and DDT. From the mid-1980s, with the growth of cities and industry (especially chemicals and distilleries) following economic liberalisation, indexes for ammonium, COD, BOD$_5$, coliform bacteria, total nitrogen and total potassium were exceeded, and eutrophication became a regular feature of the reservoir (Table 11.1). By the 1990s, 80–90 million tons of wastewater per annum from the upper reaches constituted about one-third of the inflow into the reservoir (Yang et al. 2006:43).

A commonly cited event (e.g. Peisert and Sternfeld, 2005:34; Yang et al., 2006:43; Li and Li 2008:47) signalling the level of degradation of the water in the Guanting is that in 1997, Beijing ceased using it for urban drinking water because it had fallen below the lowest quality category, Grade V. The practical significance of this act is unclear, however, since very little of the Guanting's limited supply appears to have gone to this use, and supply to industry continued without significant change after 1997 (Fig. 11.1) (Beijing Shi Yongding He Guanlichu 2002:150). Nonetheless, following an improvement in some water quality measures (especially mercury, hexavalent chromium, iron and nitrogen), despite a low average storage of only 170 MCM, some flow was supplied to a drinking water treatment plant in 2005 on a trial basis (Li 2007).

11.6.2 Miyun

With the dwindling and befouled Guanting Reservoir ruled out for municipal water supply, the Miyun has become the primary, and for a time nearly exclusive, source of Beijing's surface water supply. From 1989, it began providing 500,000 m^3 per day to the Beijing No. 9 Water Works, and twice that after 1995. Its water was much cleaner than that of the Guanting, because its catchment is less populated, with more limited agricultural land and industry, and not as erosion-prone due to rockier land and greater forest cover (Beijing Shi Shuiliju 1999:209). At the same time, the growth of economic activities, declines in inflow and the demise of the Guanting as a source of water supply led to increasing concern about the reservoir's water quality.

In 1985, the municipal government issued a pioneering trial water protection regulation (*banfa*) to protect the Miyun, following it ten years later with a more detailed and more formal regulation (*tiaoli*) dividing the catchment area within Beijing into three water protection zones. The effect and limitations of this regulation are summed up by Peisert and Sternfeld (2005:36).

The regulation bans certain economic activities, particularly inside water protection zone 1 (which encompasses the reservoir, the area inside the lakeside road and all areas 4 km off the shoreline). This regulation has helped the authorities control tourism and industrial development in the protection zone 1, as well as enforce the closure of some mines and smaller enterprises in protection zones 2 and 3. In recent years, the Miyun protection regulation has sparked programmes to limit fishery and agricultural impact on the reservoir; however, it does not apply to 70% of the 15,788 km^2 catchment area in Hebei Province, which includes about two-thirds of the 860,000 people living in the Miyun watershed.

Needless to say, this regulation and its enforcement have sparked a number of serious conflicts. These are both between downstream Beijing and upstream Hebei and within Beijing municipality itself, between the relatively wealthy urban beneficiaries of the water supply and the economically challenged residents of Miyun County.

Hebei argues that it supplies the water, yet is poorer and uses less per capita than Beijing, so it should be compensated for the water to develop watershed protection. Beijing argues that the concentrated inflow during the rainy summer months rinses into the reservoir the often toxic wastewater (notably containing ferrum and phenol) from industries such as iron mines and a brewery, while disturbed soils from open-pit mining and deforestation to provide fuel to poor households also end up in Miyun. In addition, over 80% of pollutants in the reservoir (in volume) are phosphates and nitrates flowing from settlements and food processing plants in Hebei.

Hence, according to Beijing, Hebei should be fortunate not to be charged for water pollution (Peisert and Sternfeld 2005:37).

Provincial autonomy over water has always been strong, with few effective mechanisms available to mediate disputes. Recent strengthening of basin management organisations has been offset by increasing decentralisation, and in the case of Miyun, one cannot be entirely sanguine about the possibility of resolution of inter-provincial conflicts. One attempt at a resolution was the approval in 2001 by the State Council of a five-year 'sustainable water use plan' for the capital that sought to address a variety of inter-provincial water allocation and watershed problems (Beijing Shi Renmin Zhengfu 2001). This will be discussed later, in the section on the twenty-first century.

It is also not entirely clear that conflicts of interest *within* Beijing Municipality are any less intractable. Peisert and Sternfeld (2005:39–42) elaborate on a number of unintended consequences of regulatory enforcement, including the following:

1. Limiting land use activities for near-shore farmers has induced them to take up more harmful activities such as mining and intensive fish farming.
2. Closures of illegal small iron mines in 2,000 led to protests by miners demanding compensation and to the abandonment of pits without remediation.
3. Restriction and eventual prohibition of waste-generating fish ponds farming within the reservoir was replaced by waste-generating fish ponds along the shores.
4. A programme to convert grain lands into fruit orchards was met with scepticism by the farmers due to concern about marketability and other factors (also noted by Enders 2005:11), while the resultant reduction in agricultural water demand was probably offset by an increase in the use of chemicals.
5. In addition to the above, it appears that the ban on tourist facilities within zone 1 is sometimes honoured in the breach. A three-star 'holiday resort' with 325 rooms remains located near the shore inside Miyun very close to the reservoir outlet. While it claims to meet ISO 14000 environmental standards, the fact that it was established in 1988 by the Central Metallurgical Ministry and is still operated by central government corporations (http://www.yunhu.com.cn) may have played a role in its continuance.
6. A key approach to watershed control is to develop 'small watershed management areas', many of them in Miyun County supported by a Sino German cooperation project. Yet these areas do not cover the entire watershed, and so far have mixed results. For example, in her survey of a model area, Enders found an often low level of awareness by residents of the effects of their behaviour on the lake (2005:11).
7. Enders also found that none of the residents of the village accepted an offer of cash by the state if they would move to the city, yet almost all families have a member working outside the village (2005:9–10).

These factors are not as intractable as the inter-provincial divide, and many of them will probably be addressed through personal and institutional learning and the growing availability of distant employment opportunities pulling the economic activity of residents further offshore. The double-edged sword of tourism, developed and used by outsiders, may prove more difficult to control in the long run.

11.7 The Underground Movement

As noted, most of Beijing's water has always come from groundwater aquifers. The primary original purposes of the Guanting and Miyun Reservoirs were flood control and hydropower, with the additional water supply they provided considered a welcome but secondary benefit. The prospect of Guanting's storage stimulated the development of large, water-intensive industries in the western part of the capital. This was justified in terms of turning Beijing from an administrative centre into an economically productive city, a leading producer of steel, chemicals and textiles as well as government decrees (Nickum 1994:40). Similarly, Miyun's water was seen as a means of bringing irrigation and greater grain self-sufficiency to both Beijing and downstream areas. Both Guanting and Miyun initially provided water to Tianjin and Hebei as well as Beijing (Li and Li 2008:47).

The surface water from the Yongding was not sufficient for urban development in Beijing's west, however. By 1990 over 200 km^2 of the relatively shallow aquifer there was nearly pumped dry in areas, while a cone of depression 40 m deep appeared in the eastern plains (Beijing Shi Shuiliju 1999:14).

Then came the dry years. Beijing's average annual precipitation, over 600 mm, is actually higher than that of London or Berlin, and close to China's national average, but the evapotranspiration rate in Beijing is much higher, and the rainfall is concentrated in the summer months and is highly variable from year to year. In particular, two extended strings of dry years, in 1980–1984 and since 1997 (with an average 446 mm in precipitation between 1997 and 2004),[2] forced major reassessments of the sources and uses of Beijing's water.

The water of the Miyun Reservoir initially did not go to Beijing's urban area, but to downstream areas to the east and south, providing surface irrigation to the eastern counties of Beijing, Langfang County of Hebei wedged between Beijing and Tianjin, and to urban uses in Tianjin. In 1970, Tianjin drew off nearly half of the reservoir's water. In the 1970s, there was a significant shift in allocation towards Beijing, with a small reversal towards the end as the plains areas of Beijing grew to rely more on groundwater (Table 11.2).

For example, the irrigated area of Tong County in Beijing's southeast increased from 3,580 ha in 1957 to 47,600 ha in 1971, largely due to the availability of water from the Miyun. After 1972, the total irrigated area continued to increase gradually, but with a significant shift towards reliance on groundwater. By 1978, fewer than 12,000 ha of irrigated land out of 45,000 relied on surface water (Sternfeld 1999:150). The growing reliance on groundwater in agriculture may be explained at least in

---

[2]Beijing Shi 2005: 25. Different sources provide slightly varying dates for the duration of the drought, beginning either in 1997 or 1999 and ending either in 2006 or not yet ended.

Table 11.2 Allocation of Miyun Reservoir water (in MCM)

	1970	1975	1978	1979
Beijing	297	884	686	530
Hebei	358	88	48	116
Tianjin	596	48	93	260

Source: Sternfeld (1999:131)

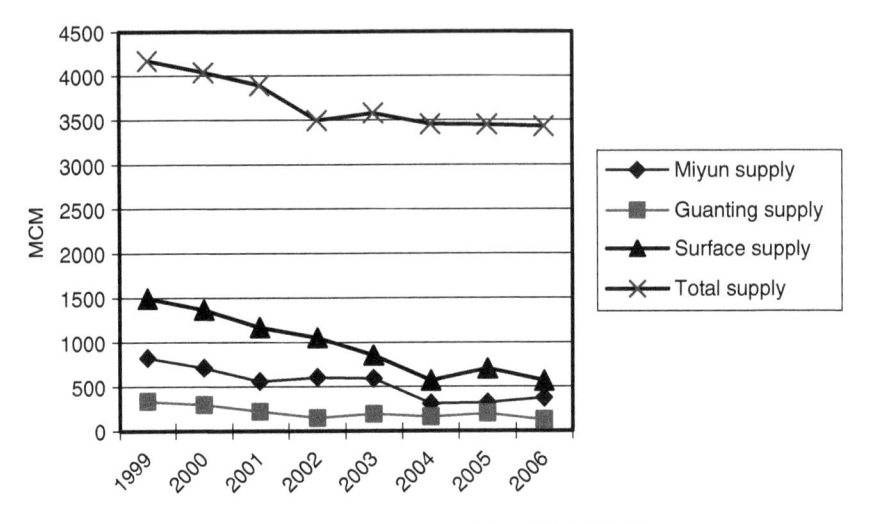

Fig. 11.2 Beijing water supply, 1999–2006 (*Source*: Li and Li, 2008:48)

part by the spread of pump-operated tube-wells throughout the plains of north China in the 1970s and the completion of the Jingmi diversion canal that made it possible for water to be transferred to the city. In addition, the State Council made a strategic policy in 1985 to shift the allocation of the Miyun water to the urban area in response to the pressures on groundwater.

In 1997, a drought year, Beijing's groundwater recharge was 1.6 km^3 (billion cubic metres, or BCM), considerably below the long-term average 2.5 km^3. The amount abstracted that year was 2.585 km^3, or 64% of the municipality's total water withdrawals. Half of this (1.3 km^3) was used by agriculture, 21% (542 MCM) by industry and 20% (517 MCM) for urban municipal purposes, mostly drinking water. The latter was provided after treatment at one of the large water plants (then eight, now ten) (Shuilibu and Nanjing 2004:74–77). In the subsequent dry years, groundwater abstraction rates remained relatively stable, indicating serious overdrafts, while surface supply fell from about one-third of the total to under one-fifth. The water provided by the two reservoirs did not decline quite as rapidly as the dwindling residual from other sources, so their share in the total surface supply climbed, reaching 92% in 2003 (Fig. 11.2). With the decline of the Guanting, the Miyun became the primary surface source for the city, while irrigation was supplied exclusively by groundwater (Beijing Shi Shuiwuju 2005:25).

11.8 A New Century and an Olympic City

The dominance of the Miyun was tenuous, due to a lack of resolution of the watershed issues noted previously, human-induced declines in the quantity and quality of water entering the reservoir, and a dry spell that was unusual even in this dry region and may be a symptom of global warming. During the first 6 years of the twenty-first century, in the midst of the most prolonged drought in memory, Beijing's population increased by 14%, from 13.8 million to 15.8 million while its total water supply fell by a quarter, from 4.05 billion m^3 to 3.43 billion m^3 (Jiao 2007:31). The supply from Guanting and Miyun, and surface sources as a whole, declined even more rapidly (Fig. 11.2).

Even before the drought and the awarding of the 2008 summer Olympic Games to Beijing in July 2001 added to the potential stresses on the city's water supply, concerns were growing about the municipality's pollution and shortage of water resources and about the consequent ecological disruption. These concerns helped to take plans for the 1267-km long Middle Route of the South-North Water Transfer Project, studied and stalled for decades (e.g. Biswas et al. 1983) off the shelf in the late 1990s. Construction on the gravity diversion from the Danjiangkou Reservoir on the Han River, a tributary of the Chang Jiang (Yangtze), began in 2002 and was expected to begin delivering up to 1.0 BCM/ann to Beijing from 2010 (since postponed to 2014), rising to 1.4 BCM in 2020 (Li and Li 2008:48). These figures constitute just over 10% of the planned headworks diversion (9.5 and 13 billion m^3/ann respectively) (Zhu 2006).

With the awarding of the 2008 Olympics, due to take place two years before the expected arrival of southern water, stresses on the capital's water sources intensified. It was estimated that the increase in demand for water by the influx of population for the 15 days of the games alone would be 275,000 m^3, while water events, river purification and greenery would add another 2 MCM in demand (Lu et al. 2002:75). The Olympics helped catalyse a new set of policies and approaches, mostly aimed at securing additional sources of surface water and protecting the watersheds of Guanting and Miyun.

As noted previously, in May 2001, the State Council officially approved a 5-year sustainable water use plan for the capital that had been drawn up over the previous two years by the Beijing Municipal government in collaboration with the Ministry of Water Resources (Beijing Shi Renmin Zhengfu 2001). One of the most significant features of this seminal document was its recognition of the use of economic measures to reduce conflicts between upstream and downstream, and its allocation of responsibilities for compensation, with the central government picking up much of the tab for projects outside Beijing's boundaries.

An example of the upstream watershed activities supported and mandated by this plan is Fengning County in the portion of Hebei north of Beijing. This county, which encompasses 26.5% of the watershed of Miyun's Chaobai River and provides it with 56.7% of its water resource, is one of the poorest counties in China with a large pastoral population. Its frost-free period of 100 days, together with a seasonally concentrated and highly variable precipitation, makes it difficult for tree and grass cover to grow quickly. It is caught in a vicious cycle of poverty and environmental degradation, with few non-destructive options for overcoming poverty.

Since the beginning of this century, Fengning County has sought to regenerate land cover, using the small watershed control approach as in Miyun, but also limiting the allowed number of livestock, banning free range grazing and providing fuel alternatives to the wood scavenged by poor farmers. The county has also promoted 'water-saving agriculture' measures, including piped, drip and sprinkler irrigation, the use of plastic ground cover to reduce evaporation, and conversion of rice fields to upland crops. Highly polluting industries were closed.

At the same time, promised funds have not been forthcoming. The plan called for a total investment of 430 million yuan in 17 water control projects to manage erosion on over 1,400 km^2 over the 2001–2005 period. In actuality, 70 million yuan was committed to four projects on over 200 km^2. Small watershed control areas have similarly been inhibited by lack of resources, aggravated by the relaxation of the corvee system (Qu et al. 2006).

From 2003, Beijing established an expensive (22 billion yuan, nearly US$ 3 billion) contingency supply arrangement with Hebei and Shanxi Provinces. Under this arrangement, a number of reservoirs upstream of the Guanting Reservoir are to release up to 240 MCM of supplemental water per year to cover shortages until the completion of the trans-basin diversion. The first instalment, in September 2003, was a 50 MCM release for ten days from the Cetian Reservoir in Shanxi, 157 km upstream from Guanting. Half of this water was lost along the way (Peisert and Sternfeld 2005:42). In October 2004, 93.9 MCM was transferred from three upstream reservoirs (50.6 MCM reaching Guanting). In 2005, this increased to 117 MCM from six reservoirs (67 MCM reaching Guanting) and in 2006, it fell to a more modest 47 MCM from seven reservoirs (11.7 MCM reaching Guanting) (Ministry of Water Resources 2003–2006).

Two problems surfaced in practice with these transfers. One was that the quality of the diverted water was far from ideal, only marginally improving Guanting's water (Yang et al. 2006). Thus, they could play a limited role in diluting concentrations, providing quantity more than quality. It was still necessary to invest in end-point adaptive water treatment measures. Below the reservoir outlet, this tends to involve a water treatment plant. At the other end, artificial wetlands, developed in cooperation with Brandenburg state in Germany, were created at the inlets to the reservoir to improve the quality of inflowing water to the still degraded levels of Grades III to V. Initial results on the Heituwa Artificial Wetland after its first year of operation (2004–2005) indicated good results in reducing BOD, COD and phosphates, but mediocre effects in the much harder task of reducing nitrogen (Li and Yuan 2006).

The other problem had to do with establishing clear inter-provincial water rights to establish a basis for compensated transfer. These were provided for in principle in the twenty-first century plan but not specified. Finally, on the last day of 2007, the State Council issued an allocation of the water of the Yongding River, establishing cross-border release quotas on Shanxi and Hebei Provinces, to be enforced by the Hai River Basin Commission (Guowuyuan 2007).

11.8.1 Inter-basin Diversion

The intra-basin transfers from the upstream impoundments on the Yongding River have somewhat revitalised the Guanting Reservoir, but are seen as emergency measures (Nickum and Lee 2006) and insufficient. With the continuing drought and the pressure of the Olympics, a number of temporary local inter-basin transfers were negotiated under force majeure from elsewhere in Hebei. These were hotly contested, according to press reports (e.g. Buckley 2008).

With the arrival of a major third surface source from the south, the roles of Guanting and Miyun are once again expected to shift (Li and Li 2008:48).

On the one hand, the South-North Water Transfer promises to guarantee the safety and water supply and greatly reduce the pressure on the water source systems of the Guanting and Miyun. It will play a critical role in increasing water storage in both reservoirs, further improving the quality of their water bodies, while restoring the ecologies and environmental conditions of the reservoir districts. At the same time, the water transfer will reduce the water supply function of the two major reservoirs, so that they can assume a greater role in water supply to the ecology (including environmental water use in rivers and lakes and replenishment of groundwater) and in water resource storage. Therefore, the main functions of the Guanting will shift from 'flood control, water supply and hydropower' to 'flood control, water resource reserve and ecological water supply', with hydropower declining in tandem with water supply for municipal and industrial uses. The Miyun will add water resource reserve to its principal functions. At the moment, however, due to its cost, water from the long-delayed SNWT may not be used for all these purposes over the long-run.

11.9 Conclusion

Over time, and sometimes not all that long a period, multiple purpose dams serve different mixes of purposes in a historical, political, cultural, developmental, environmental and economic context that is in constant flux. Unforeseen opportunities and unanticipated consequences can often be the norm. This is especially the situation in Beijing, where the two major reservoirs have played shifting roles as alternative sources for various purposes come and go, the metropolis expands in population, extent and especially economic level, relatively poorer watersheds upstream develop and become the objects of control, negotiations across administrative boundaries with neighbouring provinces and within the municipality itself become ever more imperative yet contentious, diversion projects become increasingly problematic, and climate change rears an ever uglier head.

References

Beijing Shi Renmin Zhengfu, Zhonghua Renmin Gongheguo Shuilibu (The People's Government of Beijing and the Ministry of Water Resources of the People's Republic of China) (2001) 21 Shiji Chuchi (2001–2005 nian) Shoudu Shui Ziyuan Kechixu Liyong Guihua Zong Baogao

(Summary report of the sustainable water resources use plan for the capital in the initial period (2001–2005) of the 21st century). Water and Power Press, Beijing, China, in Chinese

Beijing Shi Shuiliju (Beijing Municipal Water Bureau) (1999) Beijing Shuihan Caigai (Flood and drought disasters in Beijing). China Water and Power Press, Beijing, in Chinese

Beijing Shi Shuiwuju (Beijing Municipal Water Service Bureau) (2005) Jianshe Xunhuan Shuiwu tuijin Beijing Jiaoqu Shuili Xiandaihua (Establish circulatory water service to promote the modernisation of water resources in Beijing's suburban areas). Zhongguo Shuili (China Water Resources), 15 (August 12):25–27, in Chinese

Beijing Shi Yongding He Guanlichu (Beijing Municipality, Yongding River Management Office) (2002) Yongding He Shuihan Caigai (Flood and drought disasters on the Yongding River). China Water and Power Press, Beijing, in Chinese

Biswas A, Zuo D, Nickum J, Liu CM (eds) (1983) Long distance water transfer in China. Tycooly International, Dublin

Buckley C (2008) Beijing Olympic water scheme drains parched farmers. http://english.h2o-china.com/k/2008-1/20081231726249031.shtml. Accessed 3 Feb 2008

Enders S (2005) Miyun: integrated watershed management for the protection of Beijing's drinking water. http://forestry.msu.edu/China/New%20Folder/S.Enders-Miyun.pdf. Accessed 15 Aug 2008

Guowuyuan (State Council) (2007) Guowuyuan guanyu Yongding He ganliu shuiliang fenpei fang'an di pifu (Official reply of the State Council on the Water Quantity Allocation Plan for the Mainstem of the Yongding River) (in Chinese) 31 December (SC Letter No. 135, 2007). http://202.123.110.3/zwgk/2008-01/08/content_852552.htm. Accessed 6 Apr 2008

Jiao Z (2007) Ying dui shui ziyuan jinque jianshe Beijing xunhuan shuiwu (We should establish a Cycle-oriented water affairs [system] for Beijing to address the urgent shortage of water resources) Zhongguo Shuili (China Water Resources) 8 (April 30):30–31, in Chinese

Li H (2005) Lishishang senlin bianqian dui Yongding He di yingxiang (The impact of historical changes in the forest on the Yongding River). Zhongguo Shuili (China Water Resources) 18 (September 30):56–58, in Chinese

Li Y (2007) Guanting Shuiku shuizhi qushi fenxi ji duice yanjiu (Trend analysis of water quality of Guanting Reservoir and [proposal of] measures). Zhongguo Shuili (China Water Resources) 3 (February 12):56–57, in Chinese

Li G, Li Y (2008) Nanshui Beidiao dui Beijing dibiao shui gongshui tixidi yingxiang ji duice (Impacts of south to north water transfer on the surface water supply system of Beijing and measures for addressing them). Zhongguo Shuili (China Water Resources) 5 (March 12): 47–49, in Chinese

Li Y, Yuan B (2006) Guanting Shuiku rengong shidi xitong shuizhi jinghua xiaoguo yanjiu (A study of the effectiveness of water purification by an artificial wetland of guanting reservoir). Zhongguo Shuili (China Water Resources) 21 (November 12):53–54, 20, in Chinese

Lu Z, Sun J, Lu L (eds) (2002) Beijing Hanqing yu Duice (Beijing's drought conditions and measures to address them). Qixiang Press, Beijing, in Chinese

Ministry of Water Resources (2003–2006) Quanguo shuiqing nianbao (Annual Report of Water Conditions in China). Posted on the Ministry of Water Resources website: http://www.mwr.gov.cn/xygb/sqnb/20050711000000354824.aspx; http://www.mwr.gov.cn/xygb/sqnb/20070814163332165742.aspx; http://www.mwr.gov.cn/xygb/sqnb/20070814416375373c8ef.aspx; http://www.mwr.gov.cn/xygb/sqnb/20070814164109e87791.aspx. Accessed 14 August 2007

Miyun Xian Shui Ziyuanju (Miyun County Water Resources Bureau) (2003) Miyun Shuihan Caigai (Flood and drought disasters in Miyun). China Water and Power Press, Beijing, in Chinese

Nickum J (1978) Labour accumulation in rural China and its role since the Cultural Revolution. Cambridge J Econ 2(5):273–286

Nickum J (1994) Beijing's maturing socialist water economy. In: Nickum JE, Easter KW (eds) Metropolitan water use conflicts in Asia and the Pacific. Westview, Boulder, pp 37–60

Nickum J, Lee YF (2006) Same longitude, different latitudes: institutional change in urban water in China, North and South. Environ Politics 15(2):231–247

Peisert C, Sternfeld E (2005) Quenching Beijing's thirst: the need for integrated management for the endangered Miyun Reservoir. China Environment Series Issue 7:33–46

Qu Z, Liu H, Li Z, Ji Z (2006) Jingjin shuiyuandi shengtai yu shui ziyuan buchang wenti (The problem of ecological and water resources compensation in a water source area of Beijing and Tianjin). Zhongguo Shuili (China Water Resources) 22 (November 30):39–41, 56, in Chinese

Shapiro J (2001) Mao's war against nature. Cambridge University Press, Cambridge

Shuilibu Shui Ziyuan Si (Water Resources Office, Ministry of Water Resources) and Nanjing Shuili Kexue Yanjiuyuan (Nanjing Hydraulic Research Institute) (2004) 21 Shiji Chuji Zhongguo Dixiashui Ziyuan Kaifa Liyong (The development and use of groundwater resources in China in the early 21st century). China Water and Power Press, Beijing

Sternfeld E (1999) Beijing: Stadtentwicklung und Wasserwirtschaft (Beijing: urban development and water economy). Technische Universitaet Berlin, Berlin

Yang D, Xie J, Zhang Y, Liu W (2006) 2005 nian shushui dui Guanting Shuiku shuizhidi yingxiang, fenxi ji duice (An analysis and countermeasures of the impact on water quality of the 2005 Water Diversion to the Guanting Reservoir). Zhongguo Shuili (China Water Resources) 19 (October 12):43–46, in Chinese

Yi M (2007) Guanting shuiku shuixi shui huanjing wuran fenxi ji duice lunwen (A paper on the analysis of environmental pollution in the water systems of Guanting Reservoir and countermeasures). Posted on www.lunwentianxia.com/product.free.3842271.1/ on 25 November. Accessed 13 Aug 2008

Zhu R (2006) China's south-north water transfer project and its impacts on economic and social development. Posted on www.mwr.gov.cn/english1/20060110/20060110104100XDENTE.pdf. Accessed 4 May 2008

Chapter 12
Resettlement due to Sardar Sarovar Dam, India

C.D. Thatte

12.1 Introduction

Water resources have to be developed to overcome their spatial and temporal variability and availability, particularly for countries in the tropics. To this end, India, like other developing countries, launched a massive programme after independence to build storages on its large rivers and soon attained the required degree of development from its potential for development. However, resistance to the building of new dams emerged over time due to the perceived neglect of adequate resettlement and rehabilitation for the displaced.

The Narmada Project became the focal point for not only those who wanted proper resettlement and rehabilitation but also those who opposed dams worldwide. Nearly three decades after construction of this dam was begun, a silver lining has become visible in the clouded course of the project during the last few years. The project is now close to completion.

This chapter reviews in totality the case of resettlement and rehabilitation for the Narmada Dam, spanning almost four decades, and provides a rare insight which could be helpful for different stakeholders in drawing important lessons for the future, not only for the water sector but also for other infrastructure development programmes.

12.2 Water Resources Development and Displacement

As the Second World War drew to a close in the mid-1940s, many countries under colonial rule were waging a struggle for self-rule. Many of these nations were populous and poverty stricken, as a result of exploitation and long-term neglect. A small

---

C.D. Thatte
International Commission on Irrigation and Drainage (ICID), Pune, India

minority in these countries was well to do, but the large majority had remained economically weak, saddled with malnutrition and lack of employment. The gap between the haves and have-nots was significant, but populations were growing at a rate higher than the growth rate of their economies. Natural economic resources had been neglected, two world wars had caused further degradation and hardly any developmental efforts were being made. Disparities between the rich and the poor were widening and there was great deprivation. At the same time, the availability of natural resources per capita was dwindling.

The newly independent countries therefore launched massive programmes for economic revival, growth and development on a fast track, including water resources development, during the second half of the twentieth century. Due to high population density in these countries, every development effort required acquisition of land and involved the displacement of people, or project-affected persons or families, from their habitats. Despite the fact that legal provisions for judicious resettlement were non-existent, the affected people easily accepted relocation in their eagerness for development. They believed in the welfare proclamations of the new governments and joined the development effort with great enthusiasm. They knew that for each person facing adversity due to displacement, hundreds would benefit. It was clear to them that on balance, positive impacts would outweigh negative ones. It had been their experience that unless development efforts were undertaken, the adverse conditions of life would not diminish but rather grow at an exponential rate. The people believed that the positive impacts of water resource development would be crucial, especially when compared with degradation due to no development at all. The decades after independence therefore saw unprecedented activity on the water resource development front.

The fact is that development alone does not displace people. The absence of development displaces perhaps far larger numbers due to chronic droughts and floods, rural agriculture below subsistence levels, landlessness and the lack of employment and better opportunities. All over the South Asian subcontinent, including India, better-off regions host nearly millions of migrants from relatively poorer regions, for precisely these reasons. The Narmada River Basin—the subject of this chapter—was no exception. Distress migration that took place in the absence of infrastructure projects such as the Sardar Sarovar Project (so named after Sardar Patel, a national leader) probably exceeded the numbers of project-affected persons now seen. But those displaced by dams can be better resettled with a well thought out and sensitively implemented package for resettlement and rehabilitation. The process can also be evaluated and monitored by independent watchdog agencies supported by governments, law courts, river basin authorities such as the Narmada Control Authority (NCA), grievances redressal authorities, other participatory democratic public–private institutions and non-governmental organisations (NGOs) from civil society to ensure that project-affected families get their dues, and that injustice and lapses are avoided and expeditiously corrected where they occur.

Since indigenous productivity and production were low in the years after independence, these countries were dependent on imports from the western world of a large proportion of their food, energy and other essential requirements. Development was therefore focused on irrigated agriculture and multipurpose water resource development to achieve self-sufficiency. Fortunately, these countries had abundant untapped water resources flowing down their large river systems fed by dependable monsoons. They also had good engineering cadres trained in water resource development. Institutions were soon set up to assess water resources in these river basins. Several programmes to carry out water resources development already identified by popular demand were undertaken. Water resource development of the Narmada River Basin (between 72.32° to 81.45° longitude and 21.20° to 23.45° latitude) was one such programme. In view of the great spatial and temporal variability of precipitation, capturing river flow in reservoirs behind dams had become a major component of India's water resources development. For instance, India had built several dams, diversion structures and canal systems in past centuries. It was well known that taller dams imply greater submergence of agricultural lands and habitats, and that farmers stood to lose farmland that they tilled to eke out a livelihood. The agricultural community was aware that their houses and lands would be needed for the water resources development being planned. In the colonial period, unfortunately, legal provisions related only to acquisition of land for public purposes (Ministry of Law and Justice 1894), with financial compensation at market rates. There was no mention of relocation, resettlement or rehabilitation. These concepts gradually evolved as water resources development took shape on an unprecedented scale.

There is no doubt that displacement of population has several impacts: resource base for production systems is dismantled; productive assets and income sources are lost; people are relocated to new environments where their productive skills may be less in demand and competition for resources greater; community structures and social networks are weakened; means of livelihood are disrupted; kin groups are dispersed; pauperisation and psychological stress can occur and cultural identity, traditional authority, leadership and capacity for community support are diminished. As such, a resettlement and rehabilitation policy for the displaced is needed from the early stages of project planning. Guidelines based on a holistic approach are required to evolve an appropriate policy. For instance, a 'Risk Impoverishment Model' categorises risk: landlessness, homelessness, joblessness, marginalisation, food insecurity, increased morbidity, loss of access to crops and common property resources and social disarticulation. A resettlement and rehabilitation plan must address these risks according to priorities that account for local situations.

India's experience with water resources development had indicated that different sizes of dams are required in a river basin: Taller dams can be built economically in hilly regions, while shorter dams are economical in plains where river depth is less. Depending on the geomorphology of the river, submergence will include hills as well as plains. People inhabiting hilly terrain tilled undulating land in smallholdings. In the plains, landholdings were large. Villages in the hills were small while those in the plains were large. But in both regions, given unabated population growth, landholdings were being fragmented. Many of the hill people are descendants of hunters and gatherers and are still dependent on forest produce. The lifestyles of people in the two areas are considerably different. Tribal groups residing in hills, for instance, have a distinct way of life. As one descends from the hills, life is more centralised and urbanised. All people who need to be displaced from the dam submergence area can either be relocated in the reservoir rim area or moved downstream to the plains. In case of the former, relocation is easier because the displaced people and the local community have a similar culture. In the latter case, the host community is usually distinctly different from the guests. The integration of the two needs to be carefully achieved. The lives of the socio-economically weak amongst the displaced, including tribal peoples, have to be brought closer to the mainstream and on par with their brothers in the plains, through the economic growth engendered by water resources development.

Several international organisations have been studying tribal peoples around the world. For instance, it is asserted that in addition to displacement, several aspects of the life and culture of tribal peoples are impacted by dams: livelihood, beliefs, myths, rituals, festivals, songs, dances, hills, woods and streams. It has been found that tribal folk often become apathetic and become disoriented in harsh new surroundings. Some have drawn the attention of the World Bank to Convention 107 of the International Labour Organisation relating to tribal groups and indigenous populations, which has already been ratified by India, and asked for those terms to be incorporated into resettlement and rehabilitation policies. The Indian Constitution had, however, included all such provisions under article 244 for tribal groups. According to these provisions, a compliance report is to be submitted to the President of India by state governors every year. Notwithstanding these provisions, a great deal remains to be done. Unfortunately, the majority of tribal people cultivate landholdings that are rain fed, undulating, small and relatively unproductive.

It is of course necessary to aim at justice and equity for tribal peoples and other weak sections of society; eminent thinkers and social leaders in India raise the question of whether the country will move towards social emancipation, equity and development, or continue to grapple with poverty and superstitions which tend to keep such communities in a sort of 'anthropological museum'. Tribal groups do not take to agriculture easily, since some do not believe in ploughing the land and follow a four-year rotational cultivation cycle, by and large degrading the soils through shifting cultivation. In addition to deforestation, these practices reduce productivity of the land they till, in contrast to the popular romantic notion of tribal groups conserving nature, forests and ecology. These groups struggle to survive; their survival must be the focus of any development effort and it is necessary to bring them into the mainstream.

12.3 Water Resources Development in the Narmada River Basin and the Sardar Sarovar Project

The River Narmada rises at an elevation of over 1000 m in central India and travels about 1312 km towards the west coast to meet the Arabian Sea. It flows for 1077 km through the state of Madhya Pradesh, between Madhya Pradesh and Maharashtra state for 35 km, between Maharashtra and Gujarat for 39 km and finally in Gujarat for about 161 km in the deltaic river plains. It drains about 98,800 km^2 of area with an average annual rainfall varying from 1620 mm in the east to 820 mm in the west. The spread of the catchment area is relatively narrow. Planning for water resources development in the Narmada Basin was begun in 1950 through three major projects mostly involving the downstream stretches. The downstream most amongst these was initially named after the city of Bharuch at the head of the Narmada delta, and subsequently renamed the Sardar Sarovar Project after Sardar Patel. The preliminary proposal was revised in 1957–1959 to build the reservoir in two stages, going to Full Reservoir Level (FRL) 49 m from riverbed level of about 15 m in the first stage and to FRL 97.6 m in the second. What is now the state of Gujarat was at the time a part of the erstwhile Bombay state. After the bifurcation of Bombay state in 1960, the first stage of construction was approved in February 1961. Jawaharlal Nehru, the first Prime Minister of independent India, laid the foundation stone of the dam.

The planning for Narmada water resources development was reviewed afresh by the new states of Gujarat, and Maharashtra and Madhya Pradesh in the upstream. Alternative proposals were made for Gujarat to have a taller terminal dam, and the upstream states proposed alternatives within their territories, in addition to a series of upstream dams. In the absence of consensus, an Expert Committee chaired by the renowned technocrat A.N. Khosla considered various proposals and in 1964 recommended in its Report a high dam with FRL 152.4 m in place of the earlier proposal[1]. The Committee recommended irrigation of over 1.8 Mha through a right bank canal in the northern region of Gujarat, also to cover parts of the state's western peninsulas of Saurashtra and Kutch. Extension of the canal into the state of Rajasthan was also recommended. Again, there was lack of unanimity between the states on these recommendations.

Gujarat filed a complaint with the Government of India under India's Inter-State Water Dispute Act, 1956 (Ministry of Water Resources 1956), and requested the appointment of a tribunal for adjudication. The Narmada Water Dispute Tribunal was accordingly set up. After prolonged deliberations, the tribunal issued its award in August 1978 further modified in December 1979. The Narmada Water Dispute Tribunal Award (Narmada Water Disputes Tribunal 1979) mostly reiterated Khosla's recommendations and favoured a high dam at the same site, irrigating over 2.1 Mha and generating hydropower through an installed capacity of about 1200 MW in the riverbed powerhouse and 300 MW in the canal head powerhouse. The award specified allocation of the unanimously agreed 75% dependable available yield of about 35.8 BCM (billion cubic metres) for irrigation between the three riparian states and to the fourth water-deficit state of Rajasthan to the north of Gujarat. The hydropower to be generated was allotted only to the three riparian states (Table 12.1).

A detailed project report for the Sardar Sarovar Project was then prepared by Gujarat state. The NCA was established to implement the tribunal's decisions.

---

[1] See the relevant dates prior to the constitution of the tribunal, available at http://nca.gov.in/imp_date.htm

Table 12.1 Narmada Water Dispute Tribunal Award allocation of available water and hydropower between party states

State	Water (million acre-feet)	Water (BCM)	Hydropower (%)
Madhya Pradesh	18.25	23.7	57
Maharashtra	0.25	11.7	16
Gujarat	9.00	0.3	27
Rajasthan	0.50	0.1	Nil

Source: NWDT, December 1979

The authority set up separate subgroups to look into rehabilitation and resettlement, environment and other aspects of the award (Narmada Waters Dispute Tribunal 1979). A Sardar Sarovar Construction Advisory Committee (SSCAC) was formed to advise the Government of Gujarat on its construction. The anticipated benefits of the Sardar Sarovar Project cover almost 75% of the population of Gujarat. Existing, ongoing and planned schemes for water resources development in Gujarat can irrigate only 2.55 Mha, whereas the Sardar Sarovar Project alone can irrigate 1.8 Mha (75% of which is drought prone) in Gujarat, in addition to about 0.2 Mha in the totally drought- and desert-prone state of Rajasthan. On completion, it will provide over a million jobs. Construction of preliminary works for the dam was begun in 1986. The salient features of the Sardar Sarovar Dam are as follows: lowest foundation Elevation (El) 5 m, riverbed El 15 m, Minimum Draw Down Level (MDDL) 110.64 m, dead storage 2.97 MAF (million acre-feet), spillway crest El 121.9 m, FRL 138.6 m with storage of 7.7 MAF, full reservoir length 200 km, Maximum Water Level (MWL) 140.21 m, dam top El 146.50 m.[2]

Land use in the reservoir submergence area, the villages submerged and the numbers of project-affected families in the year 2001 are indicated in Table 12.2. During the last 26 years since the tribunal award, land use patterns have changed. The demographic features of the submergence area have also undergone significant change, due to population growth. In some locations, there was in-migration as people moved to the area because employment opportunities grew, and also to avail compensation, as resettlement and rehabilitation packages evolved.

There are varying figures for project-affected families—the Narmada Water Dispute Tribunal indicated 6147 (with project-affected persons numbering 39,700), the World Bank Morse Commission puts the figure at 12,000 (project-affected persons 60,000) (Morse and Berger 1992) and the states currently offer the figure of 41,000 (project-affected persons 205,000) (Narmada Control Authority 2001). These numbers will continue to rise with population growth, till completion of the project. It is instructive to examine the proportion of submergence area to irrigation area, which is 1.65 ha to 100 ha (about 1:60). For the Hirakud Reservoir, one of the country's larg-

---

[2]See Narmada Control Authority, available at http://nca.gov.in and Sardar Sarovar Narmada Nigam, Ltd, available at http://www.sardarsarovardam.org/

Table 12.2 Land, villages, project-affected families in Sardar Sarovar Project submergence (October 2001)

State	Cultivable land (ha)	Forest land (ha)	Other uses (ha)	Total land (ha)	No. of villages	Project-affected families
Madhya Pradesh	7,883	2,731	10,208	20,822	192	33,014
Maharashtra	1,518	6,288	1,592	9,398	33	3,221
Gujarat	1,877	4,523	1,069	7,469	19	4,684
Total	11,278	13,542	12,869	37,689	244	40,919

Source: Narmada Control Authority, November 2001

est, this ratio was about 1:30. Similarly, the submergence for 1 MW powerhouse installed capacity is 24 ha; it was 1875 ha for the Tilayya Project of the Damodar Valley Corporation, 3313 ha for the Maithon Project and 1266 ha for the Sriram Sagar Project. These figures show that the geographical features of the Narmada River require relatively much smaller submergence for potential benefits. Secondly, it needs to be clarified that of the 192 villages submerged in Madhya Pradesh, 116 account for only 613 ha of privately owned land, of which only 2 ha is agricultural (with total 12,894 project-affected families in the 116 villages). About 77 villages have 6087 ha of privately owned land, of which 25.3 ha are under agriculture (Narmada Control Authority 2001).

12.4 Package for Resettlement and Rehabilitation of the Displaced

The Narmada Water Dispute Tribunal in its award used the word 'oustee' to describe displaced or project-affected person. (And a family unit of such persons was referred to as a project-affected family.) The term related to any person who for 1 year prior to notice being issued under Section 4 of the Land Acquisition Act of 1894 had been ordinarily residing, cultivating or carrying on trade or occupation in the area likely to be submerged permanently or temporarily. All sons who were not minor were treated as separate families. The definition of a project-affected family has since been made more inclusive by the Narmada Water Dispute Tribunal Award: Such a family now includes the husband and wife, minor children and dependent persons, as well as individuals holding legal titles (co-sharers). Major sons and unmarried daughters are counted as separate families. Large areas of government or common lands were historically encroached upon; those in possession of such land prior to 13 April 1987 and the landless with homesteads are also counted as project affected.

Resettlement may be defined as the final movement of project-affected persons (possibly after a transit camp) to a new location, with full compensation for lost land and property. Rehabilitation is defined as a long-term strategy to restore and improve socio-economic status. It thus includes the attempt to recover the original status for the displaced, as well as strategies and measures to make the resettlement site psychologically sustainable. It also connotes physical, social, economic, cultural and ecological rehabilitation, including that of civic amenities.

According to the Narmada Water Dispute Tribunal Award, a primary school with a playground for every 100 families, a place of worship for every 500 families, a source for drinking water for every 50 families up to 500 families and a post office are to be provided. In addition, children's park, cremation or burial sites, meeting places for women, a common threshing ground and common water pond for livestock are also to be provided. Ecological rehabilitation includes an environmental awareness and education programme, plantation programmes with social forestry, community forestry, kitchen gardens, farm boundary plantations and fuel-wood management. The award also requires that effort be made to develop other capabilities amongst rehabilitated people that encourage self-employment in a variety of vocations: shop keeping, fishing, piggery, rabbitry, goat raising, vegetable cultivation and marketing, bee keeping, sericulture, water farming, fodder farming, animal husbandry, gardening, hand-pump repairs (drinking water), bicycle repairs, electrical repairs, basket weaving, etc.

A resettlement and rehabilitation plan must be integrated with ongoing rural development programmes such as integrated rural development, training of rural youth, employment guarantee scheme, social assistance, tribal development, integrated child development schemes, mother and children's care, besides agricultural extension services.

Resettlement and rehabilitation strategies should address the following issues:

a. The level of displacement should be minimised by exploring viable alternatives for project design. The cost of land is increasing and planners need to reduce the need for land.
b. Involuntary resettlers need to be compensated in terms of replacement value prior to dislocation.
c. Assistance and support must be provided during periods of relocation and transition, and must address any loss of livelihood induced by the project.
d. Resettlers' living standards, income-earning capacity and production levels should be improved. Their quality of life should not only be protected but upgraded.
e. There should be specific features for the socio-economically weak amongst project-affected persons.
f. Land is a central aspect of resettlement and rehabilitation policy. The land provided should be cultivable and irrigable.
g. If land is not available, economic rehabilitation is necessary.
h. Project-affected persons should be integrated with their host communities. Care should be taken to avoid overcrowding and the overuse of natural resources to avoid conflict with hosts.
i. There should be flexibility rather than a fixed standardised approach. The aim is to achieve goals without undue uniformity and rigidity.
j. The resettlement and rehabilitation programme should be backed by adequate counselling services through social workers to help redress grievances.
k. The package must compensate for the loss of sustenance derived from common property resources and economic institutions such as grazing land, civic and

social services, mines, buildings, meeting halls, water sources, bathing ghats, etc. It must also compensate, to the extent possible, for unintended cultural fragmentation and the disruption of social and family ties.

The state of Gujarat was to acquire and make available irrigable land and housing sites for rehabilitation of project-affected families a year in advance of submergence of each stage. The resettlement grant was set at Rs 750 per family, including gratis transport of household belongings and dismantled materials from the submergence area. Grant in aid was Rs 500. It was prescribed as nil for compensation greater than Rs 2,000. The resettlement and rehabilitation package for each family comprised the following: (1) 2 ha of irrigable land; (2) full compensation for an existing house; (3) a free 500 m^2 residential plot, a 45 m^2 house valued at Rs 45,000 in lieu of a temporary tin shed with plinth and roof tiles, and new roof tiles to replace old ones; (4) free transportation of dismantled components to new site; (5) subsistence allowance of Rs 15 per head for 25 days per month, in the first year of settlement; (6) resettlement grant of Rs 750, increased with escalation in consumer price index at 8% from January 1980; and (7) free life insurance at Rs 6,000 per person, house insurance of Rs 5,000 and Rs 1,000 to insure household belongings.

In 1989, the Morse Commission, set up by the World Bank as a result of pressure from NGOs opposing the project in the early stages, recommended consultation with project-affected families for resettlement and rehabilitation package. Gujarat went much further and extended to such families a final say in the selection of agricultural land, rehabilitation plots, temporary facilities followed by permanent resettlement and standard welfare measures. Specified civic amenities now include a well for drinking water with a cattle trough, place of worship, drinking and domestic water supply, sanitation, distribution lines for electrical power, street lights, approach road, primary school, primary health centre, a Panchayat (village parliament) building, seeds store, children's park, village tank and vocational training centre[3]. Displaced persons now also have priority in project employment. There are, however, some minor variations between the provisions made by the three states in tune with their own policies. All provisions of cost would change with time due to inflation. If Gujarat were unable to accommodate displaced families from other states, or such families did not wish to move to Gujarat, the state of origin bears the cost of resettlement and rehabilitation.

An allotment of a minimum of 2 ha of irrigable agricultural land has been prescribed for each project-affected family where more than 25% of the family's landholding has been acquired for the project, on the principle of '(irrigable) land for land'. A project-affected family could ask the 'Land Purchase Committee' of the concerned state to facilitate purchase of the piece of land they selected from the three options provided by the Committee. Such state committees were assigned the task of compiling information of suitable plots of land for offering to the project-affected family. Resettlement and rehabilitation officials would later step in and arrange the

---

[3]See Provisions of NWDT awards and state resettlement and rehabilitation policies, available at http://nca.gov.in/forms_pdf/Rehabilitation%20%20Policy.pdf

documentation with regard to compensation and instalments. If the assessment by the resettlement and rehabilitation officials was higher than indicated earlier, the difference was borne by the state of Gujarat.

The state of Madhya Pradesh did not provide 'land for land' but offered a land allocation from 2 to 8 ha against acquired size. Mature sons of affected households were eligible only for house plots, and were not treated as separate project-affected families. The provisions of Maharashtra state are broadly similar, but only one acre is allocated for a landless person if he migrates with others. There is no stamp duty for registration of documents, and a 100% subsidy up to Rs 5000 for procuring productive assets, along with Rs 600 as cultivation assistance, insurance, etc. have also been provided. The Gujarat Industrial Training Centre has been set up for displaced youth, and the Punarvasavat (Rehabilitation) Trust was set up to assist in various tasks, including plinth construction, costing up to Rs 10,000.

In this package, key concepts such as project-affected family, cut-off date and social and economic injury from temporary versus permanent submergence need to be more closely defined. One size certainly does not fit all, and compensation tends to attract many who do not qualify. The incidence and impact of temporary submergence too needs careful study. If homesteads go under water even temporarily, they must be relocated at higher locations. However, such flooding may enrich farmlands going under water occasionally for a few days. Should uniform rates of compensation then apply? Drawdown and tank bed cultivation, and the race to occupy what are called riverine islands in eastern India, illustrates the high value farmers place on temporarily submerged lands.

12.5 Opposition to the Sardar Sarovar Project Dam

The Narmada Bachao Andolan (NBA; Save Narmada Campaign), an NGO that has stridently opposed the Sardar Sarovar Project for the last two decades, was ironically founded the same year that the NCA was set up. Initially the project authorities went along with the NBA, accepting several of its suggestions for improving the resettlement and rehabilitation package as it evolved and so far as they were reasonable. However, opposition continued and was enlarged on different fronts. Starting with concerns about resettlement and rehabilitation, the NBA quickly took up one after another all possible 'issues' for opposing the project such as: environmental concerns, technical issues, financial viability and, lastly, the development model adopted by the country. However, it never came up with an offer to assist project authorities in implementing the evolved resettlement and rehabilitation package.

These issues attracted the attention of the media, the common man and various international organisations, some with esoteric goals and some with genuine ones. As a result, the NBA came to occupy a position of significance. For instance, a NBA leader was inducted on the World Commission on Dams (WCD) of 1998–2000. The WCD received a set back when its recommendations (WCD 2000) were rejected by many as impractical and by and large not found acceptable by the developing world, as they were not in tune with the ethos of these societies. Opposition to dams like the Sardar Sarovar Project continued, unabated in intensity and without pause. The NBA suffered a series of defeats in dealing with concerned state governments, project authorities, the World Bank and finally the courts right up to the Supreme Court of India (Supreme Court Order, 17 April 2006). Every year come summer, fresh threats for fasts unto death were made and then withdrawn after extracting concessions and garnering media attention worldwide. Similarly, every monsoon sacrificial self-emersion in the rising waters of the Sardar Sarovar Reservoir was threatened and withdrawn, but cleverly used. Resettlement and rehabilitation got linked with monitoring by the highest court in the land year after year, possibly for the first time in the world, over almost two decades. This assumed responsibility by the Supreme Court seems slated to continue till the dam is completed.

In the meanwhile, India's Environmental Protection Act was passed, according to which environmental clearance was required and was obtained in June 1987, asking the states concerned to take certain steps including resettlement and rehabilitation, pari passu with the construction of the dam, with a correspondence between these ameliorative actions and the progress of dam construction. Large projects cover vast areas of influence and their implementation spans long periods of time. Taking such actions in advance of construction, as in the case of smaller projects, is simply not possible for projects of size and construction span like the Sardar Sarovar. The World Bank had entered the scene with a credit package for the Sardar Sarovar Project. The project's opponents cleverly targeted the Bank for siding with project authorities and not project-affected persons. The Bank formed the international Morse Commission to look into several issues including those brought up by the NBA. The result was a biased report in June 1992 which was, predictably, rejected by the Government of India (Morse and Berger 1992). The World Bank too was not convinced of the acceptability of this report. In spite of the rejection of the Morse report, the Bank continued to raise several issues with the Indian government. The government was therefore forced to ask the Bank to withdraw from the Sardar Sarovar Project in 1993.

In August 1993, a Five Member Group of experts to conduct review discussions was set up, by the Ministry of Water Resources in response to suggestions from protesting activists. The Five Member Group's interim and final reports became available during 1994. The Ministry of Environment and Forests also issued guidelines for Environmental Impact Assessment in 1994 (Ministry of Environment and Forestry 1994). Shortly thereafter, the NBA filed a Public Interest Litigation in the Supreme Court, seeking to halt the Sardar Sarovar Project for not acting on various conditions of environmental clearance. After prolonged hearings, in 1995 the Court ordered a stay on continued construction of the dam higher than El 80 m at which it then stood, to enable states to fulfil the pari passu condition satisfactorily and left it to the NCA to proceed accordingly (Gowariker et al. 1994; Iyer et al. 1995). The dam was raised to El 85 m by 1999 and to El 90 m by 2000. Gujarat constituted a Grievance Redressal Authority in 1999, and Maharashtra and Madhya Pradesh soon followed suit.

After prolonged hearings and stoppage of dam construction for nearly 5 years, the Supreme Court issued its verdict in October 2000 (Supreme Court of India 2000). It allowed raising the dam to El 95 m which was accomplished by 2002, and subsequently to El 100 m in 2003. In 2002, the Ministry of Environment and Forests set up a multi-disciplinary committee of experts under the Chairmanship of the Author, to review fulfilment of conditions laid down at the time of environmental clearance, and the committee reported that the environmental clearance conditions were satisfactorily fulfilled. The NCA allowed raising of the dam to El 100.64 m in 2004, and in March 2006 to El 121.92 m (Narmada Control Authority 2006).

Opposition to the raising of the dam and its resettlement and rehabilitation, however, continued haltingly stage by stage. All the concrete has been placed and the dam has been raised as permitted so far. The raising of piers and installation of 30 spillway crest gates 18 by 18 m in size is undertaken in the next stage; the spillway bridge will be constructed thereafter. About 94% of work on the dam is complete, and the remaining work is likely to be completed by 2011. The dam height is to be kept at El 146 m. The reservoir elevation, however, will be raised subject to progress on resettlement and rehabilitation.

It is not easy to win over all who are against development, especially if the stands and perspectives are varied and shift over time. Ensuring development with justice for all is an indicator of welfare aims, but dissent and the inability to resolve differences between those for and against often derails the process. The leadership of the NBA often appeared to the gullible as revolutionary in spirit, but in many respects remained outdated and did not take on board counter-arguments. An organisation concerned with the people's welfare needs to stand for a social conscience, to grasp the cost of development and the potential of science and technology, to possess a sense of democratic institutions, to comprehend the challenges in large-scale projects in terms of dealing with beneficiaries as well as those likely to be affected by development and finally to have an understanding of the ethical dimensions. Unfortunately, these qualities were not to be found in the NBA's campaign, and their leadership behaviour was far from exemplary; for instance, they were often embroiled in complex situations such as being found in contempt of court. The leadership continued to call the Supreme Court's decisions as 'anti-people', equating itself with the 'people', and repeatedly agitated in civil society, though it eventually received less and less attention. Recently, the leadership seems to have withdrawn somewhat from the Sardar Sarovar Project and to be focusing its attention on the displacement of slum dwellers and on acquisition of farmland for special economic zones or infrastructure projects such as the car production project at Singur, West Bengal where the private sector seems to be the main target.

The last struggle by the NBA in relation to the Sardar Sarovar Project was waged after the NCA's decision in 2005 to allow the dam to be raised from 110 to 121.9 m. The NBA again approached the Supreme Court, seeking a stay on the NCA's decision because of non-compliance with resettlement and rehabilitation terms as directed earlier by the Court. The Court declined to stop the raising of the dam. A Ministers' Group was hurriedly convened for visiting the resettlement and rehabilitation sites in Madhya Pradesh. Their report was cited in the Supreme Court during the next hearing by the government's advocate. The Court again refused to stop construction, and asked for the Prime Minister's intervention to investigate the matter. The Prime

Minister accordingly set up the Sardar Sarovar Project Relief & Rehabilitation Oversight Group in April 2006, and its report was submitted after a thorough investigation about the status of resettlement and rehabilitation in Madhya Pradesh (Shunglu et al. 2006). The Prime Minister accepted the findings and the government's advocate then made submissions before the Supreme Court validating, more or less, the NCA's decision. The Court finally disposed of NBA's public interest litigation which had sought a halt to dam raising on 17 April 2006, and reiterated its order to allow construction up to 121.90 m. It said that stopping dam construction work would be unwarranted and expensive, as the gains from its completion would benefit millions. NBA supporters and environmental activists reacted rather negatively. The NBA leader had gone on fast time and again to achieve their aims. While such self-inflicted suffering arouses concern and sympathy, democratic governments have a wider—ineluctable—social and political responsibility, and cannot abandon due process in favour of any one set of demands. The NCA, chaired by the Union Water Resources Secretary, had cleared raising the dam's top elevation to 121.92 m on 8 March 2006. This was only done after reports on actions taken for resettlement and rehabilitation had been vetted by the Grievance Redressal Authorities of the three states, as per the instructions of the Supreme Court, and thereafter approved by the Narmada Resettlement and Rehabilitation, and Environmental Sub-Groups of the NCA.

12.6 The Work by the Oversight Group

In response to the Supreme Court's observations in its order of 17 April 2006, the Prime Minister of India constituted the Oversight Group, comprising V.K. Shunglu (former Comptroller and Accountant General) as Chairman, G.K. Chadha (former Vice Chancellor of Jawaharlal Nehru University) as Member, and Jaiprakash Narayan (representing an NGO, Loksatta) as Member, to verify the status of resettlement and rehabilitation in Madhya Pradesh as per Action Taken Reports I and II due to the raising of the Sardar Sarovar Dam elevation from 110.64 to 121.92 m, as allowed by the NCA.

The Oversight Group was required to conduct sample surveys to ascertain progress on rehabilitation and gather related facts. The group's terms of reference included the following: (a) ascertain numbers of families deemed project affected due to submergence, (b) estimate numbers of project-affected families who had not fully received the resettlement and rehabilitation package, (c) determine whether offers of 'land for land' were made in a fair and transparent manner, (d) ascertain if project-affected families refused offers voluntarily and accepted Special Relief Package, (e) ascertain when all resettlement and rehabilitation measures will be in place and (f) recommend a system to ensure the project-affected families receive within 3 months the benefits of resettlement and rehabilitation as per orders from the Narmada Water Dispute Tribunal Award, the Supreme Court and Grievance Redressal Authorities.

It is necessary to look into the arrangements for resettlement and rehabilitation in Madhya Pradesh, before turning to the work of the Oversight Group. The Indian government had framed the Narmada Water Scheme as required by legal provisions in 1990, to implement the Narmada Water Dispute Tribunal Award. The NCA, chaired by the Secretary of the Ministry of Water Resources, administers the scheme for all states. Its Sub-Group chaired by the Secretary for Social Justice is responsible for resettlement and rehabilitation. Each one of the three states had independent setup for implementation of resettlement and rehabilitation in the state. The Narmada Valley Development Authority set up by the Government of Madhya Pradesh administers resettlement and rehabilitation for Madhya Pradesh. As per the Supreme Court's judgement (in 2000), the NCA has to allow the dam height to be raised as long as resettlement and rehabilitation continued pari passu. The Madhya Pradesh government set up a Resettlement and Rehabilitation Grievances Redressal Authority in 2000 to receive grievances, which are forwarded to the Narmada Valley Development Authority for verification and comments. The applicant has a chance to present his case in a hearing, before the Authority issued a decision on the grievance.

The process for allowing the dam height to be raised consists of the following steps: the Madhya Pradesh government submits an Action Taken Report about the pari passu condition to the resettlement and rehabilitation sub-group of the NCA and the Grievance Redressal Authority; the NCA then asks the latter to verify a sample of cases and give its opinion. Action Taken Reports I and II of 2005–2006 covered 14,061 and 5,307 project-affected families, respectively. The last consultation for the purpose was conducted in March 2006. The NCA's resettlement and rehabilitation sub-group then met, took a view and sent its report to the NCA, which allowed raising of the dam from El 110.64 to 121.92 m. This decision was challenged by the NBA. After a hearing in April 2006, as mentioned above, the Supreme Court upheld the NCA decision and allowed resumption of dam raising, but asked the Government of India to verify deficiencies if any in the resettlement and rehabilitation process and in any Action Taken Report through an independent agency.

In order to fulfil their terms of reference, the Oversight Group undertook the following massive tasks: (a) National Sample Survey Organisation teams were set up to verify the numbers of project-affected persons/families, which were validated by an independent team under a retired Deputy Comptroller and Accountant General; (b) Six eminent resource persons visited all 86 sites to ascertain resettlement and rehabilitation facilities and entitlements; (c) A team of 'Institute of Public Auditors' checked the Grievance Redressal Authority records of 10,000 applicants, and enumerated, verified and reported on them. A summary of the investigation is presented below.

Over 25,000 persons (18,360 project-affected families from Action Taken Reports I and II), including those claiming exclusion from 177 villages, were interviewed and enumerated. As a sample, about 5% were validated. The Government of Madhya Pradesh Action Taken Reports of the Narmada Valley Development Authority had covered 150 and 73 villages, respectively. The massive data was tabulated in 44 columns and was verified. In the Action Taken Report I, about 32% of project-affected families with landholdings less than 25% being proposed for acquisition were considered, while the second report had only 1% such cases. The figures indicated that additional work had been accomplished in the field between the Action Taken Report 1 and 2. In addition, the Oversight Group visited, verified and monitored progress. During this time, activists questioned their work, though they did not provide any specific information or grievances to the group.

The National Sample Survey Organisation reported that 6,485 persons in all claimed to be affected by the Sardar Sarovar Project, about 99% of whom had claimed the loss of homestead. However, 3,000 were not residents of these villages as per the 2001 Census. About 3,620 (56%) were from 22 villages, which raises doubts about the veracity of their claims. About 1,367 had approached the Grievance Redressal Authority, while 5,118 (79%) had not. No reasons were given. Facilitator NGOs are indeed welcome for simplifying and aiding such work. The NBA, on the other hand, disallowed physical surveys in some villages, and visual impressions were therefore relied upon. About 1,650 persons in these villages had homesteads at elevations lower than 121.92 m. Some 200 who could be affected possibly approached the Grievance Redressal Authority but they did not mention it during field verification.

The figures of project-affected persons and villages varied across the census held every 10 years. The 1991 Census had indicated 177 villages, 24,539 houses and 143,773 persons. The 2001 Census came up with figures of 29,172 houses and 162,448 persons. For 22 villages, the respective figures in 1991 were 3,253, and 1,805, while they were 8,522 and 43,391 in the 2001 Census. Likewise for 155 villages, the 1991 figures were 21,286 and 125,722, while the 2001 figures were 20,650 and 119,057. These figures collected by the Oversight Group suggest a movement of population, attracted possibly to obtain compensation.

With regard to authenticity of entitlement receipt/disbursement, the Group verified a 44-column information base of the Action Taken Report of the Government of Madhya Pradesh. The result showed for the most part higher receipts than the Action Taken Report figures, indicating progress in work done since then. Progressive disbursement was built into the process. Project-affected families received a grant for transportation following the decision to move.

The Madhya Pradesh's Narmada Valley Development Authority procedure provides proximate available irrigated land for land to a willing project-affected family. A non-willing project-affected family can apply to Narmada Valley Development Authority for the Special Relief Package. The Group found that about 10% project-affected families had opted for irrigated land for land, the rest for Special Relief Package as they could buy land of their choice. Special Relief Package had been challenged but was considered as compliant of provisions of the Standing Committee of the National Water Development Agency by the Grievance Redressal Authority.

Project-affected families who chose the Madhya Pradesh state's Special Relief Package received their payment soon. Some small discrepancies were noticed, and were reported for correction. In rural areas, homesteads on agricultural fields were not enumerated, nor were subject to record of ownership, barring the 2001 Census. Survey teams thus faced difficulty in verification of claims thereafter.

Of 4,286 project-affected families, 407 opted for land with government land, and others for Special Relief Packages. About 38% found the special package attractive, 26% followed relatives or others in their decisions and 3% were not satisfied with the land offered. The Oversight Group found no reason to believe that 'coercion' had been applied. The Special Relief Package was found acceptable by people, as a project-affected family could buy land of their choice. Of the project-affected families reluctant to accept allotted land, about 21% had their own choice of land, 27% felt that the site offered was too distant and about 16% did not like the land. Though it was initially challenged, the Package had been accepted by the Grievance Redressal Authority as compliant with Supreme Court and Narmada Water Dispute Tribunal provisions.

On approval by the Narmada Valley Development Authority, the first instalment was paid to identify the land for purchase. On completing the purchase, the second instalment was paid directly to the seller. The Action Taken Report cited 652 cases, while the Oversight Group found 1137, and by 20 June, 1,650 cases of land purchase were complete. The Group found that the pace of completion had been slow but was picking up. The major reason for slow progress was escalation in the cost of land. The numbers of project-affected families receiving grants for resettlement and rehabilitation, other assets and actual transfer, were greater than in the Action Taken Reports, though for several such families a smaller amount was received.

In terms of resettlement and rehabilitation sites and the Narmada Water Dispute Tribunal Mandate, Gujarat informs Madhya Pradesh about an area to be submerged 18 months in advance. Project-affected families can occupy their lands till a notified date. In December 2005, the Narmada Valley Development Authority so notified the project-affected families in all 177 villages at El less than 121.92 m; they were expected to move in 2006 during the monsoon season. Yet as submergence was to be temporary, the families did not expect any problems. They planned to stay or perhaps move to higher ground and return. In the past, there was no legal requirement to move, and some families waited for higher compensation while others awaited improvement of their plots. In the preceding 3 years, there had been no inundation. In June 2004, at dam El 110.64 m, backwater reached El 114–115 m; thus, no immediate threat of submergence was felt.

The terms of reference of the Oversight Group had referred merely to the lack of facilities at the resettlement sites as per the Narmada Water Dispute Tribunal Award. Six teams were constituted to assess facilities for 86 sites comprising 79 up to El 122 m out of which 37 were found in good, 23 in average, and 19 below average/poor status. They did not pose a problem for moving, and it was felt that deficiencies could be removed by March 2007. About 15,000 project-affected families needed land allotment, about 14,391 had been allotted their respective lands, 1,451 houses had been constructed and 757 such families had moved. Many waited as they wanted to shift in a large group, by caste.

The Oversight Group found that most sites were connected to the Tehsil town (several Tehsils make a District), located on or close to roads with approach roads. Inner roads within sites were constructed but needed repairs. The private or government land acquired was levelled and plots demarcated, and some plots needed filling and improvement. Most lands were free from litigation; however, some discrimination was alleged. Electrification was done satisfactorily, and hand pumps were installed and were working. Some sections of pipe network were yet to be installed. Some sites needed better drainage or sanitation. Other institutional infrastructure was found to be in place. It was felt that all that remained to be done could be completed by the end of March 2007.

The Government of Gujarat stated before the Supreme Court that the dam height could reach El 121.9 m by the end of July. Raising the dam to 121.92 m, raising of piers and erection of gates was to begin after the 2006 monsoon and could continue up to July 2008. The Oversight Group concluded that: (a) There was no substantial deviation in numbers of project-affected families, and claims to this effect made by activists were not substantiated; (b) Settlement should be carried out within 3 months, and over 4,000 cases were referred to the Narmada Valley Development Authority by the Grievance Redressal Authority; (c) The actual payments made for entitlements corresponded to the figures in the Action Taken Reports; (d) The Grievance Redressal Authority felt that the Special Relief Package was a legitimate substitute to the land for land deal stipulated by the Narmada Water Dispute Tribunal Award. Disbursement of the second instalment had picked up pace; (e) The numbers of developed plots, and facilities created and made available at resettlement and rehabilitation sites exceeded the requirements of project-affected families at the time. Remaining deficiencies could be removed by March 2007, and plans could be made for additional 19,000 project-affected families up to 122 m and another 15,000 project-affected families beyond; (f) Voluntary moving from submergence locations was slow due to Narmada Water Dispute Tribunal Award provisions that allowed them to remain still preferred to move; (g) The district administration should be fully involved, rather than performing only a supportive role to the Narmada Valley Development Authority; (h) The Ministry of Water Resources should strengthen the NCA with greater support for the success of future action plans.

As mentioned earlier, the Oversight Group's report and the statement by the Government of India in the Supreme Court convinced the Court of the adequacy of the resettlement and rehabilitation package, and the institutional arrangements in place for successful accomplishment of resettlement and rehabilitation as per guidelines laid down by the rule of law. The water resources sector has to consolidate these gains and move expeditiously to complete the project as early as possible.

12.7 Conclusions

The whole process of resettlement and rehabilitation for the displaced of Sardar Sarovar Project has been tested and proven. The dam has been built up to the spillway crest elevation. The fabrication and installation of crest gates has been entrusted to an agency that can complete the task soon, provided the state governments act quickly and in line with the procedures laid down. They have to strive to remove the deficiencies noticed so far and ensure that the displacement and resettlement and rehabilitation of the people likely to be affected up to the top of dam are accomplished in the next 2 years. Fresh attempts can be made to win over the dam's opponents, and to aim collectively for a win-win situation for the much needed water resource development for the country.

References

Gowariker V, Iyer RR, Jain LC, Kulandaiswami VC, Patil J (1994) Report of the Five Member Group set up by the Ministry of Water Resources to Discuss Various Issues Relating to the Sardar Sarovar Project, New Delhi

Iyer RR, Kulandaiswami VC, Jain LC, Gowariker V (1995) Further Report on the Five Member Group on Certain Issues Relating to the Sardar Sarovar Project, New Delhi

Ministry of Environment and Forest (1994) Notification on Environmental Impact Assessment of Development Projects, New Delhi, 27 January

Ministry of Water Resources (1956) Inter-State Water Disputes Act. Government of India, New Delhi

Ministry of Law and Justice (1894) Land Acquisition Act of India. Government of India, New Delhi

Morse B, Berger TR (1992) Sardar Sarovar – report of the independent review. Resources Future International, Ottawa

Narmada Control Authority (2001) Twenty-second Annual Report and Accounts of the Narmada Control Authority, Indore

Narmada Control Authority (2006) Clearance for construction up to 121.92 metres. Narmada Control Authority, Minutes of the Seventy-Six (Emergency) Meeting, Item No. LXXVI (Emer.) - 1 (756), 8 March

Narmada Water Disputes Tribunal (1979) Final order and decision of the tribunal, Gazette of India, New Delhi, 12 December

Shunglu VK, Chadha GK, Narayan J (2006) Report of the Sardar Sarovar Project Relief and Rehabilitation Oversight Group on the status of rehabilitation of project affected families in Madhya Pradesh, New Delhi

Supreme Court of India, Civil Original Jurisdiction, Writ Petition (C) No. 319 of 1994. Judgement of 18 October 2000, AIR 2000, SC 3751

Supreme Court of India, Record of Proceedings, I.A. NOS. 18–22 in writ petition (civil) No. 328 of 2002. Order of 17 April 2006

World Commission on Dams (2000) Dams and development: a new framework for decision-making. Report of the World Commission on Dams. Earthscan, London

Chapter 13
Impacts of Kangsabati Project, India

R.P. Saxena

13.1 Introduction

Water is the most vital element that supports life. It is also the prime mover of economic development and symbolises survival, growth and prosperity, so much so that it has long been the focus of attention and the centre of all future planning and strategy. Over the centuries, however, it became clear that in order to attain the objective of overall development there was a need for large dams or, for that matter, large multi-purpose projects, to spread benefits widely, which were strictly localised in earlier times. In post-independence India, several dams and multi-purpose projects have been constructed to increase food production, energy generation, drinking water supply, fisheries development, employment generation, etc. In the long run, most of these projects have not only been successful in delivering the benefits that were expected prior to their construction, but over the years, they have radically transformed the economic scenario of their command areas. In spite of this, for the last decade or so, there has been a scathing attack on the feasibility of large dams within the domain of 'development'. Numerous questions have been raised, ranging from environmental to ecological and social issues, some of which are rather trivial and seem deliberately posed to create confusion among the masses (notwithstanding the fact that most people have been direct beneficiaries of those large projects).

A glance at India's water resources scenario will further substantiate the need for large storages in decades to come. The rising population of India is already a matter of concern and the projections made for 2025 as well as 2050 make matters even more grave. The per capita water availability has plummeted by about two-thirds

R.P. Saxena
Central Water Commission, Ministry of Water Resources,
Government of India, New Delhi, India

over the last 50 years; this downward trend will continue until the population stabilises. There is a limit on annual availability of water in the country and there is also looming threat over being able to maintain the existing share of irrigation in water use, relative to other value-added sectors like industry, domestic use, power, etc. India is predominantly an agriculture-based country. About 68% of its population earns its livelihood from agriculture. The continuous decline in created irrigation capacity has thus been a perpetual concern. Future estimates of food requirements project a monumental figure of 500 million tons by the year 2050, and regardless of the criticisms levelled against large storages it is a foregone conclusion that large water resource projects will continue to occupy a pivotal role in all future strategies and planning. In fact, to achieve such Herculean objectives it is imperative to adopt the twin strategies of development and management by which more and more area has to be brought under assured irrigation on the one hand, and the gap between irrigation capacity created and utilised has to be reduced on the other. The impact assessment of projects through performance evaluation studies is therefore of paramount importance in firming up future strategies. In the following sections, the focus will be on the Kangsabati Project, and the vital aspects of pre-project assumptions and the consequent post-project socio-economic and environmental impacts will be addressed with a view to determining ways to maximise benefits and simultaneously minimise negative impacts as far as possible in future projects.

13.2 Objectives of Impact Assessment of Projects

Like all other developmental projects, water resources projects are conceived, formulated and implemented with certain objectives. On completion of such projects, it is necessary to conduct performance evaluations of these projects from time to time, in order to (i) assess the efficiency of projects, (ii) learn whether projects are performing as expected and (iii) devise and implement remedial measures to improve efficiency of the system wherever warranted.

Growing apprehensions about the utility of water resources projects have been voiced by a group of people in the recent past, especially in the context of adverse environmental impacts. These critics tend to underplay the contributions of water resources projects towards India's development, and such a perspective is bound to undermine the progress of water resources projects in the country. Post-project impact or evaluation studies can yield factual findings on such apprehensions and the utility of water resources projects in economic progress.

Reservoir projects inevitably involve large-scale submergence of lands and other immovable properties and lead to the displacement of people from their habitations and places of work. Post-project impact assessment studies can bring to light the economic upliftment of project-affected persons after project implementation.

13.3 Present Status in India

The need for periodical performance evaluation studies of completed projects in order to evaluate the impacts of projects, seek reasons for their inefficient performance and identify ameliorative measures has been repeatedly expressed at various platforms. The National Water Policy (Ministry of Water Resources 2002) of India also recommends that the study of the impact of a project on human lives, settlements, occupations, as well as socio-economic, environmental and other conditions, both during construction as well as later, should form an essential part of project planning. The Central Water Commission (2002) also brought guidelines for Performance Evaluation of Irrigation Systems.

The practice of conducting performance evaluation studies of irrigation projects was initiated in India in the 1970s. Since then such studies, covering more than 100 irrigation projects located in different parts of the country, have been carried out by various central, state and other agencies such as the Ministry of Water Resources, Central Water Commission and irrigation departments of states. During the ninth plan period (1997–2002), the Central Water Commission successfully completed post-project performance evaluation studies for five commissioned irrigation projects in the country. The Kangsabati Project in the state of West Bengal was one of them.

13.4 Kangsabati Project

13.4.1 Project Description

The Kangsabati Project is located in Bankura District, a drought-prone area in west-central West Bengal. The project comprises two dams, one on the Kangsabati River and the other on the Kumari River, a tributary of Kangsabati. A link channel connects the valleys of the two rivers. The reservoir has a total live storage of 97,560 hectare-metre (ha·m). Apart from 72,890 ha·m earmarked for irrigation and domestic needs, the project also provides 24,670 ha·m for flood moderation.

The gravity-type earthen dam with concrete saddle spillway has two head regulators supplying water to left bank and right bank canals. The total length of main and branch canals is 620 km. Construction of the project began in 1956–1957. It was partly commissioned in the year 1965 and started providing irrigation benefits. The gross command area and cultivable command area under the project are 617,409 ha and 396,050 ha respectively. The project has an irrigable command area of 340,809 ha and an annual irrigated area of 401,890 ha. The flood reserve was intended for moderation of flood discharge. In addition, the supply of drinking water to areas where scarcity was acute, pisciculture and provision of recreational facilities were important project aims. Extending irrigation facilities to about 17,000 ha of wasteland, where no cultivation was possible otherwise, was also planned (Government of West Bengal, 1953).

Table 13.1 Targeted growth of irrigation potential

Year	Targeted growth of irrigation potential (in ha)		
	Kharif[a]	Rabi	Total
1960–1961	20,234	–	20,234
1961–1962	101,171	–	101,171
1962–1963	242,811	16,187	258,998
1963–1964	323,748	30,351	354,099
1964–1965	323,748	53,621	377,369
1965–1966	323,748	60,703	384,451

[a]The crops sown during the monsoon season are referred as 'Kharif', while the term 'Rabi' is used for winter crops
Source: Unpublished collection from consultancy work

13.4.2 Impact Assessment of Kangsabati Project

The project was begun as early as 1956 and no base-line survey of the socio-economic, agro-economic and environmental conditions in the project area were conducted at that time. The impact assessment of Kangsabati Project was carried out through performance evaluation studies to ascertain benefits from the project and to compare pre- and post-project scenarios (Water and Power Consultancy Services (India) 2003). The following five dimensions were considered:

1. System performance
2. Agro-economic impacts
3. Socio-economic impacts
4. Environmental impacts
5. Impact of command area development programme

13.5 System Performance

13.5.1 Irrigation Potential

The growth of irrigation potential envisaged in the project is indicated in Table 13.1.

Against the planned total annual irrigation potential of 384,451 ha for 1965–1966, the annual irrigation potential achieved (in area irrigated) is indicated in Table 13.2.

The table shows that irrigation coverage of the Kharif crop was greatest in the year 1990–1991 (84.92% of irrigation capacity created) and the maximum Rabi area irrigated was 75.11% in the year 1999–2000. There has always been an appreciable gap (more than 15%) between irrigation potential created and irrigation potential utilised. The main reasons for the lag in irrigation potential achieved are

13 Impacts of Kangsabati Project, India

Table 13.2 Annual irrigation potential achieved, 1965–2000

Year	Kharif area (ha)	Rabi area (ha)	Boro/hot weather paddy area[a] (ha)	Total area (ha)
1965–1966	29,119	–	–	29,119
1966–1967	42,788	–	–	42,788
1967–1968	50,608	–	–	50,608
1968–1969	74,862	–	–	74,862
1969–1970	100,790	–	–	100,790
1970–1971	117,787	–	–	117,787
1971–1972	101,387	–	–	101,387
1972–1973	140,309	–	–	140,309
1973–1974	175,010	15,650	–	190,660
1974–1975	196,021	24,819	–	220,840
1975–1976	203,881	41,948	–	245,829
1976–1977	212,671	–	–	212,671
1977–1978	223,760	20,349	–	244,109
1978–1979	238,911	29,100	–	268,011
1979–1980	197,437	–	–	197,437
1980–1981	240,338	21,428	–	261,766
1981–1982	147,106	1,626	–	148,732
1982–1983	183,832	2,732	–	187,564
1983–1984	221,268	21,007	–	242,275
1984–1985	261,670	7,515	–	269,185
1985–1986	263,531	23,186	7,838	294,555
1986–1987	256,079	21,072	18,925	296,076
1987–1988	267,991	3,345	–	271,336
1988–1989	273,570	–	–	273,570
1989–1990	271,616	18,343	17,704	307,663
1990–1991	274,939	36,983	25,000	336,922
1991–1992	269,598	40,998	–	310,596
1992–1993	267,325	6,909	–	274,234
1993–1994	259,221	40,878	20,921	321,020
1994–1995	262,387	41,490	1,302	305,179
1995–1996	249,623	45,117	26,440	321,180
1996–1997	248,866	–	–	248,866
1997–1998	247,892	44,826	16,307	309,025
1998–1999	167,126	40,223	20,095	227,444
1999–2000	253,506	45,593	27,936	327,035

[a]The term 'Boro' is a local term in West Bengal state and technically it is used for 'Hot weather paddy'
Source: Unpublished collections from consultancy work

overestimation of net irrigable area and irrigation potential, and excessive seepage loss due to porous nature of the soil through which the canals have been aligned, especially in the head reaches.

13.5.2 Irrigation Achievement

Irrigation of the Boro crop (hot weather paddy) was not planned in the project. This was begun in the year 1985–1986, after Rabi irrigation commenced. The irrigation benefit accruing from the project, including the Boro crop, was greatest (336,922 ha) in the year 1990–1991. The net irrigated area was 277,219 ha. Thus, the intensity of irrigation for the year 1990–1991 comes to 121.54%.[1]

13.5.3 Groundwater Recharge

The overall development of groundwater due to the project has been found to be 14.40%. In the hard rock terrain and its adjoining areas (the western fringe of platform zone), the development of groundwater varies from 0.16% to 6.15%. In the eastern and north-eastern regions, which basically consist of alluvial soil, it varied from 13.10% to 41.91%. In the platform area, it varied from 7.93% to 56.08%.

13.5.4 Flood Mitigation

Heavy rainfall causes a sudden rise of water level in the Kangsabati River, and the water spills over the banks, causing widespread inundation. Floods occur on average once every 3 years. With the construction and operation of the reservoir, the severity of floods has been significantly reduced. It has been observed that with flood moderation technique, the peak flow could be reduced to 54% on an average. Flood moderation performance in respect of typical floods in the Kangsabati Reservoir is shown in Table 13.3.

13.5.5 Ancillary Benefits

The Kangsabati Project was conceived primarily as an irrigation project and flood moderation was the second major objective of the project. The other project benefits are listed below.

13.5.5.1 Drinking Water

There were many areas, especially in Bankura District, where scarcity of drinking water was acute. The innumerable tanks in and around the villages were dry in the summer when water demand for human and animal use was at its peak. Shortage of

---

[1] The sources for these figures are the project report and unpublished collection from consultancy work.

Table 13.3 Flood moderation performance

Year	Peak inflow (m³/s)	Inflow (ham/h)	Peak outflow (m³/s)	Outflow (ham/h)	Amount of moderation (%)
1978	12,998	4,684	4,661	1,678	64
1984	2,203	794	1,150	414	48
1984	1,951	703	779	280	60
1984	2,386	860	851	306	64
1990	4,212	1,518	2,302	829	45
1993	3,643	1,313	1,711	616	53
1995	2,811	1,013	1,700	612	40
1997	3,438	1,239	1,431	515	58

Source: Unpublished collections from consultancy work

drinking water and the consequent use of polluted water for drinking caused epidemics and high percentages of water-borne diseases among inhabitants in these areas.

Since the Kangsabati Project was commissioned, the situation has improved. Water from the reservoir is released during the dry season to meet the water needs of human beings and cattle.

13.5.5.2 Drought Mitigation

In the post-project situation, the project area suffered drought and scarcity conditions in only 2 years across a span of 35 years (1965–1966 to 2000–2001), as compared to 15 years in a period of 43 years (1902–1944) in the pre-project era. On completion of the project and development of the irrigation system, the area under stable irrigation increased. Even in years of low rainfall, the irrigated area during the Kharif season varied from 62% to 84%. This shows the significant role played by the Kangsabati Project in drought mitigation.

13.5.5.3 Pisciculture

Fish is an integral part of the local cuisine and pisciculture is thus very common in this region. People are quite familiar with the art of fish farming and practise it in nearly all ponds, tanks, bunds, etc., in the area. Before the project was commissioned, pisciculture was a seasonal activity, generally conducted only during the monsoon months. The impoundments related to the dam have provided a much greater scope for fish production in the region. Increased production of fish has contributed to enhanced supply of high quality food, rich in protein, vitamins, calcium, phosphorus and other nutrients necessary for human health.

13.5.5.4 Recreational Facilities

The reservoir is located in a dense forest and is highly suitable for recreational facilities. Facilities such as boating, fishing and picnic spots have been developed.

For instance, there is a 'Deer Park' at Banpukuria, and Jhilmili is an alternative picnic spot where the local 'Tusu' festival is held each year with great fervour. Tourist lodges, holiday homes, youth hostels and government rest houses of the Irrigation and Water Drainage Department, the Department of Tourism and the Directorate of Youth Services with provisions for refreshments, viewing spots, etc. have been made available. The area is well connected by rail and road.

13.5.5.5 Conjunctive Use of Surface and Groundwater

Within the command area, there are patches of high land areas for which it is difficult to arrange surface irrigation. Such areas, particularly lands lying in the middle and tail reaches, are irrigated by extracting groundwater during the Kharif and Rabi seasons, as well as for perennial crops. With additional recharge of groundwater, on account of the impounding of water in the reservoir as well as the application of surface and groundwater irrigation, there exists ample scope for future development of groundwater throughout the eastern and north-eastern sections of the command area.

13.6 Agro-Economic Impacts

13.6.1 Cropping Pattern

Table 13.4 shows the pre-project and post-project cropping patterns for the command area of the Kangsabati Project. It is evident from the table that the cropping intensity increased by 19.59% in the post-project era.

13.6.2 Mechanisation and Use of Fertilisers

During the sample survey conducted in the command area of the project, the farmers reported that farming was being done mostly with manual labour using mainly the organic manure with nominal use of fertilisers during the pre-project period. The sample survey indicated that on the average, the cost of fertilisers used was 11.78% of total cost of cultivation. Similarly, the cost of using machineries such as tractors, etc., was about 16.53% of the cost of manual labour inputs. It was thus concluded that the cultivation techniques of the pre-project period, with practically no mechanisation and with nominal use of fertilisers, developed into moderately mechanised farming with moderate use of fertilisers during the post-project era.

13 Impacts of Kangsabati Project, India

Table 13.4 Pre- and post-project cropping patterns

Crops	Cropping pattern as percentage of net cultivable area	
	Pre-project (1960–1961)	Post-project (1996–1997)
I Kharif		
Rice Aus or Aman, includeing high yielding variety	83.10	89.00
Jute	1.60	–
Groundnut		11.00
Total Kharif	84.70	100.00
II Rabi		
Wheat	0.40	2.90
Potato	–	1.40
Oil seeds	0.30	4.40
Pulses	–	1.50
Vegetables	1.60	3.80
Other Rabi crops	7.00	–
Total Rabi	9.30	14.00
III Hot weather (Boro)		
Hot weather rice	0.01	Varying season to season
IV Perennial		
Sugarcane	0.40	–
Cropping intensity	94.41	114.00

Source: Unpublished collections from consultancy work

Table 13.5 Agro-industries in the post-project era

Industries	Midnapore District	Bankura District	Total
Agro-based	45	43	88
Textile	6	–	6
Storage and Warehousing	17	4	21

Source: Unpublished collections from consultancy work

13.6.3 Agro-Based Industries

There has been considerable growth of agro-industries in Midnapore and Bankura districts in the post-project era.[2] The post-project status of agro-based industries is indicated in the Table 13.5.

---

[2]This is a very old project so specific pre-project data on agro-based industries could not be obtained. However, the 1961 Census Report of the districts indicated that no cold storage or warehouse was located in the districts. The District Statistical Handbooks for the years 1996 and 1997 indicated the position of agro-based industries under the post-project situation as given in the Table 13.5. It was therefore concluded that there had been considerable growth of agro-industries in Midnapore and Bankura districts in the post-project era.

Table 13.6 Livestock in the command area

Livestock	Status (percentage increase)
Cattle	49% by the year 1989
Buffaloes	11% by the year 1989
Sheep	112% by the year 1989
Goat	98% by the year 1981
Pigs	365% by the year 1981
Fowls	246% by the year 1989

Table 13.7 Yield of principal crops, pre- and post-project

	Yield (kg/ha)			Increase in yield (%)	
	Pre-project	Post-project			
Crop	1960–1961	1996–1997	2001–2002	1996–1997	2001–2002
Rice Aus	795	2,151	3,094	171	289
Rice Aman	985	2,004	3,313	103	236
Rice Boro (hot weather)	889	3,035	3,594	241	304
Wheat	524	2,275	2,300	334	339
Rapeseed and mustard	171	847	1,095	395	540
Til (sesame)	429	719	–	68	–

Source: Unpublished collections from consultancy work

13.6.4 Livestock

Table 13.6 shows that livestock in the command area has increased considerably in the post-project era.

Animal husbandry and dairy farming were found to be subsidiary occupations for about 90% of households in the area, but poultry was a subsidiary occupation for only about 10% of households while the corresponding percentage for fisheries was nominal. Animal husbandry became the major subsidiary occupation of cultivators in the project command area under the post-project situation.

13.6.5 Productivity of Crops

Table 13.7 shows the yields of principal crops sown in the command area during pre-project and post-project situations.

A comparison of crop outputs of the pre-project period with those of the post-project period shows that crop yields have gone up manifold in the post-project period.

Table 13.8 Increase in farm and non-farm employment

Category of employment	Numbers as per 1961 census (pre-project)	Numbers as per 1991 census (post-project)	Percentage increase
Cultivator	285,818	386,068	35.07
Agricultural labour	169,251	308,648	82.36
Total agricultural workforce	455,069	694,716	52.66
Non-farm worker	91,195	198,784	117.97
Total workers	546,264	893,500	63.57

Source: Unpublished collections from consultancy work

13.7 Socio-economic Impacts

Socio-economic impact studies examine project impacts on agricultural production, economic condition of beneficiaries, social status and overall living standards of inhabitants of the project command area. Studies on the socio-economic impacts of the Kangsabati Project on the people in the command and nearby areas have been undertaken by comparing the socio-economic status of inhabitants in the post-project period with that of pre-project period. In the event of non-availability of relevant data for the pre-project period, data for similar nearby areas with no project benefits (or control areas) were used for the analysis. Primary as well as secondary data were used to carry out socio-economic impacts studies for the head, the middle and the tail reaches. Farmers were categorised according to their landholdings as (a) marginal (up to 1 ha), (b) small (more than 1 ha and up to 2 ha), (c) medium (more than 2 ha and up to 4 ha) and (d) large (more than 4 ha). Several parameters listed below were taken into account for the analysis of socio-economic impacts.

13.7.1 Farm and Non-farm Employment

As per the census of 1991, 76% of the population in the command area was dependent on agriculture, while at the state level only 68% of the population was dependent on agriculture for livelihood. The higher percentage of command area inhabitants choosing agriculture may be attributed to favourable agricultural infrastructure, scope for multiple cropping and improved crop yields, and hence enhanced crop production.

The extension of irrigation facilities and increased intensity of irrigation have led to an expansion of secondary and tertiary activities in the command area. This has resulted in greater work opportunities and employment for farmers, labourers and industrial workers. Increase in farm and non-farm employment in the project area in the pre-project and post-project situations is given in Table 13.8.

The farm and non-farm employment as per pre-project 1961 census and post-project 1991 census indicated that there had been reductions in the percentages of the population working in the agriculture sector, in spite of the substantial increases in the total number of workers in the agriculture from pre-project situation to post-project situation. It can therefore be concluded that population growth has surpassed the growth rate of employment in the agricultural sector. This may be due to the fact that agricultural land in the command area, unlike the population, is finite. Further, there may be a saturation point for absorbing labour for agricultural activities in the command area. In future, with improvements in and mechanisation of agricultural technology labour force needed for the command may be further reduced.

13.7.2 Poverty Alleviation

A comparison of the net income per household and per capita within the project command area (representing the 'with project' situation) and the corresponding income per household and per capita income in the control area (representing the 'without project' situation) indicated that the project had contributed considerably towards poverty alleviation by substantially increasing incomes.

13.7.2.1 Annual Income

The economic condition of households has improved after the project was commissioned, and this is true for all types of households such as farm households, agricultural labour households and non-agricultural households. Agriculture is the principal source of income for farm households, and the income from agriculture ranges between 55% and 63% of their total annual income. Other sources of income include dairy, poultry, small business and other miscellaneous sources. The study further reveals that per household and per capita income for the households of all categories of farm size in the command area ('with project' situation) are considerably higher in comparison with those of households in the 'control area'.[3]

Table 13.9 provides the annual per household and per capita income of farmers in the command and control areas.

13.7.2.2 Dairy Income

Animals are an integral part of Indian agriculture. Apart from providing milk, oxen and buffaloes are used for ploughing agricultural lands, carrying agricultural

---

[3]The study refers to unpublished collections from consultancy work.

Table 13.9 Per household and per capita income (US $1 = approx. Rs 46)

Farmer category	Area	Annual income (in Rs) Per household	Per capita
Marginal	Command	18,347	3,675
	CADA Command[a]	22,334	4,408
	Control	8,019	1,514
Small	Command area	27,589	4,715
	CADA Command	29,861	5,246
	Control	11,933	1,964
Medium	Command	36,995	6,313
	CADA Command	34,594	6,696
	Control	17,129	2,477
Large	Command	59,752	10,670
	CADA Command	72,771	14,554
	Control	–	–
Overall	Command	35,670	6,343
	CADA Command	39,890	7,726
	Control	12,360	1,985

[a]CADA Command is the area of the project command being managed by the Command Area Development Authority
Source: Unpublished collections from consultancy work

produce and other miscellaneous farm work. Crop stems are used as fodder for animals and their solid wastes are in turn used as manure. As elsewhere in India, dairy farming is an important economic activity in the command area. With an increase in green cover in the post-project era, availability of high quality food for animals has improved and this has resulted in enhanced annual production of milk per mulching animal. The number of animals per household has increased after commissioning of the Kangsabati Project. Table 13.10 shows the annual milk production per animal and annual income from dairy farming per household. It can be seen that income from dairy farming has increased by 150–175% in the post-project scenario.[4]

13.7.3 Cultivation Cost

As expected, the total annual cultivation cost per hectare is substantially higher in the post-project period. The cultivation cost per item, as a percentage of the total annual cultivation cost, has been provided in Table 13.11.

---

[4]The available data is from the Project report and from the unpublished collections from consultancy work.

Table 13.10 Annual milk output per animal and annual dairy income per household

Farmer category	Area	Milk production per animal per annum (in litres)	Net annual income per household (in Rs)
Marginal	Command	871	10,336
	CADA Command[a]	893	11,760
	Control	703	5,490
Small	Command	837	6,509
	CADA Command	841	7,174
	Control	791	5,590
Medium	Command	963	7,643
	CADA Command	956	6,332
	Control	804	4,430
Large	Command	924	4,386
	CADA Command	949	9,688
	Control	–	–
Overall	Command	899	7,219
	CADA Command	910	8,739
	Control	766	5,170

[a]CADA Command is the area of the project command being managed by the Command Area Development Authority
Source: Unpublished collections from consultancy work

Table 13.11 Itemised annual cultivation cost

	Annual cultivation cost as % of total cultivation cost in		
Item	Command area	CADA Command	Control area
Seed	4.68	4.78	7.18
Manure	4.77	4.60	4.82
Fertiliser	11.78	12.18	10.14
Irrigation	15.17	16.25	–
Pesticide	11.81	11.77	–
Family human labour	21.00	20.65	33.39
Hired human Labour	12.03	11.65	11.29
Hired bullock charges	3.78	4.59	6.52
Owned bullock charges	4.75	2.69	7.94
Machinery charges	5.46	5.96	–
Irrigation charges	15.17	16.25	0.00
Miscellaneous charges	4.75	4.88	1.70

Note: The sample survey was carried out in the post-project period. Being a very old project some inferences were drawn on the basis of interview of farmers and census report to find out the impact of the project
Source: Unpublished collections from consultancy work

13.7.4 Consumption Patterns

Table 13.12 provides the annual expenditure per household for various categories of farmers, under 'with project' and 'without project' situations, for food and non-food

Table 13.12 Annual expenditure per household

Farmer category	Area	Food items	Non-food items	Total
Marginal	Command	9,112.18	1,562.82	10,675
	CADA Command[a]	9,315.55	2,311.45	11,627
	Control	9,463.98	517.02	9,981
Small	Command	12,977.69	2,413.31	15,391
	CADA Command	12,818.54	4,236.46	17,055
	Control	11,336.46	1,035.54	12,372
Medium	Command	15,740.53	6,133.47	21,874
	CADA Command	15,900.29	6,235.71	22,136
	Control	12,280.72	2,901.28	15,182
Large	Command	23,880.88	9,773.12	33,654
	CADA Command	21,864.52	10,096.48	31,961
	Control	–	–	–

Annual expenditure per household (in Rs)

[a]CADA Command is the area of the project command being managed by the Command Area Development Authority
Source: Unpublished collections from consultancy work

items. The food items include cereals, pulses, vegetables, milk and milk products, non-vegetarian items (meat, fish and eggs) while non-food items consist of clothing, footwear, medicine, education, tea, tobacco, liquor and other miscellaneous items.

13.7.5 Assets

The assets of the farming community in this region include residential houses, cattle sheds, commercial establishments, agricultural lands and other immovable properties. Houses in the command area and CADA command area are of three types, i.e. kutcha, pucca and mixed, while most of houses in the control area are of the kutcha type. It can safely be concluded that houses in the command area and CADA command area are better built than those in the control area.

Similarly, the availability of consumer durables per household for each of the four categories of household is higher in the command area and CADA command area as compared to the corresponding categories in the control area. Consumer durables include furniture, light and heat appliances, modern durables (radio, TV, cassette player, etc.), vehicles and other miscellaneous durables. Table 13.13 on housing property and consumer durables per household substantiates these conclusions.

13.7.6 Immigration and Out-Migration

As shown in Table 13.14 on immigration and out-migration of people in the command area during the pre- and post-project periods, immigration to the command

Table 13.13 Housing and consumer durables per household (in Indian Rupees)

Farmer category	Area	Per household Housing property	Consumer durables
Marginal	Command	17,685	4,314
	CADA Command[a]	17,931	4,542
	Control	9,121	2,316
Small	Command	20,316	4,781
	CADA Command	21,494	4,987
	Control	9,908	3,253
Medium	Command	27,612	6,012
	CADA Command	28,241	6,302
	Control	14,167	4,819
Large	Command	35,403	6,896
	CADA Command	55,000	7,418
	Control	–	–

[a]CADA Command is the area of the project command being managed by the Command Area Development Authority
Source: Unpublished collections from consultancy work

Table 13.14 Immigration and out-migration in the command area

Period	Immigration as % of total population	Out-migration as % of total population
Pre-project (1961 census)	4.22	7.82
Post-project (1991 census)	4.85	6.77

Source: Unpublished collections from consultancy work

area has increased while out-migration from the command area has fallen at the same time. This, in itself, is an indication of improved avenues for earning a livelihood in the command area in the post-project period.

13.7.7 Impact on Literacy

Table 13.15 details the literacy level during the pre-project and post-project periods in the command area as well as the state as a whole. The percentage of the population that is literate has improved considerably in the post-project era. Literacy among females has increased fourfold while literacy among males has doubled. Literacy in the project area was lagging substantially behind the corresponding figure for the state during the pre-project situation. In the post-project period, literacy in the command area has surpassed the corresponding state figure for males and is lagging marginally behind the state figure for females.

Table 13.15 Literacy level in command and state, pre- and post-project

	Literacy percentage			
	Pre-project (1961 census)		Post-project (1991 census)	
Category	Command	State	Command	State
Male	34.90	46.57	68.80	67.81
Female	10.52	20.27	41.61	46.56
Total	22.85	34.46	56.41	57.70

Source: Unpublished collections from consultancy work

Table 13.16 Infrastructure and institutional facilities, pre- and post-project

	Number of facilities in the project area during	
Facilities	Pre-project period (1961 census)	Post-project period (1996–1997 census)
Post office	205	477
Bus route	61	167
Commercial bank	6	162
Rural/cooperative bank	6	47
Fertiliser depot	N.A.	2,043
Villages electrified	N.A.	2,670

13.7.8 *Infrastructural and Institutional Facilities*

The overall condition of postal communications, road networks, marketing facilities, banking institutions and electrification are listed in Table 13.16.

It is apparent from the table that infrastructural and institutional facilities have improved considerably in the post-project period.

13.7.9 *Benefit–Cost Ratio*

Considering the annual benefits from irrigation alone, the annual benefit–cost ratio works out to 1.53.

13.8 Social and Environmental Impacts

The Kangsabati Project was started nearly 50 years ago, in 1956, at a time when socio-environmental impact assessments were not a standard procedure. Thus, a systematic socio-environmental impact assessment was not undertaken at the pre-project stage.

Even during the post-project period, no serious attempts have been made to monitor socio-environmental impacts of the project on its surroundings. However, such impacts, as observed in and around the project area, are discussed in qualitative terms below.

13.8.1 Displacement of Human Population

It is a fact that a large number of families were displaced due to the execution of the Kangsabati irrigation project. The land required for the project was acquired in the late 1950s and early 1960s and the work began in 1956. The reservoir was impounded in 1964. According to information obtained from project authorities more than 40 years later, people were displaced only from the reservoir site and there was no displacement from the canal area. From the old office records of the project, it appears that 175 villages were partly or fully submerged and the number of persons displaced was in the order of 35,000—this is an estimated figure and not based on actual records. A detailed survey of the displaced and host populations was not carried out, since it was not necessary to obtain clearance for these aspects of development projects at that time. As such, no attempt was made to resettle and rehabilitate displaced persons and all those displaced opted for cash compensation (a total of about Rs 63,000,000) and moved out peacefully.

13.8.2 Loss of Forest

The construction of the Kangsabati Reservoir submerged an area of 14,284 ha, of which 352 ha (2.5%) was forestland. No reserved forest area was submerged. These forestlands were transferred to the project during the 1950s and 1960s, that is, before laws regarding compensation for forestland were enacted; as a result, no land in compensation for this transfer of forestland was demanded. However, new forests were planted in wastelands or other government lands under various afforestation schemes and programmes each year. Further, some forestlands were converted to agricultural lands, and the extension of irrigation to these areas resulted in the loss of forestlands.

13.8.3 Impact on Soils

The area towards the north and the west of the project area were susceptible to erosion. Overgrazing, indiscriminate felling of forest and a preference for rice cultivation even in sub-marginal uplands with very coarse soils susceptible to erosion, made the situation critical in some locations.

13.8.4 Flora, Fauna and Wildlife

Fortunately, the forests lost due to construction of the Kangsabati Project were not home to any endangered plants or wildlife. It cannot, however, be denied that there was some disturbance of wildlife due to reservoir impoundment. There was also loss of habitat for some species. Nevertheless, there was no effect on the sizeable wildlife population in the area. The plants growing in this area were mostly common species with widespread distribution outside the reservoir area as well. The risk of losses to the plant gene pool was, therefore, remote.

13.8.5 Ecosystem

There was no encroachment on existing or potential wildlife sanctuaries or other sites important for the life cycles of local birds and animals. There was also no encroachment on the nature reserve or other environmentally sensitive areas.

13.8.6 Erosion and Siltation

Erosion in the catchment area of the Kangsabati Reservoir was rather severe, resulting partly from deforestation and adverse soil conditions. This eroded soil progressively silted up the reservoir. The Kangsabati Reservoir was designed for a silt inflow of 590 m^3/km^2/year, but subsequent studies had found rates as high as 1055 m^3/km^2/year. Following impoundment of the reservoir in 1964–1965, a 1993–1994 sedimentation survey found that dead storage was silted up by about 21.5% and live storage by 5.7%. (Unpublished collections from consultancy work.)

   Problems were encountered in keeping the banks of the canals stable in the lateritic red soil, and some bank erosion was common. In cutting zones in the lateritic areas, where the depth of cutting intercepted the bottom of kaolin clays, cavities developed giving rise to slippages and bank collapse.

13.8.7 Water Logging Within Project Area

The records available indicate a temporary rise in the groundwater level during the post-monsoon periods. However, there was no trend indicating a permanent rise in groundwater levels over the years. Even in the post-monsoon stage, groundwater rarely reached the root zone of crops. Thus, the variation in groundwater level was mostly due to rainfall and not caused by seepage leaks and percolation of irrigation water.

13.8.8 Change in Water Quality

Long-term data on water quality of both surface water and groundwater were not available. However, a one-time study of water quality had shown low levels of salinity as compared to permissible limits, and indicated that salinity had not increased as a result of irrigation in the post-project period. No effect of eutrophication was felt in the reservoir area.

13.8.9 Loss of Religious, Historical and Cultural Monuments

Under a special statute of the Government of India, the Antiquities and Art Treasuries Act of 1952, the Archaeological Survey of India preserves and maintains historical and cultural monuments of regional and international importance. There was no loss of religious, historical or cultural monuments due to submergence caused by the project reservoir.

13.8.10 Transportation and Communication

Transport infrastructure in the project area was developed by providing approach roads, bridges and inspection paths over and along the canal banks. Sufficient crossings for vehicles, people and livestock were also provided across the main and branch canals, distributaries, etc. Bridges were built over regulators and falls, in addition to other road crossings where no structures were necessary. Communication and transport in the project area was thus improved rather than disrupted.

13.8.11 Incidence of Water-Related Diseases

Water-related diseases that required attention in the area were (i) vector-borne disease like malaria, filaria, etc., caused by mosquito vectors breeding in impounded water bodies, and (ii) diseases such as diarrhoea (including gastroenteritis and bacillary dysentery) related to poor quality of drinking water and inadequate sanitation. The reported incidence of malaria and filaria in the project area was not significantly different from that in surrounding areas and the state as a whole.

However, small tanks were used in villages for fish culture, domestic washing, cleaning of cattle, etc. This mix of uses made the water unhygienic for bathing and cleaning of utensils, causing diarrhoeal diseases, skin irritation and other infections. Unhygienic habits such as defecation in the fields and walking barefoot were also responsible for health problems.

13.9 Impacts of Command Area Development Programme

The Command Area Development Programme was introduced during the fourth five-year plan, with the objective of accelerating utilisation of the irrigation potential and increasing the productivity per unit of land and water, through specific land and water management schemes. The Kangsabati Command Area Development Authority was established in 1974 and continues on-farm development works and other related activities in the command area of the project. Implementation of this programme generated awareness about the utility of the scheme. People of the area perceived a sense of security regarding assured availability of water through field channels. This also motivated farmers to foray into commercial crops. The programme helped reduce losses and increase irrigation efficiency. Proper maintenance of on-farm development works undertaken by the Kangsabati Command Area Development Authority yielded substantial benefits which could not be achieved earlier.

13.10 Conclusions

An analysis of the socio-economic impacts of the Kangsabati Project (within the command area) as compared with corresponding impacts in the control area, representing the 'without project' situation, indicated that implementation of the project has led to increases in income and assets, improvement in living conditions and, thus, alleviation of poverty in the command area of the project. The consumption patterns for food and non-food items, quantity of livestock and production of milk per mulching animal per household have shown improvement under the 'with project' situation as compared to the 'without project' situation.

The construction and operation of the project has led in general to both negative and positive impacts on social life in and around the project. Some of the negative effects of the project were the displacement of population from the reservoir area, the loss of forest and effects on wildlife, fauna and flora due to submergence. However, the fact that the presence of the reservoir provided habitat for a significantly large variety of flora and fauna cannot be overlooked altogether. While the construction activity led to soil erosion in the upper catchment area and contributed to sedimentation in the reservoir, adverse effects such as significant soil degradation due to water logging, or a permanent rise in ground water level were not noticed.

The more important positive impacts of the project included stability in crop production with enhanced crop yield. Extension of irrigation facilities and increased intensity of irrigation led to the expansion of secondary and tertiary activities of the command area. On-farm development works and associated activities undertaken as part of the project through the Command Area Development Programme created awareness and assurance amongst the farmers. An increase in infrastructure and institutional facilities in the post-project period also provided greater work opportunities and employment to farmers, labourers and industrial workers.

Significant moderation in flood peaks was achieved through implementation of the Kangsabati Project and its regulated operation. Water releases from the project provided easy access to safe drinking water for humans as well as livestock, and thus led to an appreciable improvement in general health. The rise in groundwater level, as observed in the project area, supplemented the surface water for irrigation and other needs. A substantial rise in levels of education and literacy are visible in project area during the post-project period. The growth of recreation facilities in the command and adjoining areas, and considerable improvement in road and transport communication through the construction of bridges, approach roads over and along canal banks have promoted tourism in the area.

It is apparent from the post-project performance evaluation of the Project that no serious adverse impacts were noticeable in the project area. Instead, the agro-economic, socio-economic and socio-environmental impacts discussed above suggest that there was greater prosperity in the command area, or 'with project' situation, than the 'control area', 'without project' situation, in the case of the Kangsabati Project.

References

Central Water Commission (2002) Guidelines for performance evaluation of irrigation systems, New Delhi

Government of West Bengal (1953) Report of Kangsabati Irrigation Project, West Bengal

Ministry of Water Resources (2002) National Water Policy, New Delhi

Water and Power Consultancy Services (India) Ltd (2003) Performance evaluation studies of Kangsabati Irrigation Project. Central Water Commission, New Delhi

Chapter 14
Regional and National Impacts of the Bhakra-Nangal Project, India

R. Rangachari

14.1 Introduction

The Golden Jubilee of the successful operation of the Bhakra-Nangal Project was celebrated in July 2004. This multipurpose project has been in the service of the Indian Nation for the last five decades and has yielded immense benefits, far beyond expectations. At the opening ceremony of the Nangal Hydel Channel on 8 July 1954, India's first Prime Minister Jawaharlal Nehru made a speech at Nangal and paid tribute to the good work done by the project. For him, nothing was more encouraging than the sight of people trying to capture India's dreams and giving them real shape. He felt that Bhakra-Nangal was such a place, a landmark that had become the symbol of the Nation's will to march forward with strength, determination and courage. He told the audience that Bhakra-Nangal was 'not only for our times but for coming generations and future times' (Central Water Commission 1989: 3). He eloquently added, 'Which place can be greater than this, this Bhakra-Nangal, where thousands and lakhs of men have shed their blood and sweat and laid down their lives as well?' Where can be a greater and holier place than this, which we can regard as higher! (Central Water Commission 1989: 5). Again, while dedicating the project to the Nation on 22 October 1963, Nehru said 'Bhakra-Nangal Project is something tremendous, something stupendous, something which shakes you up when you see it'. He added, 'This dam has been built up with the unrelenting toil of man for the benefit of mankind and therefore is worthy of worship. May you call it a Temple or a Gurdwara or a Mosque, it inspires our admiration and reverence'.[1]

---

[1] From the transcript of Nehru's speech available with Bhakra and Beas Management Board (BBMB) office at Bhakra Dam.

R. Rangachari
Centre for Policy Research, Delhi, India

India became free from colonial rule six decades ago and since then has been making determined attempts at development. It needs to redouble its efforts at a fast pace because of its large population, which is already over a billion, and still growing. India's present population is thrice its size at independence in 1947. Making the Nation self-sufficient in food grain production has always been an important goal. While countries in the temperate zone receive rains throughout the year, India, under the influence of the monsoon, gets its annual rainfall within a limited period of three months. Since India's fresh water resources are limited and very unevenly distributed over space and time, storage dams are essential for meeting the water requirements round the year. Independent India made a significant start in this direction immediately after 1947. The Bhakra-Nangal, Hirakud, Damodar Valley, and Chambal Valley projects were actively pursued in the 1950s.

All over the world and for centuries, dams have been built in order to provide water for irrigated agriculture, domestic and industrial use, to generate power and to help control floods. Dam-based projects were often undertaken to serve some desired optimal combination of a number of these purposes. Dams and water resource development schemes have, thus, played a key role in economic and social development and made important contributions to the all-round development of project regions as well as nations. Affluent countries undertook the construction of dams and embarked on industrialisation on a large scale from an early date. Many of them already have in place the infrastructure necessary for water resource management. In contrast, most 'developing' and 'least developed' countries, including India, have been late entrants in these efforts and still have a long way to go.

It will also be relevant to recall that there have also been those who tended to take a dark view of these issues. Certain groups in developed countries, even while themselves enjoying the comfortable lifestyle enabled by decades of dam construction, irrigation and industrialisation, began to argue that the projected economic benefits from schemes involving large dams were not actually being realised, and that the environmental and social costs remained unaccounted for. Their sympathisers and cohorts in the developing world, too, then picked up this refrain. In recent decades, opposition to such schemes has become organised and vocal. In polemical debates, all large dams are portrayed as having a bleak record. The real issues involved tend to be obfuscated and arguments became polarised.

The tendency to denigrate all engineering projects and engineers as exploitative and venal, and to blame the contractors who build, the engineers who plan and design and the government that authorises a dam project has become fashionable in India too. Presenting cases in a highly exaggerated manner in order to draw attention is also common. For instance, this is how one well-known Indian critic of large dams put it: 'The fact that they do more harm than good is no longer just conjecture. Big dams are obsolete. They're uncool. They're undemocratic. They are a government's way of accumulating authority…… They are a brazen means of taking water, land and irrigation away from the poor and gifting it to the rich' (Roy 1999: 56).

The anti-dam lobby and its supporters have made unsupported hyperbolic assertions and alleged that dams had failed to deliver what they promised. The critic Arundhati Roy said in July 1999 that, 'Not a single big dam in India has delivered

14 Regional and National Impacts of the Bhakra-Nangal Project, India

what it promised. Not the power, not the irrigation, not the flood control, not the drought-proofing' (Roy 1999). Unfortunately, there have been precious few, if any, comprehensive independent analyses on how dam projects were selected for execution, how they performed over time and on the returns from these investments. Issues relating to dams and their impacts became the battleground in the arena of sustainable development.

The setting up of the World Commission on Dams (WCD) and its Report, which was released in November 2000, did not result in a balanced review but only accentuated the controversy. The WCD Report in fact questioned the very utility of dams and generated acrimonious debates regarding their impacts. The WCD Report has been critical of large dams. While it grudgingly admitted that 'dams have made an important and significant contribution to human development' it added that 'in too many cases an unacceptable and often unnecessary price has been paid to secure those benefits, especially in social and environmental terms, by people displaced, by communities downstream, by tax payers and by the natural environment' (WCD 2000: xxviii). The report highlighted the likely negative impacts of some dams while remaining almost silent about the beneficial effects. Much of its 'knowledge-base' comprised submissions by interested parties and lobby forces, and these had not been analysed and checked for their accuracy. Moreover, it made selective use of this material.

A large part of the WCD work involved a broad review of the experience with large dams. Its crosscheck survey of 125 dams around the world was stated to be part of the 'knowledge base' that enabled its review. The WCD Report stated that its evaluation was based on the targets set for the large dams by its proponents: the criteria that provided the basis for government approval (WCD 2000: xxx). WCD concluded from its 'knowledge base' that 'shortfalls in technical, financial and economic performance have occurred and are compounded by significant social and environmental impacts, the costs of which are disproportionately borne by poor people, indigenous people and other vulnerable groups' (WCD 2000: xxxi).

The general conclusion drawn by the WCD is serious enough to merit a close scrutiny based on scientific and critical performance analyses of existing dams over a sufficiently long period of service.

The Third World Centre for Water Management, assisted by the Nippon Foundation, initiated a performance analysis of some completed dams. The aim was to scientifically assess how far the dams had contributed to the social, economic and environmental development. The Bhakra-Nangal Project of India was considered appropriate for such a study. The Centre for Policy Research (CPR), New Delhi, was entrusted with this study, and the author was in charge of it. A study with regard to the positive and negative social, economic and environmental impacts of the Bhakra-Nangal Project was accordingly made. The CPR conducted this study during 2003–2004. The tentative findings were peer reviewed and discussed in a workshop and modifications, as necessary, made in the Study Report (Rangachari 2004). The findings of the CPR study are detailed in a book by the author published by the Oxford University Press (Rangachari 2006). The present chapter draws heavily from this study.

This chapter analyses the planning, execution and performance of the Bhakra Dam of the Bhakra-Nangal Project with an emphasis on the handling of its many inundation-related issues and impacts and an examination of how these compare with contemporary principles and procedures. The views expressed by those impacted by the project, ascertained by random field interviews, are also presented. The principal lessons learnt are also indicated.

14.2 Bhakra-Nangal Project

As mentioned above, the Bhakra-Nangal Project (BNP) is among the earliest water resource development projects of independent India. However, the origins of the project can be traced to 1908, when the Lieutenant-Governor of Punjab, Sir Louis Dane, directed the Irrigation Department of the province to investigate the scheme with a dam in the Bhakra gorge on the Sutlej River of the Indus system. A project report was prepared in 1910, but it was not immediately taken up for construction. The British government went on delaying the approval to the scheme due to their differing priorities that envisaged the development of 'crown waste' lands in West Punjab as a high priority.[2] However, further investigations continued sporadically and the Bhakra Project proposal underwent many revisions. The project was finally approved on the eve of the partition of British India and the emergence of India and Pakistan as free nations, which also involved the division of the Indus River system and the Punjab Province. Independent India had to revise the project after 1947 in light of the changed situation and construction forged ahead from the early 1950s.

The head-works of the BNP are located in the undivided state of Punjab in India. There are two integral parts of the project, namely, the Bhakra Dam and the Nangal Barrage. The former is a storage dam at Bhakra across the Sutlej including two powerhouses. The latter is a barrage across the same river at Nangal, enabling the diversion of water into the Nangal Hydel Channel, which also feeds the new irrigation canal system. The project features are briefly described below.

14.2.1 The Nangal Barrage

About 13 km downstream of Bhakra, the Sutlej emerges out of the Siwalik hills at Nangal and enters the plains. The Nangal Barrage is built at this point. It is a concrete structure 291 m long between abutments and 29 m high with a small storage capacity of 30 million cubic metres (MCM) to take care of the diurnal variations in the releases from the dam upstream.

---

[2]In late nineteenth century, British India initiated development of canal irrigation in arid wastelands. These canals mostly irrigated 'crown waste lands', that is uncultivated lands owned by the Government. This led in due course to the development of some notable 'canal colonies' in west Punjab and Sind, all of which presently lie in Pakistan.

The Nangal Hydel Channel (NHC), with a carrying capacity of 354 m^3/s, takes off just upstream of the Nangal Dam. The Ganguwal and Kotla powerhouses are built at the 19th and 30th km., respectively, from the head on the NHC. They had an original installed capacity of 77.65 MW each and commenced power generation in 1955 and 1956, respectively. The present uprated capacity is about 84 MW each.

The Bhakra Main Canal takes off from the Nangal Barrage system for extending irrigation to new areas. The project also involved remodelling of the existing Ropar headworks, enlarging the Sirhind canal, as also extension and improvement of irrigation.

The irrigation system comprises some 1,110 km of main and branch canals, and 3,379 km of distributary channels. The total area benefited is 4 million ha, of which the new area covered is 2.4 million ha. The barrage and canal system became operational from 8 July 1954.

14.2.2 Bhakra Dam

The Bhakra Dam is a concrete gravity structure across the Sutlej River. The maximum height of the dam over the deepest foundation level is 225.5 m. Its length at the crest is 518.16 m. The normal reservoir level at 513.58 m corresponds to a gross storage capacity of 9.62 billion cubic metres (BCM). The effective storage is 7.19 BCM. There are two powerhouses at the foot of the dam, one on either bank. The Right Bank Powerhouse had an installed capacity of 600 MW and the Left Bank Powerhouse of 450 MW. These have since been uprated to 785 MW and 540 MW, respectively.

Initial work for the dam commenced in 1948. River diversion and concreting in the foundation started in 1955. The construction was done through a government agency and completed in 1963. The total cost of the project was Rs. 245.28 crores (1 crore = 10 million).

Though the barrage and canal system became operational in 1954 and the dam itself was completed in 1963, the storage and use of Sutlej waters as well as development of irrigation had to be governed by the provisions of the Indus Waters Treaty.

14.2.3 Construction

There are many comprehensive accounts of the construction of the Bhakra-Nangal Project. Those interested in the intricacies and details of the planning and construction of the different components of the project should refer to these published accounts. The Bhakra and Beas Management Board[3] (BBMB) itself published the *History of Bhakra Nangal Project* in October 1988 (BBMB 1988).

---

[3]The Government of India set up in 1967 the Bhakra Management Board for the administration, maintenance and operation of the Bhakra-Nangal Project. After completion of the Beas Project works, in 1976, it was renamed BBMB and entrusted with the management of both the Bhakra-Nangal and Beas Projects as well as the regulation of the waters of the Sutlej, Ravi and Beas rivers in India.

Prime Minister Nehru placed the first bucket of concrete (in the spillway apron) on 17 November 1955. The first stage of dam concreting commenced in January 1956. After the shifting of the batching and mixing plants to higher locations, the second stage of concreting started in October 1958. Concreting in the main dam was practically complete by October 1961, except for the spillway bridge piers and towers and visitors' balcony. These were completed by October 1963.

With regard to the dam powerhouses, the first bucket of concrete was placed in the foundations of the Left Bank Powerhouse in March 1957. The first power unit was commissioned on 14 November 1960. Work on Power Plant-II on the Right Bank commenced later, only in 1963 and was completed in 1968.

The pioneering work of the Indian engineers and other specialists in successfully completing the project after overcoming obstacles on the way is laudable. To count them all will create a long list indeed and include many unsung warriors. They came up with many innovations and adaptations to suit the Indian conditions. No account of the design and construction of the project can be completed without due recognition to a small number of expatriate specialists and experts in different fields, whose services were utilised by India.

14.3 Indus Waters Treaty

Almost immediately after 1947, differences arose between India and Pakistan on the use of the waters of the Indus River system and on various development proposals for the future, including the Bhakra Project. Difficulties mounted over the years; irreconcilable positions were taken and threatening statements made by both sides.

The Indus Waters Treaty, that became effective from April 1960, resolved the water-related problems of the Indus Basin[4] (Ministry of Water Resources 1999: 44–154). Simply stated, following a transition period of ten years, this Treaty awarded the waters of the Eastern Rivers, namely, the Ravi, the Beas and the Sutlej rivers, for the unrestricted use of India, except for some very limited reservations for use by Pakistan. Similarly, it allocated the waters of the Western Rivers namely, the Indus, the Jhelum and the Chenab rivers, for use by Pakistan and for certain uses by India. There was to be no quantitative limit on India's use of the Western Rivers for domestic and industrial purposes. Provisions to allow for their non-consumptive use in India were also made. A permanent Indus Commission was established for the implementation of the Treaty and to serve as a regular channel of communication and contact between the two governments. Stringent and elaborate provisions for

---

[4]For fuller details, see the text of *The Indus Waters Treaty*, 1960, along with all its annexures. These have been published by both governments of India and Pakistan. Some of those connected with the negotiations have also written their accounts of the evolution of the Treaty. For instance, see— *Indus Waters Treaty—An Exercise in International Mediation* by N.D. Gulhati, Allied Publishers, New Delhi, 1973.

the resolution of differences and disputes were also included. The Treaty was to continue in force until terminated by a duly ratified treaty for that purpose between the two governments.

The Treaty stipulated that during a 'transition period', India should limit its withdrawals for agricultural use, limit abstractions for storages and make deliveries to Pakistan from the 'Eastern Rivers' in a specified manner. The limits for withdrawals from the Sutlej and Beas rivers at Bhakra-Nangal, Ropar, Harike and Ferozepur (including abstractions for storage by the Bhakra Dam and for the ponds at Nangal and Harike) were specified by sub-periods of the year. The transition period was to end in March 1970 (but was extendable up to 1973 at Pakistan's request).

The three 'Eastern Rivers', the waters of which were allocated for use by India, carry, on an average, about 40.5 BCM (32.8 MAF). Of this, the Sutlej accounts for 16.8 BCM (13.6 MAF). The Bhakra-Nangal Project was to utilise, fully and beneficially, the available flows of the Sutlej River allocated to India.

The Treaty was signed in September 1960 by Field Marshal Mohammad Ayub Khan, President of Pakistan, and Jawaharlal Nehru, Prime Minister of India, and 'for the purposes specified' in some articles of the Treaty[5] by William Iliff of the World Bank. It was ratified by both countries in December 1960 and became effective from the first of April 1960 (Ministry of Water Resources 1999).

The Treaty did not seek to optimise the use of the waters of the Indus system as a whole for the benefit of both India and Pakistan, but resolved the irritants between the two nations at that time. Hopefully, the two nations will some day find the ways and means for optimisation of the system's potential for their mutual benefit.

After the Indus Treaty, water requirements in Pakistan were to be taken care of by that country through appropriate links from the Western Rivers. Pakistan undertook the construction of the Mangala and Tarbela dams. In addition, it executed 'replacement works' to convey the waters of the Indus, Jhelum and Chenab rivers to the Sutlej—at Suleimanki and below Islam—for replacing the waters of the Ravi, Beas and Sutlej rivers which were 'reallocated under the Indus Waters Treaty'. Pakistan undertook the construction of some new barrages, as well as the re-modelling of existing ones, so that the requisite flows to the new link canals could be diverted.

14.4 The Pre-project Scene

The province of Punjab was split into the new provinces of East Punjab and West Punjab at the time of independence in 1947. East Punjab remained in India and West Punjab became part of the newly carved state of Pakistan. The Punjab Boundary

---

[5] The Treaty states after the signatures of the Plenipotentiaries of India and Pakistan that it was also signed by W. A. B. Iliff of the International Bank for Reconstruction and Development 'for the purposes specified in Articles V and X and annexures F, G and H'.

Commission Award given by Sir Cyril Radcliffe laid down the boundary between India and Pakistan. Delineated in August 1947 on religious considerations, it cut across the Indus basin, its river system, as also the vital canal irrigation systems therein, which had all been developed only under the conception of a single administration. This left everyone in the region unhappy.

After Partition, while nearly half the cultivable land in the Indus Basin remained with India, only 20% of the basin's irrigated area remained within its territory. This left several million hectares of arid but fertile land in the Indian part of the Indus Basin literally high and dry. Moreover, the allocation of a major part of the highly developed canal-irrigated areas to West Punjab worsened the food problem for India. Between the Sutlej and Yamuna rivers lay the vast, alluvial, gently sloping plains, which formed the watershed between the Indus and the Ganga. The Bhakra-Nangal Project aimed at bringing these areas under assured canal irrigation to protect the region from recurring droughts and famines.

The communal holocaust that accompanied the partition led to the uprooting of millions of people from their homes. Practically, the entire Hindu and Sikh population in the area that became Pakistan migrated to India. They needed land for cultivation, power for setting up industries and opportunities to restart their livelihoods. This sudden, mass arrival into India of millions of refugees aggravated India's problems. The Government of India was compelled to act very quickly to ameliorate the situation. Bhakra-Nangal Project seemed ideally suited to provide the solution.

14.4.1 Famines

Southwest Punjab and the adjoining areas of the state of Rajasthan had for centuries suffered from many devastating famines caused by severe droughts, till the very end of the colonial rule. Records speak of the Mughul Emperor Akbar ordaining in 1568 that 'this jungle (Hisar District[6] which now lies in the command of the Bhakra-Nangal Project) in which subsistence is obtained with thirst, be converted into a place of comfort' (Central Board of Irrigation and Power 1965: 42). The efforts undertaken obviously did not stop famines from recurring. The arrival of the British on the scene and their rule for two centuries, too, did not make much difference with regard to droughts and famines in this region. However, the colonial rulers appointed three successive Famine Commissions in the last three decades of the nineteenth century to report on the dimensions of the droughts, their causes and relief to the affected people. Detailed codes for providing relief were drawn up. However, Hisar District featured again and again in famine distress records. The District Gazetteer of Hisar states that the first such case for which authentic accounts existed relates to the year 1783 when thousands died. Similar accounts are available for the famines

---

[6]Hisar was originally part of Punjab province, but after the province was bifurcated, it became part of Haryana state.

of 1860–1961, 1868–1969, 1876–1978, 1899–1900, 1918, 1929–1930, 1932–1933 and 1938–1940. Likewise, accounts are available in the District Gazetteers relating to other districts of Punjab such as Ferozepur and Amritsar and those of Rajputana (now Rajasthan) such as Sri Ganganagar.

14.4.2 Drought-Prone Area

The vulnerability of any area to drought depends on the extent to which physical and climatic conditions play an adverse role in creating an unstable agriculture. The Irrigation Commission set up by Independent India in its report submitted in 1972 (Ministry of Irrigation and Power 1972: 400) defined drought-affected areas as those where the annual rainfall was less than 100 cm and further where 75% of this rainfall was not received in 20% or more of the years. Over large areas of Punjab (including Haryana) as also Rajasthan, the average annual rainfall is below 60 cm, and the western parts of Rajasthan and adjoining areas in Punjab receive hardly 10 to 30 cm rain. The Irrigation Commission pointed out that irrigation, to the extent it could be provided, would give protection to many of the drought-prone areas. After adequate irrigation coverage is extended to these areas, they need no more be classified as 'drought affected'.

Annual rainfall in the Punjab plains varies between 40 and 80 cm. The rainfall in Haryana is scanty and erratic. In the southern part of the state, it is as low as 25–40 cm. Rajasthan is the driest state of India. The annual rainfall is as low as 13 cm in the Thar Desert. The districts of Bikaner, Ganganagar, Barmer, Jodhpur, etc., fall in the arid and semi-arid zones. The Bhakra Dam was planned to provide assured irrigation to the plains districts of Punjab and Haryana and the Ganganagar District of northern Rajasthan.

As long as crop cultivation was dependant only on rainfall, the crop pattern was curtailed and crop yields were low. The average yields of food crops like wheat, rice, maize, etc., were extremely low. The average food grain yield in Punjab for the 5-year period 1950–1951 to 1955–1956 was 730 kg/ha. While undivided Punjab was the granary of India, post-partition Punjab in India became deficient in food grains.

14.4.3 Drinking Water

Water sustains life and, hence, is a basic need and right. The fundamental right to life is thought to include the right to drinking water. Unfortunately, large parts of the project region completely lacked this basic amenity till the BNP was developed and completed. In particular, the areas in Rajasthan and adjoining parts of Haryana and Punjab suffered acutely due to this problem. It was a typical sight in this area to see women and children walking miles to fetch potable water for domestic use, deftly balancing a number of pots on their heads and in their hands. Similarly, at many places, the menfolk had to go for miles in camel-drawn vehicles to fetch water from far-off sources.

14.4.4 Floods

While droughts and famines took their toll, Punjab was also subject to flood devastation during 'wet' years. Soon after the Sutlej River and its many tributaries emerge from the hills into the plains, they spread out and start meandering. Their steep, and sudden, changes in gradient aggravate silt and erosion problems. The swollen rivers often inundated fertile cultivated lands in the plains and deposited silt over them. Crops and properties were, thus, lost and post flood cultivation rendered difficult.

14.4.5 Communications

Hilly regions in India had always been remote and inaccessible. Roads and railways were non-existent. The BNP region was no exception. The Sutlej catchment area upstream of the Sirhind canal headworks in Ropar remained mostly inaccessible. In the absence of communications, these remote areas were deprived of development and even the minimum standards of quality of life. Bhakra remained an inaccessible spot even in 1947. It took two days for the project investigating team and inspecting officers to reach the Bhakra Dam site at that time. Conducting surveys upstream in the reservoir area meant going into even more inaccessible territory.

Prior to the decision to implement the BNP, the nearest rail terminus was at Ropar, 69 km (43 miles) away from the project site. The decision to extend the railway line along the left bank of the river till Nangal and beyond was taken to facilitate the movement of workers and materials. Construction of the railway line was entrusted to Northern Railways, part of Indian Railways and covering the northern zone. The route passed through very difficult terrain and involved crossing over 60 torrents by bridges.

The Ropar-Nangal rail link was opened to traffic in October 1948. A rail-cum road bridge was built across the Sutlej at Olinda in 1953 to provide a link to the other bank. Till then, the only communication to the right side of the river from the left bank approach road was by two suspension bridges built by the project. The railway line on the right side was laid in 1953.

Construction of the highway from Ropar to Nangal was also undertaken around the same time. The 58-km (36 miles) long highway from Ropar to Nangal was constructed by the Public Works Department, Punjab, and was completed by 1947. The road from Nangal to Bhakra, about 11-km (7 miles) long, was constructed on the left bank of the Sutlej by the Bhakra Project administration. This was a 7.6-m (25 ft) wide hill road.

14.4.6 Medical Facilities and Health

The Bhakra-Nangal Project report (1955) prepared by Punjab government does not give details with regard to medical facilities and health conditions in the project area. Preexisting nutritional and health conditions had to be surmised. Waterborne diseases like cholera, diarrhoea, dysentery, etc., were known to be prevalent in all areas with drinking water of poor quality. As in other hilly and remote locations in India, there were hardly any modern hospitals or health facilities in the project region and upstream areas. In the absence of road or rail facilities, if someone fell seriously ill, it was a long, hazardous journey over many days to take the patient to the nearest hospital with some modern health care facilities. Even during the days of investigations and the initial days of the project construction, these facilities were unavailable to the project staff or other people of the township till 1951.

14.5 Submergence and Land Acquisition

14.5.1 Maximum Inundation Area

The Bhakra Project investigation and planning group studied different storage possibilities and other considerations over the years. The early project report of 1919 envisaged a 121-m (395 ft) high dam with a storage of 3.18 BCM (2.58 MAF or million acre-feet), mainly for irrigation. This was later raised to 152.4 m (500 ft) high dam to provide storage of 5.86 BCM (4.75 MAF) in the project report of 1934–1935. In 1940, the question of raising the dam to a height of 750 ft was considered by the government in order to provide for greater water and power benefits. This increase was, however, subject to the consideration of two issues: suitability of the foundation and abutments, and the political problem of the submergence of Bilaspur town. The foundation conditions suited a much higher dam, but the Raja of Bilaspur was agreeable only to a lower level, in order to reduce the land area that would be submerged.

The position changed drastically after partition and independence in 1947. The Raja of Bilaspur now agreed to a higher extent of submersion in his state, including Bilaspur town. In 1948, a decision was made to build a higher dam that was safe as determined by geological studies and foundation conditions, and to fully exploit the irrigation and power potential of the Sutlej River. The dam, as finally planned and constructed, is 225.55-m (740 ft) high to provide a gross storage of 9.62 BCM (7.8 MAF). The reservoir covers a maximum area of 168.35 km^2 (65 miles^2 or 41,600 acres) when full and extends to about hundred km (60 miles) from Bhakra. The submergence area per unit volume of storage (BCM or MAF) was found to be the lowest among all the storage dams completed in India since the start of the twentieth century.

Table 14.1 Details of land acquisition for Bhakra submergence area

District	No. of villages affected	Total land acquired (ha)	Privately owned land (ha)	No. of families owning private land
Kangra (Una)	110	5,483	5,483	3,333
Bilaspur	256	12,313	5,611	3,838
Mandi	5	162	15	35
Solan	4	26	26	3
Total	375	17,984	11,135	7,209

Source: BBMB (communication to the author based on unpublished records)

14.5.2 Land Acquisition

Land to the extent of 17,984 ha (44,440 acres) was acquired by the project in the districts of Kangra (presently Una), Bilaspur, Mandi and Solan of Punjab and Himachal Pradesh. In addition to the small town of Bilaspur, 375 villages were involved. Only 48 villages of Kangra and 14 villages of Bilaspur were completely submerged, while the remaining villages were affected in varying degrees. About 30% of the land acquired was classified as forest-land even though much of it was really degraded forest.

Of the total land required, 6,849 ha (16,924 acres) were already government owned and only the balance 11,135 ha (27,516 acres) of privately owned land had to be acquired. The land was acquired as per the provisions of the Land Acquisition Act, 1894, as it then existed. This involved the migration of 7,209 families or a population of 36,000 people. The salient details are given in the Table 14.1.

For the urban displaced, a new town of Bilaspur was built just 2 km away on high land overlooking the old town, and a thousand urban families resettled there.

14.5.3 Bhakra Rehabilitation Committee

In view of the number of persons to be affected, and the desire of the government to deal with the matter in an interactive and comprehensive manner, the Bhakra Control Board set up the Bhakra Rehabilitation Committee to examine the inundation-related issues under the chairmanship of the Secretary to the Government, Public Works Department. The Committee was asked to advise the government on the following matters:

a. Principles and methods of rehabilitation with particular reference to:

i. Basis of rehabilitation on vis-à-vis land for land, cash compensation, etc.

ii. Places of resettlement, after ascertaining public opinion, both among the population to be displaced as well as among the people of the area where the displaced persons would be rehabilitated

iii. Fixing responsibilities of the government/authorities for rehabilitation

b. Procedure for determining compensation to the displaced persons
c. Procedure for determining compensation in individual cases
d. Rough estimates of cost and recommendations regarding its incidence
e. Construction of new town in lieu of Bilaspur

The General Manager of Bhakra Dam, Joint Secretary of Revenue in Himachal Pradesh, Deputy Commissioner of Bilaspur and Deputy Commissioner, Resettlement, were the official members of the Committee. The three members of the Himachal Pradesh Territorial Council and the Member representing Kangra in the Legislative Assembly of Punjab were non-official members.

The Bhakra Rehabilitation Committee decided, at the very outset, that in respect of the lands to be submerged, the displaced persons would be compensated, as far as possible, in the form of land. In cases where the displaced preferred not to receive compensation in the form of land, they would receive cash or part land and part cash. Compensation for houses, trees, etc., was to be provided in cash. Liberal compensation was paid for land, houses, trees, *gharats*[7] and other property to be submerged. The acquired lands were even leased out to the erstwhile landowners on a temporary basis till the actual submergence, with the proviso that they could not evict existing tenants. Owners of houses in the submerged area were permitted to make free use of material from their houses, regardless of the acquisition value.

Since the resettlement of those likely to be affected by the rise of water up to R.L. 1,280 ft[8] was an immediate issue, the resettlement was carried out in two stages, namely, first those affected by inundation up to R.L. 1,280 and later, those affected beyond that level. The Government of Himachal Pradesh framed its own rules and principles for the first phase of resettlement.

In this first phase, displaced persons/families who were either landowners or tenants who had cultivated land in the reservoir area were given twice the area cultivated. In case of uncultivated land, an equal area of land was given. All others were granted up to 10 *bighas* of land.[9] People displaced due to submergence between R.L. 1,200 ft and 1,700 ft were offered resettlement in Hisar District.

Displaced persons in the second phase who could not move to Hisar because of a small amount of compensation (up to Rs. 500) were also granted: 10 *bighas* of land in Himachal Pradesh. Any other landless displaced person who was not able to resettle elsewhere was also entitled to 5 *bighas* of land for his dwelling/homestead.

14.5.4 Resettlement

As per the resettlement policy, stakeholders were consulted and their views and desires regarding resettlement were fully considered. Nearly a third of the affected

---

[7]Garats are indigenous small water mills, mostly for grinding wheat.
[8]R.L. stands for Reduced Level, commonly used technical term denoting the elevation over a bench mark.
[9]Roughly, a bigha equals five-eighths of an acre.

Table 14.2 Rehabilitation of displaced families

	Number of families
Resettled within Himachal Pradesh by the Himachal Pradesh government	2,398
Those who preferred cash compensation and resettlement on their own	2,632
Families that preferred resettlement in the irrigation command	2,179
Total	7,209

Source: BBMB and Deputy Commissioner, Resettlement, BNP (communication to the author based on unpublished records)

families were resettled in areas of Himachal Pradesh in the vicinity. A little over a third of the families preferred to receive cash compensation, so that they could make their own resettlement arrangements. A little less than a third of the families wished to resettle in land within the irrigation command of the project. The details are provided in Table 14.2.

In addition to resettlement, the following actions were taken to provide means of livelihood and more amenities for families resettled in Himachal Pradesh:

- Free fishing licences in the reservoir for three years
- Gainful employment for local people in project construction

Surveys for locating suitable land in the Sirmour, Mandi and Mahasu districts of Himachal Pradesh, as also in the Hoshiarpur and Ambala districts of Punjab, were carried out, but very little land was available. However, a similar survey in Hisar District within the irrigation command showed that sufficient land in compact blocks was available at cheap rates. The displaced people, too, expressed their preference for lands within the irrigation command in Hisar. About 5,342 ha (13,200 acres) of land in the BNP command was acquired in compact blocks in 30 villages of Hisar District.

Compared to the price paid for the land to be submerged, the cost of land to be purchased for resettlement in Hisar was low. It was thus found that the full value of compensation owed for land under submergence could not be ensured solely through land in the command. Therefore, it was determined that compensation would be given partly in the form of land and partly in cash, subject to two overriding considerations: No displaced person would be allotted more than 25 acres (about 10 ha) of land or get less than the extent of his cultivated land acquired for submergence, if the compensation amount was adequate to meet its cost.

In order to help small landowners among those displaced, compensation up to the first Rs. 1,000 was made fully in the form of land. For larger compensations, a sliding scale was adopted, the effect of which was to progressively reduce the fraction of the total compensation that will be given in the form of land. For instance, one who became eligible for a total compensation of Rs. 2,000 got land valued at Rs. 1,400 and cash of Rs 600. If the total compensation was Rs 4,000, the land component was equal to Rs. 2,050 and cash Rs 1,950. If the application of the sliding scale resulted in the land component being less than five acres, then a minimum of five acres was permissible. As the project had acquired some extra land in Hisar,

14 Regional and National Impacts of the Bhakra-Nangal Project, India

those who were eligible for less than 5 acres were given an option to increase their allotment up to 5 acres.

Landless tenants in the submerged area were also declared eligible for allotment of land in the command. They were given land to match the acreage of their tenancy (as recorded in revenue records for the 1957 *kharif*[10] season), subject to a maximum of 5 acres (about 2 ha), the cost of which was to be paid by them in 20 equal half-yearly instalments, beginning after a grace period of two years. Even artisans and labourers affected by the project, who did not own or cultivate land as tenants but wished to shift and settle in the Hisar District, were each allotted a half-acre of land free of cost.

Allotments of *abadi*[11] plots for the construction of residences and shops in the resettlement villages were also done. *Abadi* sites were on the basis of four allottees per acre. Model layouts for *abadi* sites were planned in each resettlement village.

The following additional amenities were provided:

- Temporary shelter accommodation
- Easy loans for construction of houses and supply and transport of building materials
- Repairs to old wells and new wells for drinking water supply, where necessary
- Bhakra canal water supply through new canals, minors and water courses
- Bridges on canals, new ferry services, new roads or village paths
- New primary schools, medical facilities, cattle inoculation and security arrangements
- Rail/bus fare plus a lump sum rehabilitation grant

In order to resettle the displaced as close to their old *biradari* (social and cultural group setting) of neighbouring villages as far as possible, allotment of land in the new area was made on the basis of *had bast*[12] numbers in the erstwhile villages in the districts of Kangra, Bilaspur and Mandi.

The office of the Deputy Commissioner, Resettlement, with its headquarters at Hisar, was started in 1961 to deal with all these issues. Even after this district was split into two or more administrative units in later years, this single centralised resettlement office at Hisar monitors the condition of people resettled in the command and deals with all issues and problems raised by them including court cases carrying on from earlier periods. Deeds transferring proprietary rights over the land to the allottees have been executed, after they fulfilled all the conditions of their allotment and cleared all pending dues. Of the 2,285 allottees, 2,212 or 97 % have completed all formalities and the ownership rights have been transferred to them.

Some of the land acquired in Hisar remained surplus, and a part of the common land reserved for other purposes such as roads, parks, rest houses, places of worship also remained unutilised. These lands were kept under *Purusharthi*[13] committees for the common benefit of the resettled people.

---

[10] *Kharif* is the crop harvested in the autumn season.

[11] *Abadi* in local vernacular stands for habitation.

[12] *Had bast* can be loosely translated as the list of property owners in the government record.

[13] *Purusharthi* literally means energetic or industrious. Here it has been used to denote the standing committee representing the displaced people that deal with common material benefits.

14.5.5 Random Survey by Centre for Policy Research

The CPR study (Rangachari 2004) aimed to ascertain the perceptions of the people who were benefited or impacted in any manner through a random sample survey and this was made an integral part of the comprehensive CPR study. Field interviews took place between July 2003 and January 2004 to gather the people's views regarding various impacts. The results are detailed in the final report of the study (Rangachari 2004). The survey generally confirmed that most of those displaced who were contacted did not have any real serious adverse comment on the resettlement and rehabilitation aspects of the project. It also revealed that poor farmers and the common men and women below the poverty level benefited as much if not more than the others. Similarly, the BBMB carried out a quick field study in 2005 to determine the views and present perceptions of the Bhakra-displaced (Central Board of Irrigation and Power 2005). This study too revealed that the resettlement and rehabilitation policy for those displaced by Bhakra was considered quite reasonable and framed with good intentions. The policy received the people's full cooperation and the resettlement process was free from corruption.

14.6 Project Objectives

14.6.1 Main Objectives

The declared objectives of the project were irrigation and hydropower development. Incidental and indirect benefits included increased production of food grains and cash crops, flood control, industrial development, reclamation of wasteland and refugee rehabilitation. The Bhakra-Nangal Project helped stabilise, improve and extend the Sirhind as well as the Western Yamuna Canal systems in addition to providing new irrigation to vast areas. The project also realised greater power generation than was planned for. The details of these and other achievements are briefly discussed below.

14.6.2 Direct Benefits

14.6.2.1 Irrigation

With a view to providing irrigation facilities to long neglected arid and semi-arid tracts of lands in East Punjab and Rajasthan, the Bhakra Project was reoriented soon after partition in 1947. The need to assure water supply to the then existing irrigation systems of Sirhind Canal, Grey Canal, Sarsuti and Gaggar Canals was also kept in view during the planning of the new irrigation canal system. This necessitated the remodelling of the Ropar headworks, modifications in the Sirhind canal system and the canal system for new irrigation command taking off at the tail of the Nangal Hydel canal near Roopnagar, as also the Nangal Barrage and Bhakra Dam. This had to be done while maintaining uninterrupted flows to existing canal systems in India, and while taking into account India's commitments and understandings in respect of water supplies to downstream areas in Pakistan.

The total culturable command area under the project is 2.372 million ha and the irrigated area is 1.296 million ha. The extent of area to be covered by irrigation under the Bhakra-Nangal Project, as envisaged at the time of its approval, was fulfilled, notwithstanding the many limitations that were subsequently imposed under the Indus Waters Treaty (1960). Additional irrigation benefits, too, became possible since 1977 by the diversion of the Beas waters into the Bhakra Reservoir, enabled by the same Treaty.

The crop pattern that was recommended by the Crop Planning Committee for this project[14] was originally followed. However, the progressive farmers soon changed the crop pattern in tune with market forces. Irrigation development under the Bhakra-Nangal Project coincided with the start of the era of the 'Green Revolution'. With professional assistance from agricultural universities and government support, further changes in crop patterns became necessary due to the demands of the new strategy adopted in agriculture. Groundwater development, increased number of tubewells and the procurement and marketing strategy of the Government of India played their part in revising the crop patterns adopted by farmers in later years.

14.6.2.2 Agricultural Production

In the 1960s, there was widespread shortage of food grains in many underdeveloped and developing countries. The USA was then the main surplus producer and had sizeable accumulated stocks. It offered many countries food grains at concessional prices and on special terms under its Public Law 480. India too began to import wheat from the USA under Public Law 480. In the beginning, the imports were small, but they soon increased. At the same time, the wheat came with strings attached and political pressure was felt (Subramaniam 1979: 8–9). There were many awkward moments in ensuring the required imports. A time came when the public distribution system was so stretched that the Nation literally lived 'from ship to mouth'.

Lal Bahadur Shastri became Prime Minister of India after the demise of Jawaharlal Nehru in 1964. Soon thereafter, there were severe pressures on the Indian economy. To make matters worse, monsoon rains failed in 1965 and 1966. India had to negotiate the import of ten million tonnes of food grains from the USA for 'otherwise it would have been impossible to avert human disaster of a magnitude unparalleled in human history' (Subramaniam 1979: 8).

Many doomsayers predicted disaster for India. Drs William and Paul Paddock said that there would be 'acute famine conditions and that millions and millions of

---

[14]In 1952, the Government of Punjab constituted a Crop Planning Committee for the project to recommend the crop pattern tract wise, taking into consideration the climatic conditions, rainfall, water table, type of soil, local agricultural practices and other local factors.

people would die of starvation in the developing countries, particularly in South Asia' (Paddock 1967: 67). The Paddock brothers classified developing nations according to their capacity for escaping famine through possible advancements in agriculture. They used the principles of 'triage analysis' adopted in military circles and put India in the category of 'can't be saved' on the analogy of 'incurables who are generally not attended to and are left to die' (Subramaniam 1979: 10).

This experience was adequate warning that a country the size of India with its increasing population pressure could not sustain itself on imported food grains on the large scale witnessed at that time. Luckily, this spurred India to forge a new strategy to increase its agricultural productivity and reach self-sufficiency as early as possible.

The new Agriculture Minister attempted to introduce many bold initiatives for sustained increase in food production, now known as the 'Green Revolution'. The new strategy for agricultural production was a package of measures that included dependable irrigation and controlled water supply, along with high yielding varieties of seeds, fertilisers, pesticides and mechanical equipment. To optimise results, certain preceding operations and post-harvest initiatives were also taken. These included consolidation of land holdings, execution of command area development works, provision of agricultural credit, marketing, price support and remunerative prices, as also developing the necessary institutions for the same.

The results were soon evident. Punjab, which produced less than 2 million tonnes of wheat in 1965–1966, raised its output to over 5 million tonnes by 1970–1971. It crossed 6 million tonnes by 1976–1977 and reached 10 million tonnes by 1984–1985. By the year 2000, it had crossed 15 million tonnes. This was followed by the rice revolution. Prior to 1965–1966, rice was grown only in some pockets in Punjab. With the extension of irrigation and the advent of high yielding varieties of rice seeds, rice was cultivated in areas that could not have supported it earlier. From a production of less than half a million tonnes of rice in 1968–1969, the state crossed 3 million tonnes by 1978–1979 and touched 5 million tonnes by 1985. In the year 2000, rice production stood at 9 million tonnes. There was no significant increase in the net sown area between 1970 and 2000.

Haryana, which was carved out of Punjab in 1966 as a new state, was not far behind, though it did not have as extensive an irrigation system as Punjab. Wheat production in Haryana was a mere one million tonnes in 1966–1967. However, it had crossed 3 million tonnes by 1978–1979 and reached 5 million tonnes by 1985. Rice production in Haryana was well over 9 million tonnes by 1999–2000. In this state, too, there was hardly any increase in the net sown area from 1970 to 2000.

A study of the statistics from year to year on the area, production and yield (productivity) per hectare for the two important food grains, wheat and rice, in Punjab and Haryana, has been made (Rangachari 2004). This has brought out many interesting results. There has been a continuing upward trend in the production and yield of rice and wheat, both in Punjab and Haryana. Within this rising trend, the average rice yields have fluctuated from year to year, with variations in the quantum and distribution of rainfall. In both states, almost the entire rice and wheat crop is now irrigated. The average wheat yield in Punjab is the highest for any state in India, and Haryana comes next. These two states are the only ones in India to have surpassed

Table 14.3 Productivity of rice and wheat in Punjab and Haryana

	Kilograms per hectare					
	Rice			Wheat		
Year	India	Punjab	Haryana	India	Punjab	Haryana
1966–1967	863	1,186	1,161	887	1,520	1,425
1970–1971	1,123	1,765	1,710	1,307	2,238	2,074
1980–1981	1,336	2,733	2,602	1,630	2,730	2,359
1990–1991	1,740	3,227	2,775	2,281	3,715	3,479
1999–2000	1,994	3,347	2,385[a]	2,778	4,696	4,166

[a]The productivity of rice in Haryana had been 2,964 kg and 2,800 kg in 1996–1997 and 1997–1998 respectively. But later monsoon rains in 1998 and 1999 were less than normal, and hence, the productivity was relatively lower in these two years. The productivity again went up after 2000–2001

Source: Statistical abstracts of Punjab and Haryana

a yield of over four tonnes per hectare. The average yield of rice is also the highest in Punjab. The traditional rice-growing states, Tamil Nadu and Andhra, follow closely and Haryana comes thereafter.

The progressive improvement in rice and wheat productivity in Haryana and Punjab is detailed in Table 14.3.

The large-scale production of rice in this region, which is not traditionally an area of rice consumption, proved to be a blessing for the farmers and the millions of Indians depending on the Public Distribution System (PDS). Punjab and Haryana contribute the maximum to India's PDS, which seeks to enhance food-security for the economically weaker sections of society. These states contributed half to three-fourths of the wheat and rice that came into the central pool in the last three decades. Another positive outcome of the rural prosperity triggered by the 'Green Revolution' is the low level of rural outmigration from Punjab and Haryana since 1966.

Punjab and Haryana have effectively tackled the groundwater rise noticed in the initial period, soon after large-scale irrigation was introduced. The encouragement given by the government for the conjunctive use of ground and surface water by offering incentives for the development of tubewells has paid off. Furthermore, the cultivable area subject to salinity has continually decreased. Certain agrarian issues have risen in recent years in this region, but these go far beyond the objectives of the present study, namely, the promise and performance of the Bhakra-Nangal Project.

14.6.2.3 Hydropower

After the Nangal Hydel channel became functional in 1955 and till the Bhakra Dam powerhouses were completed in 1966, the average annual generation in the two powerhouses on the Nangal Hydel channel was around 600 million kWh. Thereafter, the project had an annual generation of around 4,000 to 4,500 million kWh till 1975. During the 25 years thereafter, power generation stood at around 7,000 million kWh per year. This is more than what was envisaged at the time of the project's approval.

The project has also built up an expertise in managing the power system, and has constantly attempted modernisation and renovation. All five units of the Bhakra Left Bank Powerhouse were uprated, thereby increasing the installed capacity from 450 MW to 540 MW, an increase of 20 %. Similarly, all five units of the Right Bank Powerhouse were uprated, raising, in the first phase, the installed capacity from 600 to 660 MW. In the second phase, this was again raised to 785 MW. Plans are afoot to increase the installed capacity of the Left Bank Powerhouse yet again to 630 MW. The project engineers, technicians and workforce have carried out the entire renovation and uprating at a fraction of the cost that would have been incurred in new power projects.

Being the largest hydroelectric complex feeding India's Northern Regional Grid, Bhakra (along with Beas) power meets the heavy peak demand of the region. Like other regional grids of the country, the Northern Regional Grid, too, faces huge peaking power shortage, particularly during morning and evening peak hours. The Bhakra generators are then synchronised, one after the other, to pick up the load at a fast rate to match the sharp rise in load and to help in controlling the excursion of frequency to dangerous levels.

The Indian grids also experience variations in frequency. Only a small number of machines in India operate in free governor mode for primary regulation and there are no designated machines for secondary control or Automatic Generation Control (AGC). However, all the ten machines at Bhakra operate on free governor mode, which makes a major contribution to the primary regulation of frequency. This helps stabilise the frequency of the Northern Regional Grid. During grid contingencies, grid disturbances have been averted by the timely intervention of the Bhakra powerhouses.

During grid collapses, power for 'black starts' has been provided by the Bhakra powerhouses. It is difficult to imagine the smooth operation of the Northern Regional Grid without Bhakra and the other hydro machines of the BBMB. This has been possible due to flexibility in operation offered by the Bhakra machines backed by a large reservoir.

The project has enabled the electrification of 128 towns and 13,000 villages. The other achievements by the Bhakra powerhouses include high plant availability (90–94%), high transmission serviceability (99.5–100%) and lowest possible cost of generation (Rs 0.06 /kWh).

14.6.3 Incidental Benefits

The project report[15] indicated a number of incidental benefits. These included flood control, immunity from famines, increased production of food and cash crops, industrial development, reclamation of state wastelands and rehabilitation of refugees.

---

[15]The term 'Project Report' here and later stands for the one that was approved in 1955 for execution, (Punjab Government 1955).

Many other benefits that were not explicitly stated were also derived from the project.

14.6.3.1 Flood Control

According to the project report, flood control is not a primary objective but an incidental one. The fairly large storage capacity of the Bhakra Reservoir was generally used to render flood moderation below the dam. Prior to the Bhakra Dam, the channel downstream was in the form of flood plains, including the flood channel about 7- to 8-km wide, and the adjacent marshy low land. After the dam was built, the Punjab government constricted the Sutlej flood plain by constructing embankments that were designed to cater to a carrying capacity of only 200,000 cusec and reclaimed the remaining area. This is a factor to be noted while examining the flood conditions in the lower Sutlej region.

Based on the actual operations of the BNP until now, a study was made by CPR (Rangachari 2004)[16] of the flood moderation actually achieved by the project over the period 1964–2002 for all inflows greater than 200,000 cusec. It was found that, barring two instances, one in 1978 and another in 1988, the Bhakra Dam was able to moderate all flood peaks, restricting the outflow to a level very much lower than 100,000 cusec.

Some critics made a serious charge in 1988 that the BBMB caused disastrous floods by 'opening the flood gates of Bhakra and Pong' (Verghese 1994: 23).[17] The specific circumstances of the 1988 situation were examined by the BBMB through expert consultants. It was found that in the light of the circumstances then prevalent, the accusations made against the regulating authorities were unwarranted. It is, however, true that there were serious flood losses in 1988, but the available evidence showed that the flood damage would have been much more but for the moderation provided by the dam.

In the case of Bhakra, as elsewhere, clashing interests between flood control and irrigation and/or power generation could be seen as reflected in the differing interests of the participating states. However, there cannot be any lasting advantage in creating economic prosperity in an area through the development of greater irrigation, power and industrial infrastructure, if floods are allowed to wipe out the accrued gains every few years.

The central and state governments, who are participants in the BBMB, have made a review of the project's flood management, mainly in the light of the 1988 experience. The BBMB, even while awaiting the emergence of a consensus among these states, decided in July/August 1990 on some fresh guidelines for the Bhakra Reservoir, including a modification of the rule curve and a lower full reservoir level. The maximum level at Bhakra will be kept at 1,680 ft for storage purposes and the space above

---

[16]See Table 10.3, p. 127.
[17]Pong stands for the dam located at Pong across the Beas River.

will be used for flood moderation. Improved flood forecasting and increased carrying capacity in the river channel below Ropar were also recommended.

The reaches of the Sutlej upstream of Bhakra are prone to flash floods. Sixty-five percent of the catchment area upstream of Bhakra lies in the Tibet region of China. Often, information about flash floods does not reach downstream areas till the flood itself arrives and causes havoc. In recent years, such flash floods had caused serious disruption and losses in locations upstream of Bhakra but all of them were fully absorbed and evened up at Bhakra. In the absence of large storage at Bhakra, such flash floods would have travelled downstream to the Punjab plains and caused devastation. Thus, Bhakra has been effective in mitigating the effects of upstream flash floods, too.

14.6.3.2 Immunity from Famines

Southern Punjab/Haryana and the adjoining Rajasthan area—now irrigated by the BNP—were frequently subject to failure of rainfall and resulting droughts and famines in the past. The First Irrigation Commission (1901–1903) stated in its Report (Government of India 1903, paragraph 15, cited also in the Bhakra-Nangal Project Report 1955: 83) that in a period of 50 years, southeast Punjab was subject to 13 dry years, which included five years of severe drought. The government was forced to spend vast sums of money on famine relief measures. When the rains failed, scarcity of food grains and drinking water, as also migration of people and animals, was common. High production of food crops enabled by the Bhakra-Nangal Project has erased these miseries.

14.6.3.3 Industrial Development

Through its large storage capacity and the vast network of its canal distribution system, the project has made water available perennially to many areas for the first time. The generation and distribution of a large quantum of power made available abundant and reliable electric energy at an economical cost to a vast region. The hinterland of the project, comprising Punjab, Haryana, Rajasthan, Delhi, Chandigarh, and Himachal Pradesh, was, thus, advantageously placed for the setting-up and development of industries. The BNP region is now home to some of the largest industries of India. There are also a large number of small and medium industries in this area. A significant trend in Punjab is the conscious effort to establish industry in rural areas. Many small and large industries moved to rural locations, giving the benefit of increased employment to people from these areas.

14.6.3.4 Poverty Alleviation

Water resource development and management activities are essential for growth as well as poverty reduction. A high level of poverty is synonymous with poor quality of

Table 14.4 Poverty ratio (%)

State/country	Rural 1973–1974	Rural 1993–1994	Rural 1999–2000	Urban 1973–1974	Urban 1993–1994	Urban 1999–2000	Combined 1973–1974	Combined 1993–1994	Combined 1999–2000
India	56.44	37.27	27.09	49.01	32.36	23.62	54.88	35.97	26.10
Haryana	34.23	28.02	8.27	40.18	16.38	9.99	35.36	25.05	8.74
Punjab	28.21	11.95	6.35	27.96	11.35	5.75	28.15	11.77	6.16

Source: Economic Survey 2001–2002, Government of India, Ministry of Finance

life, deprivation and low human development. The Planning Commission[18] periodically estimates the incidence of poverty at the national and state levels, and records for the period 1973–1974 to 1999–2000 are available. Based on a large-scale sample survey on consumer expenditure, the poverty ratio was estimated at 27 % in rural areas and 23.6 % in urban areas, aggregating to 26.1 % for the country as a whole. At the national level, there has been a decline from 1973–1974 till the present, but rural–urban and interstate disparities are visible. Regions that experienced a faster rate of growth in output (GNP), particularly where such growth was mainly through agricultural production, exhibit a reduction in the proportion of people below the poverty line. Punjab and Haryana indicate the lowest levels of poverty, while the states of Orissa and Bihar are at the other extreme. Furthermore, there are no marked urban–rural variations in Punjab and Haryana (see Table 14.4).

The availability of assured irrigation and plentiful electricity for households, industries and agriculture, in which the Bhakra Dam played a very crucial role, had thus helped in bringing down the poverty level in the states of Punjab and Haryana. The benefits of the economic development in this region have been shared both by rural and urban areas.

14.6.3.5 Food Security for the Poor

A well-targeted and properly functioning Public Distribution System (PDS) is important for poverty alleviation. The PDS has evolved into a major instrument of the government's poverty eradication programme and is intended as a safety net for the poor. The large-scale production of rice and wheat in this region after the project enabled Punjab and Haryana to contribute the maximum to the PDS. A part of the centrally procured food grains is provided to the states during emergencies to run the 'food for work' programme, where a part of the wage payment is made through food grains. Between 1996 and 2000, Punjab's contribution to the central rice procurement was around 40 % of the total. The corresponding wheat contribution was around 60 %. Through the 1980s, the average contribution of Punjab and Haryana was around half in respect of rice and two-thirds in the case of wheat. For decades, these states have remained very important contributors to the central procurement pool of food grains.

---

[18]The Government of India established in 1950 the Planning Commission to assess the country's resources and to formulate a plan for their most effective and balanced utilisation. The Commission has so far brought out eleven successive five-year plans for this purpose.

Punjab and Haryana have, thus, been the kingpins of public distribution in India, and the BNP has played a crucial role in this.

14.6.3.6 Refugee Rehabilitation

The partition of India in 1947 and the gory aftermath of communal frenzy left a legacy of hatred between communities in Punjab and nearby areas. The sudden uprooting of people from their lands, possessions and jobs, and the unplanned migration of millions of people was one of the greatest challenges that had to be tackled by independent India. It was estimated that about 6 million people came to India from West Pakistan. Among them were the agriculturists who had developed the canal colonies and gained vast experience in converting arid wastelands to rich cultivated fields. There were industrialists as well as skilled workers who had lost their workplaces as well as jobs overnight. They needed land for cultivation, power for setting up industries and opportunities to restart their livelihood.

The Bhakra-Nangal Project was helpful in the timely and proper rehabilitation of refugee agriculturists in the new areas brought under irrigated cultivation. Similarly, the abundant and cheap power supply from the project enabled the setting-up of many small, medium and big industries that helped resettle non-agricultural refugees who had other industrial skills, training and experience.

14.6.3.7 Livestock and Milk Production

One method of raising productivity from land involves ruminants like cattle. Since they can convert roughage, which humans are unable to digest, into animal protein, they can help transform crop residues like straw and stalks into meat and milk. India has been steadily increasing her milk production over the last four decades, overtaking USA as the world's leading milk producer in 1997. Between 1961 and 2001, India's annual milk production went up from about 20 million tonnes to 85 million tonnes. Punjab, Haryana and Rajasthan have played a leading role in India's milk production. Since the early 1970s, about a sixth of the Nation's production comes from Punjab and Haryana alone. They also stand first and second, respectively, in per capita milk production since 1971.

There are two special features about India's milk production. First, these output levels have been achieved almost entirely by using farm by-products and crop residues as cattle feed, unlike some developed countries such as the USA that divert grains from humans to cattle. Second, the milk production in India has, so far, remained largely with small farmers and marketed through cooperatives run by them. Apex cooperative milk producers' federations were also set up in Punjab and Haryana in the mid-1970s. There was a similar expansion in the poultry sector, and egg production increased significantly in the two states. For millions of farmers, the integration of milk production with the crop production system is an important source of supplementary income.

14.6.3.8 Water Supply

The Bhakra-Nangal Project is the source of raw water supply to urban and rural population in the project area. The national capital, Delhi, also receives a significant part of its water supply from the Bhakra Dam. Many medium-sized cities and small towns such as Chandigarh, Patiala, Hisar, Sirsa, Ludhiana and hundreds of villages in the vast canal command have benefited from the project water supplies. In arid and semi-arid desert regions of Rajasthan, there was no local or nearby source of potable water prior to the project. Women had to trudge several miles daily to fetch water from distant sources. This scenario changed for the better after the Bhakra supplies.

14.6.4 Indirect Benefits

14.6.4.1 'Multiplier Effect' of the Project on the Economy

Many primary and incidental benefits from the Bhakra-Nangal Project have been noted earlier. There are, in addition, many indirect and induced benefits and impacts as well. These arise out of the myriad linkages between the project's direct consequences and all other sectors of the economy. For instance, hydropower from the dam not only provides electricity to urban and rural regions but also enables increased outputs of industrial commodities (e.g. fertilisers, chemicals, and machinery). This, in turn, requires inputs from other sectors of the economy, such as steel. Increased agricultural outputs encourage the setting-up of food processing industries. There are, thus, many backward and forward linkages.

The World Commission on Dams (WCD) recognised the importance of indirect impacts. Its Report stated:

> As with livelihood enhancement, the broader impacts of irrigation projects on rural and regional development were often not quantified. Dams, along with other economic investments, generate indirect economic benefits as expenditure on the project and income derived from it lead to added expenditure and income in the local or regional economy. The WCD case studies give examples of these `multiplier' benefits resulting from irrigation projects. (WCD 2000: 100)
>
> -------- a simple accounting for the direct benefits provided by large dams – the provision of irrigation water, electricity, municipal and industrial water supply, and flood control – often fails to capture the full set of social benefits associated with these services. It also misses a set of ancillary benefits and indirect economic (or multiplier) benefits of dam projects. (WCD 2000: 129)

The economic development of Punjab and Haryana in the last three to four decades reveals that a very close relationship exists between water resources development and the structure of the economy. The linkages induced by input demand and output supply generate growth impulses, which are transmitted from one sector to another of the economy. Additional employment opportunities are generated in the non-farm sector, for instance, for servicing and repairing of machinery and implements.

These effects extend within the project region as well as outside it. In other words, the major outputs of a dam project will have inter-industry linkages that result in increased demand for output in other sectors of the economy. Increased agricultural and industrial outputs lead to the generation of additional wages/income for households, in turn leading to higher consumption of goods and services. All these lead to changes in output-input of various sectors. These could be estimated and aggregated in terms of their 'multiplier value', the ratio of the total direct and indirect impacts of the project to its direct impacts alone. There are large employment opportunities created for migrant labour that come to the BNP area from other regions on a regular basis (mostly from the states of Bihar and Uttar Pradesh). These labourers can not only earn high wages, but also acquire agricultural and mechanical skills during their work in the BNP area. Similarly, the 'food for work' programme creates employment opportunities in various parts of the country.

Increase in personal disposable income will also create demand for services like transport, communication and recreation. The increasing preference of consumer durables such as scooters, cars, refrigerators, televisions and kitchen gadgets will, in turn, create the need for repair and maintenance services. This kind of overall change in the socio-economic scene and its beneficial impact on the tertiary sector has been witnessed in both Punjab and Haryana.

Analytical frameworks for estimations of the regional value-added multipliers are available. Investments in water resource development and management have significant region-wide impacts. The economic and employment multipliers have been assessed under a wide variety of settings in different parts of the world and found to vary between the values of two and six. A study of the Bhakra-Nangal Project conducted by Bhatia et al. (2008) for the World Bank (2008) indicated that the multiplier effect might lie in the range of 1.78 and 1.9. In other words, for every rupee of direct output of the project, another rupee was generated in the form of downstream or indirect benefits. This figure would have been much higher than 2.0 if industrial development, significant additions to the agricultural labour pool and remittances of the migrant seasonal labour to other regions were also taken into. A source cited in the study estimates migrant labour in Punjab and Haryana as approaching a million. The extent of their remittances to their villages elsewhere is also significant, and has helped improve the living conditions of the people there.

14.7 Some Impacts of the Reservoir

In 1985, the Government of India legally enforced comprehensive Environment Impact Assessments based on published guidelines. As the Bhakra-Nangal Project was begun some four decades earlier, there were no legal requirements or guidelines that needed to be followed at that time. Critics generally list a few common social and environmental impacts for any reservoir project. These relate mostly to resettlement and rehabilitation of the displaced, sedimentation, water logging, etc. The BNP authorities responsible for planning and executing the project had taken great pains to study these aspects and to ensure minimal adverse impacts.

The National Policy on Resettlement and Rehabilitation was not notified till 2004 (and this is already under reconsideration), whereas BNP was undertaken and completed over five decades earlier. However, the project had adopted progressive, liberal and humane provisions in respect of 'Project affected families'. As was discussed above at length, the resettlement and rehabilitation provided by the project authorities rendered adequate justice to displaced people, well beyond the legal obligations of those days.

The views of those to be displaced from their land were first ascertained. Many of them preferred to be resettled in compact blocks in an area served by the Bhakra canals. Suitable such lands were located in Hisar District in compact blocks and 13,200 acres of land in the cultivable commanded area were acquired in 30 villages. A detailed plan and guidelines for resettlement were drawn up. Even if the displaced family had land in more than one village, the policy required that she or he be allotted new land in one location. Provision was made for families who wanted to settle together in Hisar. It was ensured that those displaced from a single village in the reservoir area were allotted land in the same village in the Hisar area. This ensured that the village communities were not dispersed. In the case of land-owning families, it was decided to allot land for all those who desired it, no less than the cultivated land in the reservoir area but with a maximum limit of 10 ha on the allotment of irrigable land. Even landless tenants were eligible for land to the extent of their tenancy subject to a maximum of two hectares, the price to be recovered in 20 half-yearly instalments after a grace period of two years from the time of allotment. The artisans and labourers who did not own or cultivate land in the submergence area and wished to resettle in the Hisar area were also allotted half an acre of land free of cost. Rehabilitation grants, temporary shelter, drinking water through new wells and other facilities were also provided.

14.7.1 Sedimentation

Sedimentation in any reservoir is an unavoidable natural process. The details of investigation and planning for the Bhakra Dam show that the planners had started collecting data on silt from the outset and conducted comprehensive studies before determining the storage capacity for the reservoir. A realistic estimate of the likely rate of sedimentation had to be made for the project. A proposed reservoir should provide for silt storage so that its full and beneficial functioning will not be impaired during its 'design life'. The project studies used silt load observations for the period from 1916 to 1939, as also the experience gained from comparable reservoirs in USA. Based on these, the likely rate of silting in the Bhakra reservoir was projected as 4.29 ha m/100 km^2/year. The useful life of the project was accordingly estimated at over 300 years. However, all financial forecasts were conservatively worked out on a '100-year life' basis.

After construction, BNP conducted comprehensive sedimentation surveys and studies repeatedly, every year. Since 1959, when impounding began, and till 2000, 28 repetitive reservoir surveys have been made. These show the rate of sedimentation as

6.2 ha m/100 km^2/year, which is somewhat higher than the project assumption. If this trend continues, the life of the reservoir will be lesser than originally worked out. However, it will still be higher than the 100-year life adopted for economic and financial purposes. Catchment area treatment has been made a part of the project.

14.8 Summary of the Assessment of the Impacts

Most development projects (whatever their purpose) are criticised mainly because they displace people and could deprive them of their livelihoods. Projects involving 'large dams' are particularly targeted on this account. It is not development alone that causes displacement. The absence of development can also displace people, perhaps in larger numbers. One should also keep in mind the perennial distress migration from regions prone to drought and flood, sometimes seasonal, often permanent, and all those in search of employment and better opportunities in places far from home. What is important is that wherever unavoidable issues relating to displacement arise, they should be handled with utmost care and consideration.

The former President of South Africa, Nelson Mandela, aptly put this matter in the right perspective during his address at the launch of the WCD Report on 16 November 2000. He said:

> The problem, though is not the dams. It is the hunger. It is the thirst. It is the darkness of a township. It is townships and rural huts without running water, lights or sanitation. It is time wasted gathering water by hand. There is a real pressing need for power in every sense of the word.[19]

The Bhakra-Nangal Project has been in the service of the Nation for five decades. This assessment of its performance has shown that the project has fulfilled all its stated objectives in a sustained manner. In addition, the project has provided immense additional benefits to the region through indirect and secondary impacts. The perceptions of the people impacted were also ascertained through random interviews: It is their view that the project has brought prosperity to them.

All in all, Bhakra-Nangal is a successful story of sustained humane development with overwhelmingly beneficial impacts for the region and the Nation.

References

BBMB (1988) Bhakra and Beas Management Board, History of Bhakra Nangal Project. Government of India, Chandigarh

Bhatia R, Cestti R, Scatasta M, Malik RPS (eds) (2008) Indirect economic impacts of dams: case studies from India, Egypt and Brazil. Academic Foundation for World Bank, New Delhi

---

[19] From the Press release issued by WCD on the occasion of the launch of the Final Report of the World Commission on Dams giving the text of the address by Nelson Mandela on 16 November 2000 at London.

Central Board of Irrigation and Power (CBIP) (1965) Development of irrigation in India, publication no. 76, New Delhi
Central Board of Irrigation and Power (CBIP) (2005) Proceedings workshop on impacts of Bhakra-Nangal Project, New Delhi, August
Central Water Commission (CWC) (1989) Jawaharlal Nehru and Water Resources Development. Government of India, New Delhi, April
Government of India (1903) Irrigation Commission Report. Government of India, New Delhi
Gulhati ND (1973) Indus Waters Treaty: an exercise in international mediation. Allied Publishers, New Delhi
Ministry of Irrigation and Power (1972) Report of the Irrigation Commission. Government of India, New Delhi
Ministry of Water Resources (1999) Report of the Working Group on International Dimensions. Government of India, New Delhi, September
Paddock W (1967) Famine 1975! America's Decision: Who will survive? Little Brown, Boston
Punjab Government (1955) Bhakra Nangal Project Report, unpublished document
Rangachari R (2004) Final Report of the Study of the Bhakra Nangal Project, Centre for Policy Research, New Delhi, May (mimeograph for restricted circulation)
Rangachari R (2006) Bhakra-Nangal Project: socio-economic and environmental impacts. Oxford University Press, New Delhi
Roy A (1999) The greater common good. Outlook, New Delhi, 24 May
Subramaniam C (1979) The new strategy in Indian agriculture: the first decade and after. Vikas Publishing House, New Delhi
Verghese BG (1994) Winning the future, Konark Publishers private Ltd, Delhi
WCD (World Commission on Dams) (2000) Dams and development, report of the World Commission on Dams. Earthscan Publications, London

Chapter 15
Impacts of Koyna Dam, India

C.D. Thatte

15.1 Introduction

The Koyna Dam, which was built across and is named after the river Koyna, was undertaken for the primary purpose of hydropower generation by Maharashtra state as part of the upsurge in nation building activities after India's independence. The dam has become a symbol of achievements in water resources development during the last five decades. It is the upstream most dam in the Krishna River Basin, 103 m in height, storing close to 2.8 billion cubic metres (BCM) of water in a reservoir with a spread of over 115 km^2, near the Sahyadri Range which serves as a Continental Divide that is over 1000 m high at the head of the major peninsular Krishna River Basin. The range runs in a north-south direction, parallel and close to India's west coast. The reservoir captures run-off in a high rainfall zone at the highest elevation and drops the reservoir waters over nearly 500 m in a westerly direction across the mountainous divide, for generating hydropower at an exceptionally low cost, in the first major underground powerhouse of the country. The Koyna Project has already seen four innovative stages of development and holds promise of a fifth stage of development and beyond in the near future (Mahabal 2000).

The project has to its credit several technological firsts in India. It has provided social benefits for the people of the region, a population about 200 times the number of the adversely affected. The direct benefits to the region are indicated in an economic rate of return of more than 25%. The indirect and incidental economic benefits derived by the region are also noteworthy. At the same time, the adverse environmental effects of the project have been insignificant. The project has proven water resources development to be an eco-friendly venture, and not, as some have argued, eco-destructive. A multi-purpose project accepted by the entire community in the Krishna River Basin following the interstate Krishna Water Dispute Tribunal Award, Koyna remains a shining example of interstate cooperation within the Indian union (Popular Article 1990).

C.D. Thatte
International Commission on Irrigation and Drainage (ICID), Pune, India

15.2 South Asian Peninsula, India and Its Water Resources

The present geography of the South Asian subcontinent is largely the result of continental drift and the rising of the mighty Himalayas: the eruption of numerous Basalt lava flows over the peninsula, the formation of the western and eastern ranges called the 'ghats' running parallel to coastlines, their orographic influence on the course of monsoon and the rise of subcontinental rivers. Like other countries of the region, India's water resources are a gift of the monsoon and are estimated at about 4,000 BCM, of which about 1,200 BCM are considered to be the dynamic yearly renewable resource. A little more than half of this resource is currently 'developed' for beneficial use. Irrigation accounts for about 83% of water abstractions. In the ultimate stage of development, the share of irrigation is expected to fall to about 70%. Other abstractions, including those for generation of hydropower, are largely non-consumptive in nature. So far, only about 24% of the country's hydropower potential has been tapped. Plans are afoot to complete all sectors of water resources development by 2050 as by then India's population is likely to stabilise for which water planning could be more realistic.

The northern region of the subcontinent has two main perennial river systems: the Indus—shared by India and Pakistan, draining into the Arabian Sea in the west—and the Ganga Brahmaputra Meghna shared by China, India, Nepal, Bhutan and Bangladesh, draining into the Bay of Bengal in the east. The southern triangular peninsula of the subcontinent, the Deccan Plateau, slopes eastward and is bounded to the north by mountain ranges running across the subcontinent. The other two sides of the triangle are bordered with similar mountain ranges. Unlike the northern region, the peninsula receives no snow and its sole water source is rainfall during the monsoons. The continental divide in the west, i.e. the Sahyadri Range that rises to an elevation of over 1,000 m within about 50 km of the coastline, causes heavy rains west of and on the divide over less than 100 days per year, spread over June to September.

Several small rivers run down westward with a steep gradient from the continental divide. The plateau at a higher elevation east of the divide, however, falls in a rain shadow. It experiences extreme variability of precipitation both in space and time. Three major interstate river basins—Godavari in the north (313,000 km^2) and Krishna (257,000 km^2) and Cauvery (81,155 km^2) farther south—drain the plateau eastward with a yearly run-off at 75% dependability of about 78, 64 and 19 BCM of water, respectively. The basins are uniformly and densely populated, and per capita yearly availability is about 1,730, 1,050, 620 m^3, respectively, indicating increasing scarcity. The states of Maharashtra, Madhya Pradesh, Chhattisgarh, Andhra Pradesh, Karnataka, Kerala and Tamil Nadu share these three basins. The river basins of Mahi, Narmada and Tapi drain the northern part of the plateau from east to west, while the basins of Mahanadi, Brahmani-Vaitarani and others, similarly drain it from west toward east (Koyna Project Special Number, IJPRVD, 1962). Figure 15.1 shows the river basins of India.

15 Impacts of Koyna Dam, India

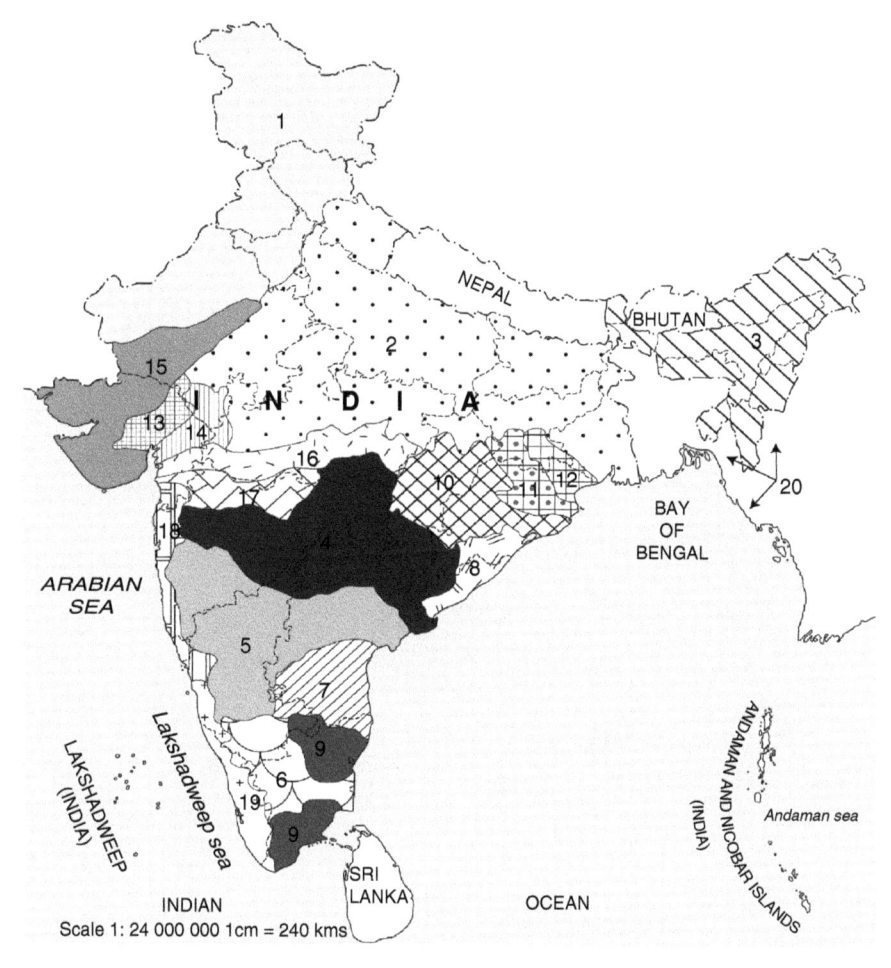

1. Indus
2. Ganga
3. Brahmaputra, Barak and others
4. Godavari
5. Krishna
6. Kaveri
7. Pennar
8. East-flowing rivers between Mahanadi and Pennar
9. East-flowing rivers between Pennar and Kanyakumari
10. Mahanadi
11. Brahmani and Baitarani
12. Subarnarekha
13. Area of inland drainage in the desert
14. Mahi
15. West - flowing rivers of Kutch
16. Narmada
17. Tapi
18. West - flowing rivers from Tapi to Tadri
18. West - flowing rivers from Tadri to Kanyakumari
18. Minor rivers draining into Myanmar and Bangladesh

Fig. 15.1 River basins of India (*Source*: Biswas et al. 2009)

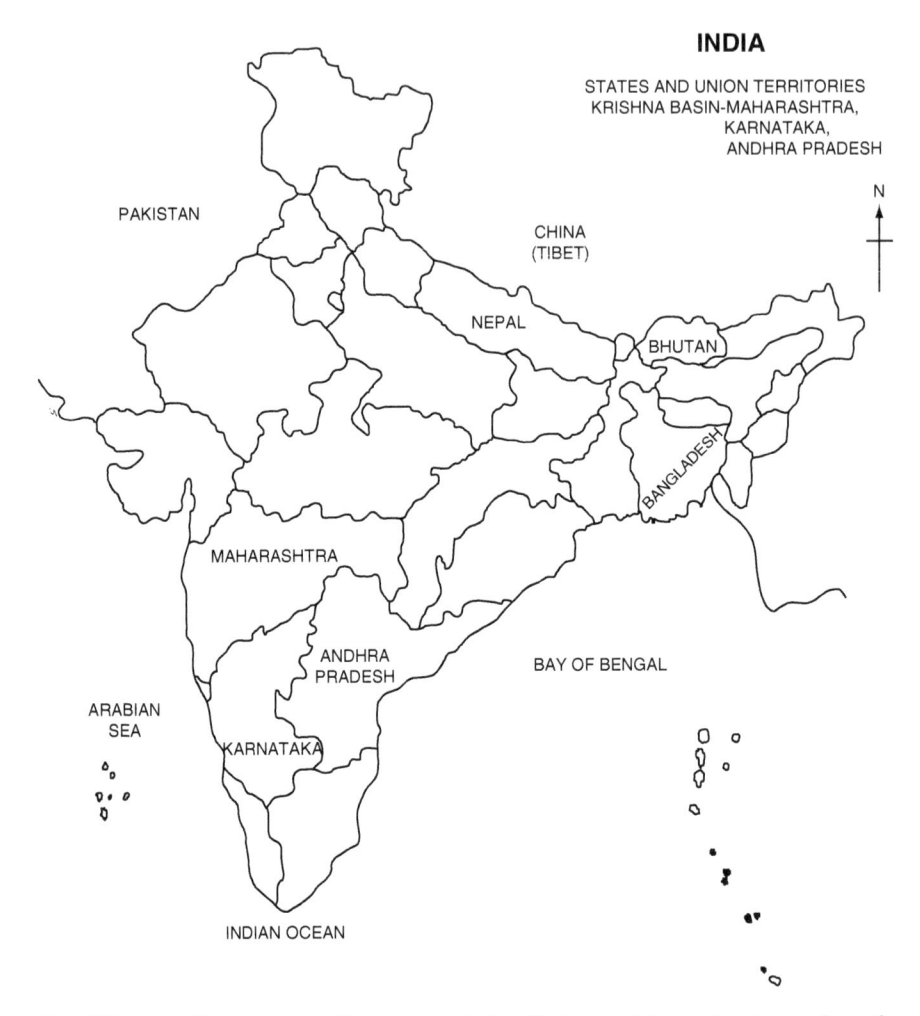

Fig. 15.2 Map of India showing Maharashtra, Andhra Pradesh and Karnataka (*Source*: Records of the Koyna Project)

Table 15.1 Salient features of Maharashtra State

Geographical area	308,000 km^2
Population	97 million
Urban population	>40%
Cultivable area	22.5 Mha
Coastal line	720 km
Administrative division	6 regions and 35 districts
Climate	Tropical
Annual rainfall	400–600 mm
Seasons	Summer, rainy and winter (rainy season of 3–4 months duration)

15.3 Maharashtra State and the Koyna Hydroelectric Project

The Koyna Hydroelectric Project was amongst the first water resource development projects built through a succession of four stages over the past 50 years in the state of Maharashtra. Its total installed generation capacity is now about 1,920 MW, and stage V is being undertaken while the scope of the project is still evolving. Maharashtra is the third largest state (31 Mha) in terms of area and the second largest in terms of population (97 million) (Fig. 15.2 and Table 15.1). The city of Mumbai (formerly Bombay), the most populous in the country (15 million), is the state capital and also the country's commercial capital. Koyna Dam is located about 240 km south of Mumbai. About 42% of the state is urbanised and, as in other major states, urbanisation and industrialisation in Maharashtra is proceeding rapidly. It is estimated that more than half the state will be urbanised by 2050. The rest of the state is dependent on agriculture. However, 35% of cultivable land holdings are less than 1 ha in size and the average holding is 2.2 ha. The fragmentation of larger holdings continues with successive generations, causing a decline in land productivity.

Large areas comprising the peninsular river basins are prone to drought. A time-tested strategy to overcome drought is the building of dams—large and small—and canals to capture and lead surface flood water of the monsoon where needed, for irrigation. The state receives about 1,300 mm of rainfall annually. About 88% is from the southwest, the remainder from the northeast monsoon. Of the state's roughly 400 rivers, four viz. Godavari, Krishna, Narmada and Tapi are shared with other states. The first two originate in Maharashtra, while the other two begin in Madhya Pradesh. Run-off at 75% dependability is about 132 BCM. Although India accounts for 4,000 large dams, Maharashtra has nearly half of them, an indication of their importance for the state's well-being. They hold about 16 BCM of water, of which almost two-thirds is used for irrigation. Due to the undulating topography, the irrigable area is in small packages, and irrigation facilities are expensive but precious and crucial. Storage is still low as compared to the yield. There are 53 major and 212 medium irrigation projects, irrigating about 3 Mha. The minor projects irrigate about 1 Mha. The state has an ultimate irrigation potential of about 13 Mha. The use of water for irrigation is about 70% of total use for all sectors, as compared to national proportion of about 83% (Koyna Hydroelectric Project Completion Report 1970).

More than 70% of India, in terms of area, is expected to remain rain fed even in the ultimate stage of irrigation development. In Maharashtra, while storage capacity and availability of groundwater is limited due to the rocky terrain, it is overdrawn because surface water for natural recharge is deficient. Surface irrigation supplies in recent times have, however, recharged groundwater significantly. The proportion of energy generated in the state that is used for pumping groundwater for irrigation is rather high. The need and demand for water and energy for irrigation continues to rise with population and economic growth, but competing demands on limited resources cause problems in the sharing of river water within and between the region's riparian states.

The unmet energy need of the region, particularly peak demand, is large and is being partially addressed through the relatively expensive thermal energy generating plants, although hydropower has been generated within the river basins at the foot of some multi-purpose dams. Prior to 1950, the installed capacity of hydropower in Maharashtra was about 280 MW. In 2005, it was over 2,700 MW of the total capacity of 14,000 MW, with three-fourths generated by the Koyna Hydroelectric Project. As per present plans, it will rise to about 11,000 MW (including output from some plants outside the state) in a state total of 27,000 MW. The hydropower component of total generation is, however, falling day by day and thus affecting peak capacity adversely, although it is recognised that even today, it is the least expensive option at about 40 paise (one hundredth of a Rupee) per unit of power against 84 paise for atomic, 100 paise for gas-based and 125 paise for thermal power. Koyna power is the most economical, due to the high drop harnessed for generation.

Maharashtra's long-term energy expansion plan includes installation of several peaking plants through hydro installations. Energy needed for industry is growing rapidly with accelerated economic growth. The continental divide, however, provides an opportunity to capture the east-flowing river water fed by heavy precipitation at the head of the plateau, and drop it westward to utilise the elevation difference of over 400 m to generate hydropower outside the parent basin, instead of using it within the basin over a small drop of the order of 100 m provided by the dam height. A similar opportunity was put in practice as early as the beginning of the twentieth century and implemented by Tata, a leading industrial house of the country, in diverting westward the waters of an adjoining northern tributary of Krishna River. The installed capacity of the Tata plant is now 440 MW. The Koyna Project was conceptualised as a similar scheme, but was twice shelved due to the two world wars. As a project, it was finally administratively approved and undertaken in 1953. Although basically for generating hydropower, like most other dam projects of the subcontinent, it also serves multiple purposes such as irrigation, domestic and industrial water supply (Gokhale Institute for Economics and Politics 1963; World Bank 1989).

15.4 The Krishna River Basin and Early Years of the Koyna Hydroelectric Project

The Krishna River rises to an elevation of 1,438 m and runs over a length of 1,392 km of which 282 km are in Maharashtra, cuts across the peninsula from west to east, before it meets the Bay of Bengal. It drains an area of over 257,000 km^2, which is 8% of the country's geographical area, home to 60 million people and 20.5 Mha of arable area. The average rainfall over the basin is 784 mm, with a peak of over 6,500 mm near the origin, yielding an average annual run-off of 73 BCM. In view of the heavy rainfall and the steep slopes, 1 mm rainfall generates as much as 7.7 million cubic metres (MCM) of run-off in the upstream. Dependable yield, at 75% reliability, is 69 BCM (Fig. 15.3).

15 Impacts of Koyna Dam, India 335

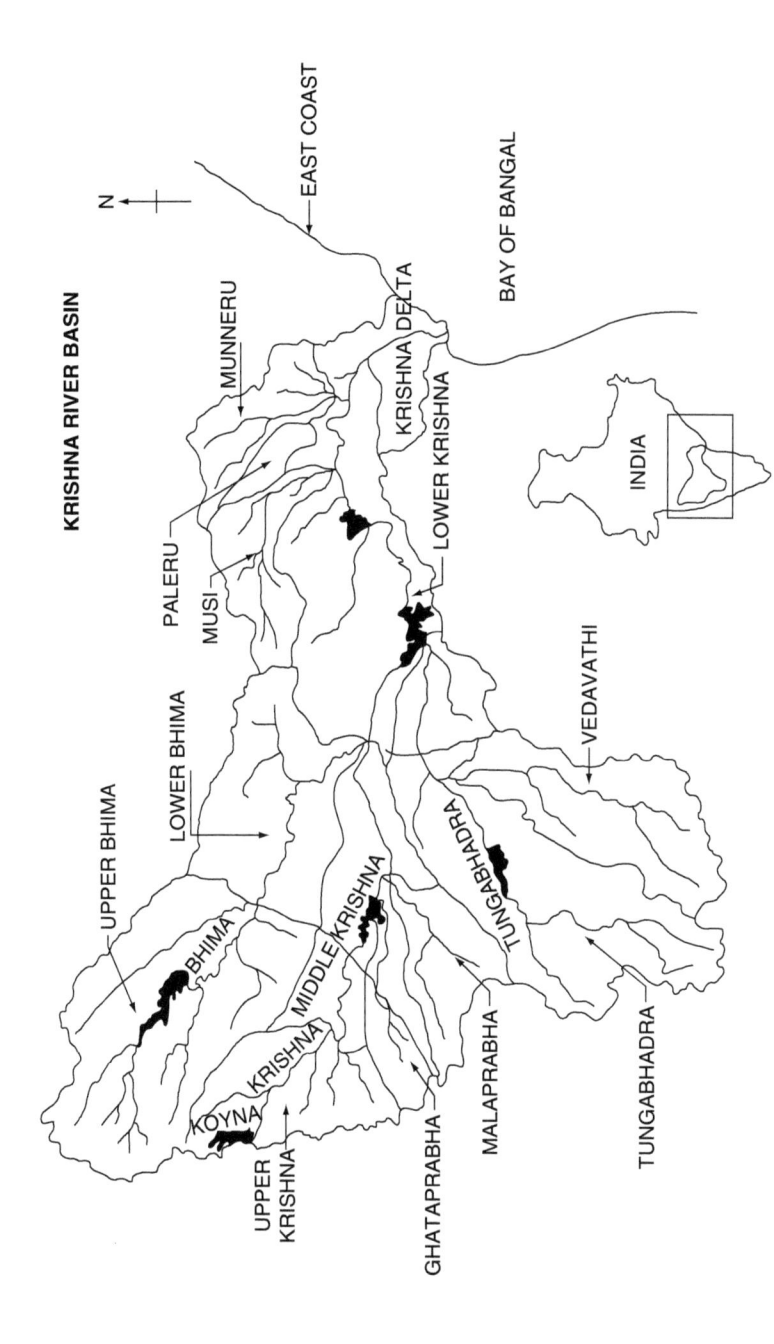

Fig. 15.3 Map of Krishna River Basin (*Source*: Koyna Project Records)

The three riparian states of the basin are Maharashtra, Karnataka (earlier Mysore) and Andhra Pradesh. The Koyna River originates at Mahabaleshwar at an altitude of about 1,600 m above Mean Sea Level (MSL), draining an area of 892 km^2 up to Koyna Dam falling entirely within Maharashtra. This is the first important right bank tributary of the main stem of the Krishna River in the head reach, draining only 0.74% of the catchment area of the Krishna Basin. Koyna runs for a length of 137 km length and steeply falls to 541 m elevation above MSL, before it meets the Krishna River. Other tributaries on the right bank are Venna and Panchganga, whereas Yerala joins Krishna on the left bank before it enters Karnataka state. Rainfall drops sharply from 6,500 mm to less than 1,000 mm just 40 km east in the rain-shadow region. At the top of the western hill, heavy winds have been harnessed to run wind generators, with generating potential of about 500 MW.

A straight gravity rubble-concrete dam across Koyna 808 m in length, 103 m in height, tapping a catchment area of 892 km^2, and storing 2.8 BCM of water is a key structure of the Koyna Hydroelectric Project as a whole, and Stages I and II in particular. The rubble-concrete technology of sinking large pieces of rock (up to 300 mm in size) in concrete was adopted from dams on the Ocker River in Germany. The dam top is at El 664.46 m (MSL), design Maximum Water Level (MWL) is at 660 m, whereas Full Reservoir Level (FRL) is kept at 658 m. The reservoir water spread is about 13% of the catchment area, which is quite high and hence the sediment trap efficiency is close to 100%. Spillway crest is at 650.29 m, dead storage level is 609.6 m and outlet sill is at 597.41 m. Gross design storage is 2.8 BCM, of which live storage is 2.68 BCM and dead storage is 0.12 BCM, which is roughly 5% of gross storage. The length of the reservoir at FRL is 55 km, of the 64 km length in the upstream.

Concreting for the first stage of the dam was begun in 1958. The second stage followed in continuation of Stage I and it comprised addition to the cross section of the dam by changing the downstream slope along the lines of a Swiss dam, and addition of four generating units to the four from the first stage. Initially, it was expected that demand for power would rise slowly, which could be met in two successive stages. The second stage, however, followed the first stage back to back. The composite dam was essentially completed in 1961 and lake filling was begun. The generating units were commissioned in 1967 (Koyna Hydroelectric Project Reports 1950, 1952).

15.5 Earthquake and Reservoir Associated Seismicity

The primary rock types in the region are basalts: vesicular, amygdaloidal and compact. There are also thin inter-trappean beds of red bole which cause some weakness in the otherwise strong rock mass. Dolerite dykes run criss-cross, in a way reinforcing the rocks and controlling the seepage paths. The region is considered to be a part of the stable continental Deccan shield with relatively low seismic activity. Despite this, the dam section was designed and constructed according to the national codes for design to account for native seismicity. Soon after impounding, however, on 11 December 1967, an earthquake of 7.5–8.0 Richter magnitude struck, causing the loss of about 200 lives; its focus was close to the reservoir and hence some attributed it to Reservoir Associated Seismicity (RAS). The earthquake was of an intra-plate nature with low return frequency, and was linked to adjustment of individual blocks of the subducting Indian plate. Peak acceleration experienced was equivalent to about 0.63 g. The possibility of RAS at Koyna was, however, discarded after detailed investigations. In addition, the epicentre has been continuously moving away from the reservoir in south and southwest directions, where the geotectonic features are different. Minor seismic activity has continued in the region since 1967.

The original structural components that were distressed due to the earthquake were initially repaired by grouting and gunite techniques for the cracks, cable anchoring the seven high monoliths, and drilling additional drainage holes in the dam body. The dam had developed leakages due to the earthquake, mainly from the distressed contraction joints. While at first leakage increased, due to continuing treatment the rate of leakage is now reduced to a low level. Further underwater treatment from upstream face of the dam is being contemplated. Some distressed equipment was repaired and additional support, anchorage, insulation, flexibility and shock dampeners were provided. Housing was strengthened to take care of native tremors as per code provisions and to preclude any future loss of life. Seismic instrumentation was augmented and a Standing Committee was set up to assess feedback and likely future earthquake activity, and to recommend further strengthening if called for. The non-overflow stretch of the dam was strengthened with concrete buttresses in 1973 to meet newly recommended design parameters. Strengthening of the overflow dam is being undertaken to incorporate additional design parameters in the aftermath of another intra-plate earthquake event, Killari (Latur), in 1993 (Pendse et al. 1975) (Fig. 15.4).

15.6 The Development Stages

The Koyna Reservoir, named Shivajisagar Lake and formed under Stages I and II of the project, taps 892 km^2 area of the river's head reach with an average annual rainfall of 3,000 mm. It has a gross volume of 2.8 BCM, about 115 km^2 of spread at FRL, 120.43 km^2 at MWL, and a maximum depth of 82 m. The average width of the catchment is 14 km while the length is 64 km, elevations falling from 1,220 to 579 m, i.e. 640 m in 64 km, indicating an average hill slope of 1 in 100. Average inflow into the reservoir for 25 years has been estimated to be of the order of 4 BCM, i.e. about 3.3 BCM at 75% dependability. It has not been possible to measure inflow and hence the estimate is based on reservoir levels and outflow. Several small tributaries join the Koyna River along its length. The catchment is elongated, hilly with steep slopes and wooded.

The storage capacity during Stage I was 1.6 BCM. The reservoir storage accounts for 85% of the 75% dependable yield at the dam, but allows 1.9 BCM of stored water to be dropped westward over the divide for hydropower generation, and a release of 0.8 BCM for downstream uses. A gated central spillway 75 m in length caters to a maximum overflow into the river downstream of nearly 5,500 m^3/s at

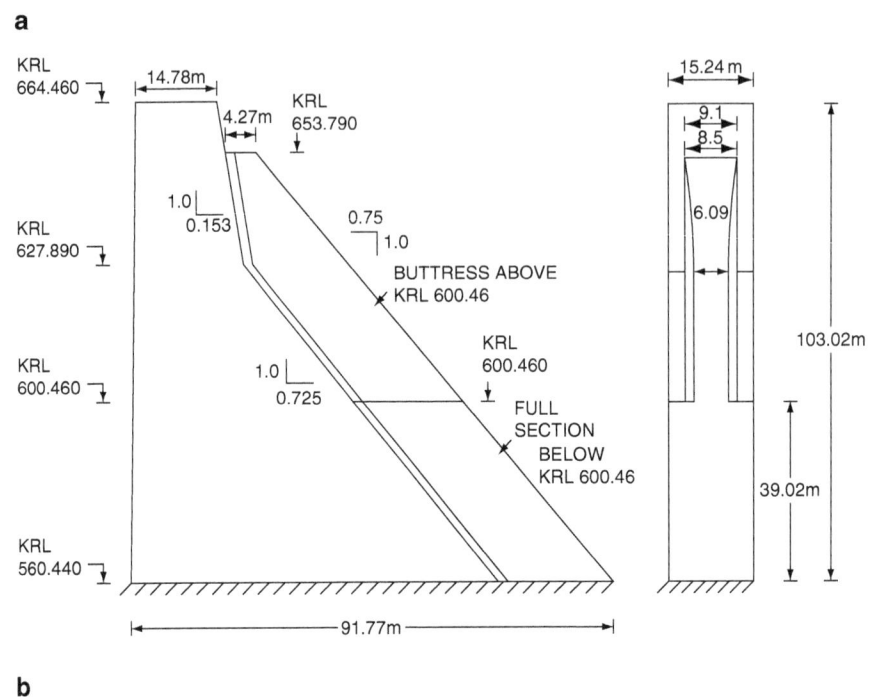

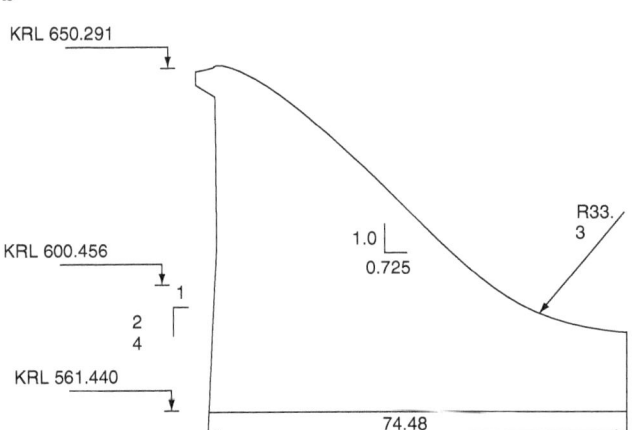

Fig. 15.4 Present cross section of (**a**) non-overflow and (**b**) overflow dam strengthened after Koyna earthquake in 1967

MWL of 652 m through six radial gates 12.5 m in width and 7.52 m in height. Designed inflow into the reservoir is 13,460 m³/s. The design maximum outflow from the six gates is 5,500 m³/s, while the observed peak flow in the river downstream is 3,245 m³/s. Flaps 1.5 m high were added on the radial gates to raise the FRL in 1995, enabling better water management for so many facilities.

The design outflow is kept low in view of limited carrying capacity of the river channel and to achieve a minimum of inundation. The overflow energy is dissipated in a stilling basin type of dissipater.

The 1.2 km long intake channel starting at the intake level of 601.98 m taps the lake upstream of the dam, carries water through the 3.7 km long concrete-lined headrace tunnel; through a surge system comprising a main surge shaft about 12 m in diameter, a lower gallery, a riser and an expansion chamber into the 616 m long steel-lined penstocks that are four in number, each with a diameter of 3 m, dropping at an incline of 45° to 2.6 m diameter. The four penstocks bifurcate into eight arms in the main powerhouse to supply water to eight units of 560 MW installed capacity at 60% load factor in all, consisting of four 65 MW French Pelton wheels commissioned in 1963 and four 75 MW Swiss Pelton wheels commissioned in 1967. The tailrace tunnel runs for 2.2 km and drops into the Vaitarani River a peak discharge of 164 m^3/s at an elevation of 123 m MSL. The river leads it down the course meeting a creek joining the Arabian Sea at about 14 m elevation.

The powerhouse at the foot of the dam on the right bank has two units of 20 MW installed capacity each from Francis turbines, passing downstream on an average 0.8 BCM of storage as regulated year-round release into Koyna River for downstream use. The powerhouse construction work was started in 1975 and completed in 1980. About 100,000 ha on the banks of the Koyna and Krishna rivers are irrigated using regulated releases accounting for 670 MCM of water. The remaining water, about 130 MCM, is used for municipal, industrial and rural water supply. Another powerhouse on left bank of the river with two units each of 40 MW is under planning and may be built in the near future.

Stage III of the Koyna Hydroelectric Project was commissioned in 1975 and comprised utilisation of the residual drop of about 120 m for generating hydropower with four units of 80 MW each, and providing an installed capacity of 320 MW, essentially for meeting peak power needs. In order to enable such additional generation, the tailrace water from Stages I and II at Pophali powerhouse was picked up and carried through a 4.5 km long and 7.47 m wide headrace tunnel to drop it into a new balancing reservoir of 11 MCM capacity, created by building a 497 m long, 66 m high masonry/concrete dam at Kolkewadi. The new reservoir's water spread is about 17 km^2 and it stores 36 MCM of water. Four 192 m long pressure shafts of 4.1 m diameter, laid in tunnels steeply inclined at 42°, bring the water to the underground powerhouse with four units of 80 MW each, providing peaking capacity of about 320 MW at 24% load factor. A collecting gallery leads about 325 m^3/s of discharge from the turbines through a 4.5 km long tailrace tunnel of 10.36 m diameter, to a tailrace channel 4.7 km in length back to the Vashisti River.

Stage IV followed in year 2000 to enable conversion of the base-load power station into one catering to peak loads (peaking station) by drawing more water from the Koyna Lake from an additional location, with new headrace tunnel, penstocks, powerhouse, tailrace tunnel, tailrace channel etc. The turbine discharge is let into the Kolkewadi Reservoir built during Stage III. The powerhouse includes four units of 250 MW each, accounting for an installed capacity of 1,000 MW. As water from the reservoir could not be drawn down for Stage IV from a new intake below lake level,

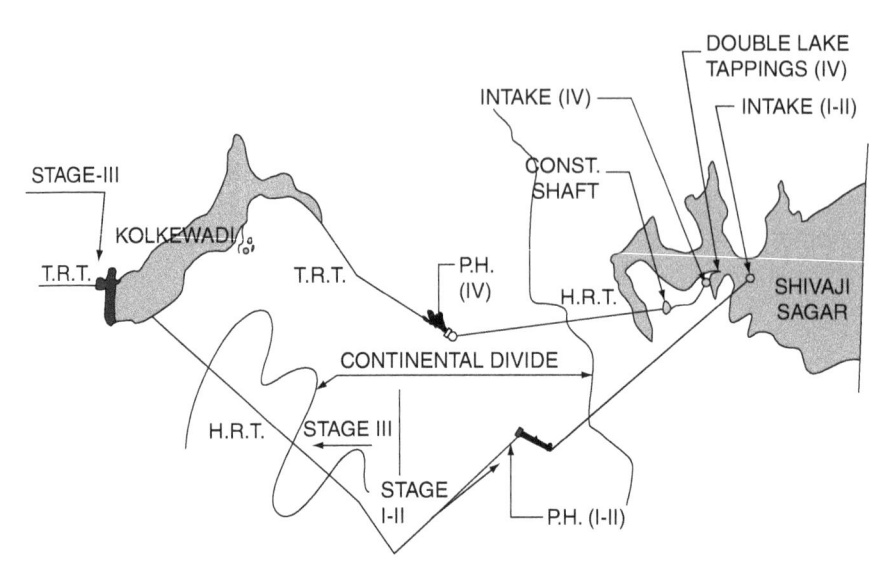

Fig. 15.5 The Koyna Hydroelectric Project (Stage I-II-III-IV) schematic layout

all facilities downstream were undertaken and completed first, leaving a small rock plug ledge 6 m wide near the lake for final clearance. The plug was carefully blasted almost 40 m below the water level of the lake, creating a unique lake tap, on 13 March 1999. An additional lake tap with headrace tunnel extended by another 4.5 km is being planned to further augment peak capacity. It will constitute Stage IV 'B' for the Koyna Hydroelectric Project and will allow, via an extended headrace tunnel, lake water 16 m deep (i.e. about 400 MCM) to be drawn from a lower level (El 593 m instead of the present El 609 m MSL) through the additional lake tap (Fig. 15.5).

Stage V is presently under planning and will consist of a pumped storage scheme upstream of Koyna Dam, enabling an installed capacity of 400 MW through two 200 MW units at Humbarli. The planned yearly generation will be about 751 million units (KWh) of energy. It is envisaged that a reservoir with 7.83 MCM storage will serve as an upper reservoir, with the existing one serving as the lower reservoir. The Vazarde, a tributary of the Koyna River at the upper dam, has a catchment area of about 285 ha with 4,946 mm rainfall, and the level of the river bed at the dam site is 931 m. A concrete dam 450 m long and 55.55 m high is planned on the stream near Humbarli village. A 320 m high natural fall in this stream into the Koyna Reservoir drew the planners' attention to the site's potential for construction of a pumped reservoir. The two reservoirs will be connected by an underground water conductor system comprising headrace tunnel of 5.8 m diameter and 320 m long, pressure shaft of 5.8 m diameter and 820 m long in total, and tailrace tunnel 8.5 m in diameter and 1,550 m long. The reversible turbines will generate power for 6 h and pump water for seven hours daily. The Benefit–Cost ratio is expected to be about 1.56 and the Internal Rate of Return about 21%.

Started as a firm-power (base-load) generating station, the Koyna Hydroelectric Project presently (2005–2006) caters to 16% of peak power demands. Each year 2,280 million units (KWh) of power are generated, 30% of power generated in the

15   Impacts of Koyna Dam, India   341

Table 15.2 Koyna Hydroelectric Project Energy Generation (MU of KWh)

	Generation in MU of KWh				
Years	Stages I & II	Stage IV	Stage III	KDPH	Total
1961–1966	6,152	–	–	–	6,152
1966–1971	14,673	–	–	–	14,673
1971–1976	16,764	–	170	–	16,934
1976–1981	19,013	–	3,977	15	23,005
1981–1986	17,126	–	3,749	278	21,153
1986–1991	13,140	–	3,340	512	16,992
1991–1996	13,088	–	3,341	714	17,1423
1996–2001	10,384	1,225	2,859	619	15,087
2001–2004	2,696	4,826	1,743	288	9,553
1961–2004	113,036	6,051	19,179	2,426	140,692

MU million units of energy in KWh

Note: Stage I generation started in 1961–1962. Stage II generation started in 1966–1967. Stage III generation started in 1975–1976. KDPH (Koyna Dam-foot Power House) generation began in 1980–1981. Stage IV generation began in 1999–2000

The Krishna Water Disputes Tribunal (KWDT) restricted westward diversion to 1.92 BCM, subject to additional diversion, till 1994 in the following stages: (a) 1974 to 1984—2.75 BCM, (b) 1984 to 1989—2.46 BCM, (c) 1989 to 1994—2.21 BCM

state of Maharashtra as a whole. The Koyna Project's scope for hydroelectric generation has steadily evolved through its four stages. Table 15.2 shows progress in installed capacity and actual units of power generated. The generation is dependent upon storage year to year but it is clear that power generation has remained more or less uniformly high, because of dependable yield in the upstream basin. Running of powerhouse to take care of peak loads (peaking) was begun only in October 1999, prior to which the stations ran as base-load units. The westward diversion is authorised and the quantum is considered as consumptive use against the share of the state of Maharashtra, the uppermost riparian state amongst the co-basin states. Further development in the form of Stage V, comprising the Humbarli pump turbine station and additional science and technology inputs, promises a more optimised output. The salient features of the evolution of the Koyna Project through five stages are indicated in Table 15.3 (Modak and Ghanekar 2002).

15.7   Sedimentation in the Reservoir

The Koyna Hydroelectric Project Reservoir stores about 2.65 BCM of water above the minimum draw-down level (MDDL) and 0.145 BCM below it. As mentioned earlier, the catchment area of the dam is 892 km^2 while the reservoir area at FRL is 115 km^2, leaving a relatively small area to contribute to the sediment load entering the reservoir. During planning, the loss of dead storage was projected at 6.50 ha·m/100 km^2/year.[1] Accordingly, dead storage was considered to have a life of

---

[1] ha·m is a unit of volume indicating one m depth over a one hectare area. 100 km^2 indicates a unit of catchment area of 100 km^2.

Table 15.3 Salient features of the Koyna Hydroelectric Project

Particulars	Koyna Dam & Powerhouse, stages I & II	Kolkewadi Dam & Powerhouse, stage III	Powerhouse, stage IV
a. Location	Koynanagar, Tal: Patan, Dist: Satara	Alore, Tal: Chiplun, Dist: Ratnagiri	Koynanagar, Tal: Patan, Dist: Satara & Alore, Tal: Chiplun, Dist: Ratnagiri
b. Latitude	17-0-25 to 17-0-30 North		
c. Longitude	73-0-38 to 73-0-45 East		
Reservoir			New reservoir not created
a. Catchment area	891.78 km²	25.40 km²	
b. Capacity			
i. Gross	2,797.4 mm³ (98.78 TMC)	36.22 mm³	
ii. Net	2,652.4 mm³ (93.6 TMC)	11.22 mm³	
iii. Dead storage	145 mm³ (5.12 TMC)	9.18 mm³	
iv. From spillway crest to top of gates	725 mm³		
c. Water spread	115.35 km²	1.67 km²	
d. Project-affected villages	98	6	
e. Project-affected families	9,069	355	
Dam			No new dam planned
a. Maximum height above river bed	85.35 m	56.80 m	
b. Maximum height above foundations	103.02 m	66.30 m	
c. Length	807.72 m	497 m	
d. Top width	10.70–14.80 m	4.8 m	
e. Spillway gates	6 (Taintor) 12.50 m * 7.62 m	3 (Taintor) 12.50 m * 6.21 m	
Dam levels			
1. FRL	657.90 m (2158.5′)	135.4 m	
2. MWL	661.337 m (2,170′)	137.16 m	
3. MDDL	609.6 m (2,000′)	130.1 m	
4. TBL	664.484 m (2,180′)	139.3 m	
5. Crest RL	650.3 m (2133.5′)	131.36 m	
6. Lowest river bed RL	579.1 m (1,990′)	82.5 m	
7. Deep foundation RL	561.73 m (1842.94′)	73.0 m	
Intake Work			Double lake tap, intake tunnels (2), intake structure

15 Impacts of Koyna Dam, India

Table 15.3 (continued)

Particulars	Koyna Dam & Powerhouse, stages I & II	Kolkewadi Dam & Powerhouse, stage III	Powerhouse, stage IV
a. Intake channel length	1,227 m		
b. Headrace tunnel	Length 3,748 m	4,351 m	4,225 m
	Diameter 6.4 m (circular)	7.41 m ('D' shaped)	7.0 m * 9.5 m (horseshoe)
c. Maximum	164 m³/s	170 m³/s	260 m³/s
Pressure shafts			
a. Number	4 converted to 8, by Y-pieces at the end	3	4
b. Diameter	3.04 m at top, 2.59 m at bottom	4.115 m each	3.90 m at top, 3.75 m at bottom
c. Length of each penstock	616 m	192 m	590 m
Powerhouse and appurtenance			
a. Approach tunnel	864 m long	780 m long	988 m long
	6.7 * 6.1 m	7.00 * 6.24 m	7.0 * 8.5 m
b. Ventilation tunnel	36.80 m	65 m	760 m
	2.4 * 2.7 m	4 * 2 m	4 * 4 m ('D' shaped)
c. Cable tunnel	3; 357 m	2	2
	390 m & 457 m	78 m	427 m & 391 m
d. Tailrace tunnel	2,215 m	4,543 m	2,154 m
	7.9 * 6.4 m	10.36 m. dia	10 * 10 m ('D' shaped)
e) Peak discharge	164 m³/s	325 m³/s	260 m³/s
f. Tailrace channel	–	4,790 m	165 m
Switch yard	(45.72 * 295) m	(340 * 137) m	(136 * 45) m
Power generation			
a. Turbine	Stage I: 4 Pelton 110,000 HP (France)	4 vertical Francis	4 vertical Francis
	Stage II: 4 Pelton 99,180 HP (Swiss)	113,000 HP each (BHEL)	233,480 KW (Neyrpic-France)
b. Design head	Stage I: 475 m	109.7	496 m
	Stage II: 490 m		
c. Generator	Stage I: 4, 65 MW each	4, 80 MW each	4, 250 MW each
	Stage II: 4, 75 MW each	–	
d. Load factor	0.6	0.24	0.18

MDDL minimum draw-down level, *Tal* Taluka, *Dist* District, *TMC* thousand million cubic feet, *TBL* top of bund level, *RL* reduced level

150 to 200 years. At that time, loss of live storage—mainly due to release of bed load or coarse sediment and consequent bay/delta formation at the entry of tributaries into the reservoir—was not considered.

Hydrographic surveys by conventional means were conducted for the reservoir about four times during the last four decades up to year 2005. Remote sensing technique was also utilised to estimate loss of live storage by getting sedimented reservoir contours above the low levels of the reservoir during the survey years. Recently, Digitised Global Positioning System (DGPS)-based automated hydrographic survey equipment has been deployed and is being used for all later surveys. The surveys so far indicate that the actual loss of dead and live storage is about 0.11% and 0.02% per year, respectively. In terms of total sediment deposited yearly in the reservoir, about 20% affects dead storage while 80% affects live storage. Loss of dead storage in the last 40 years since impoundment can thus be inferred at 4.4%, whereas that of live storage may be to the order of 0.8%. At this rate, sedimentation loss of dead storage may actually take about 1,000 years.

Another important aspect controlling the loss of live storage is the storage behind crest gates. These gates have been operated to flush deposition over the crest. Reservoir operation rules require building up the reservoir to El 667.4 m up to 22 August each year by keeping the gates closed, filling it further up to El 668 m by 30 September by opening gates as required, and then trying to keep it at El 668 m as long as possible by closure of gates. Records of reservoir operation indicate that minimum water was released in the years 1981–1982 and 2002–2003 over the crest. The elevations of the reservoir in these years were about 637.4 m in June, rising to a maximum of 667.4 m in September and then depleting to 637.4 m, almost uniformly by 2.5 m every month. Koyna Reservoir has a sizeable draw-down of about 30 m every year, possibly causing some sloughing/erosion of the deposition in draw-down heights to lower levels. When needed the six gates are opened, starting from the end gates, to clear the deposition as much as possible near shore. At closing too the end gates are closed last, allowing the erosion process to continue as long as possible (Maharashtra Engineering Research Institute Records accessed by the author in 2004 in his capacity as Chairman of an Expert Committee of the Government of Maharashtra).

15.8 The Krishna Water Dispute Tribunal

The Krishna Water Dispute Tribunal was set up by the Government of India to adjudicate on issues raised by the three riparian states—Maharashtra, Karnataka and Andhra Pradesh. Its final award of 1973 allocated the basin's 75% dependable surface water flow of about 59 BCM, assessed at Vijayawada near the head of the Krishna delta, between the states. The share of the most upstream state, Maharashtra, was fixed at 16 BCM, that of Mysore (i.e. Karnataka) at about 20 BCM, and of Andhra Pradesh, the most downstream state, at about 23 BCM. The tribunal award envisaged augmentation of river flow due to regeneration after application of abstracted water to irrigation, and allocated that too between the three states.

At the time of the tribunal, the Koyna Hydroelectric Project was already in operation, diverting westward the waters of the Krishna (Koyna) Basin for hydropower generation (about 1.9 BCM and lake loss of 0.21 BCM due to evaporation, adding up to 2.11 BCM). The tribunal considered the controversial issue of priority between the irrigation needs in the basin's scarcity regions in the downstream, and the committed diversion outside the basin for power generation in the upstream. The tribunal award upheld the westward diversion as legally valid, citing several similar cases worldwide, though the non-consumptive use was considered as consumptive, adjusting it against the allocated share for Maharashtra state. Some other out-of-basin westward diversions, including lake losses of 0.07 BCM, totalling about 1.27 BCM in other sub-basins of the Krishna were also similarly protected, subject to a 5-year diversion of about 6 BCM. The tribunal noted that such diversions comprised about 5% of the 75% dependable flow of the Krishna River.

Some concession was granted for the extra use of water for hydropower generation during specified time steps, viz. 2.75 BCM in 1974–1984, 2.46 BCM in 1984–1989 and 2.21 BCM in 1989–1994, as irrigation development in the downstream would take time and the water would otherwise be wasted. However, the Krishna Water Dispute Tribunal did not allow additional westward diversion, in order to ensure that the intra-basin needs in the downstream areas of other states could be met. The tribunal award was reopened after 30 years, as per specific provisions. A new tribunal was constituted by the Indian government in April 2004, under the chairmanship of Supreme Court Justice Brijesh Kumar and two members. Fresh submissions were made by the party states to the newly constituted tribunal, and the earlier award has continued to prevail in the meantime, as the Award of the new Tribunal is yet to be declared. Any evolution and/or consideration of future stages for the Koyna Hydroelectric Project will be subject to the parameters of the award of the new tribunal (Government of India 1973, 1976).

15.9 Social Impacts

The Shivajisagar Reservoir created behind Koyna Dam (during stages I and II) submerged 11,535 ha of land in 98 villages, affecting 9,069 families comprising about 30,000 people and 3,755 households, mainly Thakar and Katkari communities that are dependent on forest produce, fishing and cultivation of some coarse grain crops. Some 6,316 project-affected families were rehabilitated by the Koyna Hydroelectric Project, while about 2,753 families preferred to rehabilitate through their own efforts. The project settled 19,500 persons (3,310 project-affected families from 76 villages) within the catchment area in acquired forests at a distance of 8–10 km from their original locations.

The submergence covered only about 4% of government owned forest-land and the rest was privately owned terraced and/or waste lands. People from the submergence area generally practised shifting cultivation in narrow, isolated land strips and migrated to the plains in search of employment after monsoon. Since such plots have been further isolated after formation of the reservoir and remain less productive, they were acquired by the project, as desired by the owners, and handed over to the Forest

Department for afforestation. People from 30 villages near the northern edge of the reservoir preferred to stay and continue cultivating their land, of which only a small part was likely to be submerged. They could also take advantage of the silted land for cultivation after drawing down of the reservoir every year by about 30 m.

In all, 6,316 project-affected families were resettled in 180 habitats of their choice in six districts adjoining the project area. About 143 were relocated in Satara District itself. Three of these districts—Raigad, Ratnagiri and Thane—are located to the west on the coastal strip, whereas three others—Sangli, Satara and Kolhapur—are to the east within the command area of the project. About 6,971 ha of land were allotted to displaced persons who asked for alternative land. The Kolkewadi Reservoir (Stage III) submerged 1,670 ha of land in six villages, affecting 355 families. About 181 landowners and 41 ha of their agricultural land, were affected. They were resettled in Raigad, Ratnagiri and Thane Districts.

For Stage IV, about 629 ha of agricultural land were acquired, in addition to about 2 ha of forest-land. No village was fully displaced. More than 57 ha were allotted to 66 landowners. AIMS Consultancy (1998) reported that no affected family/person had gone to court against the acquisition of land for the project. Of the three villages affected in Stage IV, Nawaja was a resettled village from Stage I, Kolkewadi was partially resettled during Stage III and Kondphansawane was affected by both stages II and III but was not a resettled village. Project-affected families were offered land in a district at some distance from the original location and they preferred not to relocate. Besides, project-affected families with yearly income below Rs 20,000 in 1997, 145 in number, were identified as below the poverty line as they were eligible for special assistance such as skills training, dairy business, transport vehicles, lift irrigation facility, trading shops etc. As per 1986 policy, project-affected families were entitled to 13 different facilities including a well for drinking water, school, roads, electricity, crematorium, latrines, gutters, cattle stand, pasture, threshing floor and market.

As in 2005, the Koyna Hydroelectric Project was in possession of about 14,650 ha of land (net) of the 27,302 ha acquired—8,755, 17,413, 508 and 626 ha for Stages I through IV, respectively. About 4% of this land was held by the government through revenue and forest departments. The land acquired also includes a part of land returned to the Forest and Revenue departments (not specifically indicated), as it was no longer required.

In an earlier era, the state government acquired property for public purposes under India's Land Acquisition Act of 1894 and provided compensation in cash. The displaced settled themselves as they wished, with the compensation at their disposal. In due course, significant amendments (in 29 instances, the last one of 1984 included) were made to the Act and new procedures were established. For the first time, the concept of rehabilitation of project-affected families was woven into the procedures to be followed, and a Rehabilitation Committee was formed in 1955 at the state level. The Koyna Resettlement Advisory Committee was formed in 1957. In June 1965, a dedicated Directorate of Resettlement was set up at the State level, under a government resolution, to attend to the issue for water resources projects. The new procedures evolved over a period of time, resulting in the Maharashtra

Resettlement of Project Displaced Persons Act of 1976. A revised Act came into being in 1989, as the Maharashtra Project-Affected Persons Rehabilitation Act. It is applicable to projects for irrigation, power, public utility, or composites serving more than one function. It is further consolidated and amended in light of ground conditions from time to time.

About 1,500 landowners from stages I and II of the Koyna Project who asked for allotments of new land in proximity to their old locations, instead of the offer made in the irrigation command area of the Ujjani Project, a major irrigation project, are yet to be satisfied. The land required is to be procured from the Forest Department, and the matter is taking time. Some complaints about land boundaries and corrections to the revenue record are pending. Another complaint is related to operation and maintenance costs for facilities provided in the past. These issues are yet to be amicably settled. Other complaints pertain to assigning revenue status to new habitations, in order to grant status of full ownership to sub-owners, and issuing certificates regarding need for employment. The resettlement and rehabilitation work from different stages is still continuing, and remains incomplete due to difficulties related to verification of claims spread across a large period of time.

Since 1894, then, the focus has moved away from compulsory acquisition of private property for public purposes, to addressing developmental needs with full compensation, and finally to rehabilitation of displaced populations at higher economic levels than those preceding resettlement. At Koyna, although the full cost of acquired land was paid, the displaced were permitted to salvage as much as possible from their property, including timber for construction of their dwellings in the new settlement. Free transport was provided to the relocation site, and construction materials were supplied at subsidised rates from the project stores. For the first time effort was made in the Koyna Project to help the displaced resettle in a more organised manner. For this purpose, a Rehabilitation Advisory Committee, comprising officials of the concerned departments and some local leaders, was set up. Alternative offers of land for allotment were made and the representatives of the displaced were invited to visit and select the land of their choice for a new settlement. The selected land was surveyed, layout of public facilities prepared, and water supply provided at project cost.

For the most part, forest- or wasteland certified as suitable for cultivation with or without improvement, of value equal to the area acquired, was offered and finally allotted to the displaced. The land was made cultivable by terracing or contour bunding, where necessary. The displaced were allowed to till their original land until actual submergence after acquisition (about one year), in addition to the newly allotted land, which required time for development. Several displaced people preferred to use the compensation to buy private lands on their own. Young people were trained in different vocations and offered employment in project work. A wildlife sanctuary was carved out on the right bank of the Koyna Reservoir in the 1970s, by notifying land from 50 villages.

By and large, the resettlement and rehabilitation policy of water resources projects has become crucial to the success of all developmental work. Liberal provisions have therefore been made, as more and more projects have been initiated. In the last decade, new efforts were made to design a model that would suit different sectors of development. As resettlement and rehabilitation issues became clearer and more critical over the past few decades, several major projects in India stalled because of inadequate provisions. Some, such as the Sardar Sarovar and Tehri projects, adopted excellent packages for resettlement and rehabilitation, and were able to solve problems to their advantage. While the debate was raging in the water resources sector, resettlement and rehabilitation became a hot topic in several other socio-economic sectors. The Government of India therefore initiated an exercise at the national level in the 1990s, to formulate a national policy for resettlement and rehabilitation (Kulkarni, Land acquisition and rehabilitation, Development of policy and implementation, 2002, unpublished article).

15.10 National Policy on Resettlement and Rehabilitation for Project-Affected Families, 2003

The National Policy on Resettlement and Rehabilitation was published in February 2004 by the Department of Land Resources under the Ministry of Rural Development, with the following objectives:

a. To minimise displacement and to identify non-displacing or least-displacing alternatives
b. To plan the resettlement and rehabilitation of project-affected families including special needs of tribal groups and vulnerable sections
c. To provide better standards of living to project-affected families
d. To facilitate harmonious relationship between the agency acquiring land and project-affected families through mutual cooperation

The National Policy lays down basic norms, packages, broad guidelines and executive instructions addressing the needs of displaced people with a focus on resource-poor sectors. It provides a broad approach for effective dialogue between the acquiring agency/administration and project-affected families, leading to timely completion of projects. The provisions are applicable to projects affecting 500 or more families—en masse—in the plains or 250 and more in hilly areas.

The key components of the policy are:

Preparation of schemes/plans for resettlement and rehabilitation: Declaration of affected zone; procedure to be followed for survey and census of project-affected families, etc.; assessment of land available for resettlement and rehabilitation; declaration of resettlement zone; power to acquire land for resettlement and rehabilitation; draft scheme/plan for resettlement and rehabilitation; management of funds for resettlement and rehabilitation; final publication of scheme/plan of resettlement and rehabilitation.

Resettlement and rehabilitation benefits for project-affected families, including Scheduled Tribes.

Basic amenities to be provided at resettlement zone. Dispute redressal mechanism: Resettlement and Rehabilitation Committee at project level and Grievance Redressal Cell.
Interstate projects' monitoring mechanism. National Monitoring Committee and National Monitoring Cell.

The resettlement and rehabilitation policies being followed in the country's water sector are relatively more liberal because of the large size of the displaced population and also because of the critical nature of water resources development. It goes without saying that the actions taken in the Koyna Project largely formed the basis for drafting the guidelines of the new policy. Successive stages of the Koyna Hydroelectric Project, however, introduced a new element of secondary displacement due to the need in subsequent stages to displace some rehabilitated people.

15.11 Economic Performance and Impacts

The generation of hydropower through the westward diversion of water from 1967 to the present, barring 4 or 5 years, has remained higher than the planned quantum. Similarly, irrigation utilisation eastward has been more than 0.6 BCM for about 55,000 of the 100,000 ha area under irrigation using the combined waters of the Koyna and Krishna rivers. The total Koyna River yield is found to be fully dependable.

A benefit-cost analysis, along with determination of Internal Rate of Return, for different options of the Koyna Project was carried out in 1962 when the Krishna Water Dispute Tribunal was considering the matter. The Internal Rate of Return is the discount rate of return that is determined from the aggregate present value of net cash inflow with the aggregate present value of cash outflow. Revenue from the sale of electricity was worked out on the basis of average selling price of electric power. For the benefit–cost analysis, annual cost was computed as a sum of interest charges on investment, plus depreciation on the investment, plus maintenance charges. The benefit stream comprised revenue proceeds from sale of hydropower after allowing for transmission losses. The study indicated that the Koyna Hydroelectric Project yielded a benefit–cost ratio of 1.17 and Internal Rate of Return of 5%. These figures were 1.99 and 17.5%, respectively, for diversion of a higher quantum of 3.3 BCM of water. It was further shown that if irrigation in the lower riparian states was accorded priority over the westward diversion of water for hydropower, the benefit-cost ratio and Internal Rate of Return both were significantly reduced, to 1.1 and 2.5%.

More recently, an effort was made to assess the financial performance of the project based on Internal Rate of Return up to the year 2000. For simplification, it was assumed that the life of the project was completed in the year 2000. Actual records were available from 1982 and data for the prior period were extrapolated. Details of irrigation and non-irrigation eastward water use were available from 1977 and were used. Nearly ten check dams hold Koyna and Krishna releases for irrigating 100,000 ha of fertile but thirsty land. The Koyna releases account for about 55,000 ha of irrigation. About 27,000 ha are under perennials, while about 16,000 and 12,000 ha are under Kharif (monsoon) and Rabi (winter) crops, respectively. Lift irrigation schemes are being implemented to utilise the remaining waters and irrigate additional arable drought-prone areas. Three major lift irrigation schemes are under construction in the east (Takari, Tembhu and Mhaisal) to supply the remaining water allocated by the Krishna Water Dispute Tribunal Award for use in Maharashtra state, involving lift heights of 250 to over 400 m and to serve an area of about 190,000 ha. They may be put into operation in the coming years, depending upon several factors, at a cost of Rs 150,000 and above per hectare. The Internal Rate of Return based on direct benefits as indicated, work out to 25.81%.

The indirect benefits of irrigation have been recently assessed by MITCON Ltd, a consultancy firm, at Rs 4,470/ha as reported by Modak and Ghanekar (2002). For annual irrigation of about 50,000 ha, the benefits amounted to Rs 5,000 million/year. The indirect benefits of hydropower have not been assessed. With these added, it is expected that the Internal Rate Return will go up significantly. Water resources projects in India have a multiplier or triggering effect of as much as 2.6, as assessed by the World Bank on projects supported by them. Koyna hydropower is generated at a minimum price as compared to other hydro projects in the country. Its triggering effect is therefore expected to be much higher. The Koyna Hydroelectric Project generates about 2,850 million units of power per year, yielding annual revenues of about Rs 6,000 million. MITCON Ltd has estimated yearly indirect benefits from irrigation at Rs 220 million.

A high level of poverty is synonymous with poor quality of life, deprivation and low human development. The country's Planning Commission periodically estimates the incidence of poverty at the national and state levels, and such records are available for the period 1973–1974 to 1999–2000. Based on a large-scale sample survey of consumer expenditure, the poverty ratio was estimated at 27% in rural areas and 23.6% in urban areas, aggregating to 26.1% for the country as a whole. At the national level, there has been a decline since 1973–1974, but rural-urban and interstate disparities are visible. Regions that experienced a faster rate of growth (GNP), particularly where such growth was mainly through agricultural production, exhibit a reduction in the proportion of people below the poverty line. The impact of the Koyna Project in terms of poverty alleviation in the communities of the region has to be comprehensively studied.

The economic benefits accruing from drought- and flood-proofing through the Koyna Dam are being assessed. As with all water resources development projects, these positive impacts are significant but remain unquantified. In addition, these economic gains have supported the growth of the population itself. Since the productivity of irrigated areas is at least double that of the best rain-fed areas, every hectare brought under irrigation saves deforestation of about the same area. Therefore, bringing 100,000 ha under irrigation has avoided deforestation of at least an equivalent for agriculture.

15.12 A Study of Lift Irrigation Schemes

The Koyna River was notified in 1959 and the Krishna River in 1964 under the State Irrigation Act, which brought their regulation under the Irrigation Department of the state government. Unlike most irrigation projects with limited water supply, Koyna provided assured releases round the year, promoting perennial crops. Further, irrigation came to be based on lifting water from rivers or from ponds behind check dams constructed to hold water for limited time periods. A composite study of the performance of the lift irrigation schemes over the previous three decades was undertaken in the 1990s. Such irrigation schemes were developed along the Koyna River up to a length of 69 km from the dam, followed by the Krishna River along a length of 153 km up to the border with Karnataka. By 1992, there were in all 4,700 lift irrigation schemes using 90,000 HP of energy, with 5,000 pumpsets in 260 villages covering 180,000 ha, plus 60 small scattered schemes spread over 200 villages utilising 3,600 HP of energy. The farmland in question was mostly mortgaged, and the lift irrigation schemes were financed by National Bank for Agriculture and Rural Development of India (NABARD) and managed mostly by sugar cooperatives. Eight Kolhapur type weirs were built for enabling lift. The Irrigation Department levies a cess on individual farmers who avail the facility and the cooperatives levy a further charge. (A study of lift irrigation schemes under operation was conducted in 1994 by the Development Group, Pune.)

The irrigation command of specific lift irrigation schemes was up to 2,000 ha. Broadly, they were grouped into five categories by coverage—less than 100, 100–300, 300–600, 600–1,000, and greater than 1,000 ha. The sample schemes comprised 1,500 beneficiary farmers: 18% with a command area up to 200 ha, 24% up to 400 ha, and 58% beyond 400 ha. In terms of crops grown the area irrigated by the projects is split as follows: perennials 56%, Kharif 26% and Rabi 18%. The command area is predominantly in the sugar belt with five main sugar factories. In the sugar industry, the sugarcane crushing capacity ranges from 1,200 to 5,000 tonnes per day. Two of Asia's largest sugar factories, in terms of capacity, are in this area, namely Vasantdada, 5,000 tonnes (Asia's largest), and Rajarambapu, 4,000 tonnes. They have taken the lead in securing fair prices, credit and other services for farmers, and also in the stabilisation of irrigation systems in the area. The role of the state Irrigation Department was limited to sanction of the crop pattern and release of water from Koyna Dam for smooth functioning of lift irrigation projects.

The survey was conducted in three main blocks of Miraj (Sangli District), Walwa and Tasgaon (Kolhapur district) which have 210 major lift irrigation schemes. Of these sample schemes, 23% had lifting energy up to 25 HP, 24% up to 50 HP, 12% less than 100 HP, 14% less than 200 HP, 14% less than 400 HP, and the rest were individual farmers. For each small project a 50 HP pumpset was deployed, 600 HP for a medium project, and 800 HP pumpset for a large scheme. The scheme sample included cooperatives, some started by the government that had become inoperative due to various reasons, and a few individual schemes. In addition, 81 dry-land farmers were included to as a control group.

The schemes were mostly in the Sangli District where 86% of the area is cultivable, with 6% forest cover and the rest fallow or wasteland. The Krishna, Warna and Yerala basins spread across the district. About 20% of the cultivable land is irrigated, 40% of which is by flow and 60% by lift. Some 70% of the area is under sugarcane. Average rainfall for the region is 625 mm, and the potential use of irrigation at present consists of 73% perennials, 27% Kharif and 36% Rabi crops.

Farmers benefiting from lift irrigation derived 57% of their annual income from agriculture, 19% from service and 12% from dairy businesses; for individual farmers the annual income was split 55%, 13% and 12%, respectively, while dry-land farmers' income was 25% from agriculture, 42% service and 10% dairy. Supplemental irrigation from wells was preferred. Surprisingly, the location of the farm (whether at head/tail along the lift canal) did not make much difference, probably because of the size of each lift irrigation scheme. The schemes were financed by banks (70%), cooperatives (20%) and individual farmers (10%). The capital investment required was Rs 2 million per small scheme, Rs 4 million per medium scheme and Rs 15 million per large scheme, while the investment per hectare of irrigated land ranged from about Rs 10,000, to 15,000 and 22,000/ha, respectively. The main cost components were pipeline-distribution network (65%), electricity/motor installation (17%), jackwell/pumphouse (8%), others 10%. Annual costs included salaries (24%), interest (21%), electrical energy (16%), administration (14%), depreciation (12%), repairs and maintenance (10%), others 3%. The economic status of farmers with private schemes was seen to be much better than that of cooperative or dry-land farmers. Better-off farmers (viz. frontline, educated, those who adopt modern methods quickly) had mostly invested in such schemes. On the whole, the economic condition of farmers in the region improved greatly due to irrigation. Farmers' priority for investment of additional income was in the following order of importance: children's education, farm development, construction/repair of homes, medical treatment, savings, expansion into additional land and other occupations.

Recovery of the irrigation cess remains a major issue for sustainability of irrigation from the Koyna Dam. One of the reasons for non-recovery was the fact that the penalty for non-recovery was three times the original amount, which could not be enforced and hence ballooning the size of recovery arrears took place. The study identified the need to convert lift irrigation schemes into Water Users Associations, and for the sale of water on volumetric basis to address this lacuna and ensure recovery of charges. Suggestions for attaining sustainability included placing responsibility for recovery with the lift irrigation scheme management itself, which should obtain consent letters from beneficiaries. The need for modernisation of old, leaking and failed machinery in lift schemes was considered urgent after due technical assessment. The government will have to take the lead in raising resources, abandoning the insistence on cropping pattern imposition, attending to water logging and salinisation, removing encroachments on drains, including rotational crops such as soybeans, and limiting the use of chemical fertilizers. Smaller schemes were seen to be better managed, and thus careful selection of scheme size was recommended for the future.

Table 15.4 Salient parameters of Koyna Reservoir water

pH	7.9
Total alkalinity (CaCO$_3$ mg/l)	20
Dissolved solids (mg/l)	40
N/10 HCl to neutralise 100 ml water (mg)	0.3
Calcium (mg/l)	9
Potassium (mg/l)	7
Sodium (mg/l)	11
Total Solids (mg/l)	48
Conductivity (μmhos)	60

15.13 Environmental Impacts

Any study of the social and environmental impacts of a project in India that was completed several decades ago is handicapped by the lack of requisite baseline data. This is understandable, as there was no formal or legal requirement to conduct an Environment Impact Assessment or to assess social impacts till the 1980s. As a result, all such data, which are now considered to be relevant baseline information, are not available in the project reports for the Koyna Project. Nevertheless, an attempt has been made to reconstruct the pre-project situation of the earlier period.

The Koyna wildlife sanctuary, spread over 43,500 ha, was established in the 1970s on the western periphery of the Koyna Reservoir in the reserved forest spread that was not open to the public; thus, it is now a rich forest providing shelter to many species of flora and fauna. In contrast, because of access for local populations, there is evidence of some degradation in other forests in the area. The reserved forest comprises mainly tropical moist deciduous and partly semi-evergreen forest. The site does not include routes for migratory birds because of the relative lack of shallow waters. Noxious aquatic weeds have not been noticed. In contrast, the population of vectors such as snails and mosquitoes has not increased in the reservoir area, probably because of predators such as fish and birds. The proposed lake under Stage V with a spread of about 108 ha, of the total catchment area of 285 ha, will be close to this sanctuary. A rapid survey carried out by the Science and Technology Park of the University of Pune and reported in 2004 (Jagdale 2004) to the project indicates that there will be no endangered species in this area. In the project area of 119 ha itself, new compensatory plantation of trees has been carried out at an average rate of 400 trees per hectare. Significant fishery can be promoted in the reservoir, but fishing is not permitted since the reservoir is within the 'wildlife sanctuary'. It is necessary to plan and allow fisheries in this lake, without jeopardizing ecosystems and without allowing encroachments on the reserved sanctuary. A proposal to carve out a 'Tiger Reserve' in the sanctuary is presently under active consideration.

The quality of water in the Koyna Reservoir is regularly monitored and is found to be unaffected since impounding. The salient parameters are listed in Table 15.4. Likewise, the Central Ground Water Board monitors the quality of groundwater from wells at about 25 stations in the irrigation command.

Riparian water needs along the river in the downstream were calculated at less than 60 l/sec year-round in the past. No specific reservation was therefore made in planning for releases of water for such needs. However, the need has grown significantly since then and is met by regulated releases round the year from the dam-foot hydropower station.

15.14 Science and Technology Impacts

The Koyna Project introduced a set of new science and technology measures, probably for the first time in India, which transformed water resources development and management at the national level. Important aspects included the following: (a) indigenous technology and skills were developed for building underground hydropower station; (b) a vast network of tunnels in basalt formations was dug with locally trained manpower; (c) the lake tap procedure was adopted and further developed, and GIS was deployed underground in the powerhouse/transformer caverns; (d) natural geographical features were availed to generate high level of energy production from the same quantum of water, and optimal use of stored waters was achieved; (e) rubble-concrete technology was used for dam building for the first time, and a 1,000-ton compression testing machine was fabricated and used extensively for advance investigations and quality control; (f) suspected Reservoir Associated Seismicity was ruled out after disproving all theories; (g) compensatory afforestation was comprehensive, and a sanctuary with luxuriant forest cover has been nurtured in the reservoir periphery; (h) vastly improved river flow in the downstream was achieved, enabling rejuvenation of groundwater through Kolhapur-type weirs downstream of the dam; (i) an entire generation of engineers, scientists, contractors, skilled workers was trained, and these cadres spread across the country, benefiting the nation as a whole; (j) dam electrical equipment remained unaffected and in good order due to a very low level of suspended sediment load passing through the turbines, and state of the art dam safety measures are in place and continuously monitored.

15.15 Future

Future evolution of the Koyna Project could include complete conversion of the base-load station into a peaking station, the conversion of Pelton wheels into reaction turbines, and the introduction of pump turbines as and when the original facilities age. Another dam-foot powerhouse is being planned on the left bank of the river with two units of 40 MW each. It might be an underground station with another lake tap, this time on the left bank. The new Krishna Water Dispute Tribunal Award was delivered on December 2010 allowing further modernisation and evolution of this key hydropower station in western India.

The changing scenarios following the Koyna Dam and several stages of the development of the Koyna Hydroelectric Project show that the dam has proved to be a great boon to the growing population of the region, generating employment opportunities through industries that are run on the hydropower generated, reducing poverty and also conserving local ecosystems. The reservoir waters have enhanced the ecology around the dam, encouraged growth of forest, reduced soil loss, and made available regulated releases in the river's downstream for irrigation as well as various non-consumptive uses. With the spread of water to areas it did not reach before, the process of desertification caused by aridity and drought has been reversed. The irrigation of 100,000 ha has generated a biomass which was simply not possible without the dam. Flood and sediment flow in the downstream has been reduced. The rejuvenated river has enabled recharge for the otherwise depleting groundwater of the region.

Alongside these benefits, the technological innovations of the Koyna Dam triggered the adoption of modern science and technology inputs in an unprecedented manner across the country. The Koyna Dam has been extremely beneficial to the state, to the region and to the nation as a whole.

The pros and cons of Koyna's development through multiple stages over several decades have been extensively debated. While the most bitter critics acknowledge its contribution, a question often raised, in hindsight, is why the developments were not planned and achieved in a single instance, as it would have been much more economical. Several reasons could be cited for and against this hypothesis. Life is, of course, the art of the possible. One can learn from history and avoid making the same mistakes again. The present chapter is an attempt to chronicle as much information as possible to enable such rational decision making in future.

Acknowledgements This case study is a compilation of available information; no studies were undertaken by the author to assess impacts. The study, however, indicates the grey areas in which more detailed assessment will be useful. The project has been a resounding success and the lessons learnt will be extremely useful for posterity.

The compilation has been made possible by the unstinting cooperation extended by engineers of the state government and the Koyna Project led by Mr. S.V. Sodal, Mr. V.V. Gaikwad, Mr. S.Y. Kulkarni, Mr. R.V. Panse and Mr. R.G. Dravid. Their contribution is acknowledged with thanks.

References

AIMS Research Pvt Ltd (1998) Socio-economic survey of rehabilitation action plan for project affected persons of Koyna. Stage IV, Final report for Department of Revenue and Forests, Government of Maharashtra, Pune
Biswas AK, Rangachari R, Tortajada C (2009) Water resources of the Indian subcontinent. Springer, New Delhi
Development Group (1994) Study of lift irrigation schemes on Krishna River, Internal report submitted to the Government of Maharashtra, Pune
Gokhale Institute for Economics and Politics (1963) Economic evaluation of alternative uses of Koyna-Krishna Water for Irrigation and Power, Pune
Government of India (1973) The Krishna Water Disputes Tribunal with the Decision, vol II, New Delhi
Government of India (1976) Further report of the Krishna Water Disputes Tribunal, New Delhi
Jagdale R (2004) Science and technology park. Interim report, University of Pune, Pune
Koyna Hydro Electric Project (1950) Report and appendices, vol 1, Mumbai

Koyna Hydro Electric Project (1952) Project report, 1st stage, Mumbai
Koyna Hydro Electric Project (1970) Completion report (chapter V) Hydrology, Mumbai
Koyna in the Belly of Sahyadri (1990) A popular article, Pune
Mahabal AS (2000) Environmental impact assessment for Humbarli Pumped Storage Scheme. Zoological Survey of India, New Delhi
Mane PM, Murti NGK (1962) Koyna Project Special Number. Indian J Power River Valley Dev
Ministry of Rural Development (2004) National Policy on Resettlement and Rehabilitation for Project Affected Families, Department of Land Resources, Government of India, New Delhi. Published in the Gazette of India, Extraordinary Part-I, Section 1, No. 46, 17 February 2004
Modak DN, Ghanekar SK (2002) Evaluation of Koyna Project Stages I, II, III. Internal report, Pune
Pendse MD, Huddar SN, Kulkarni SY (1975) Rehabilitation of Koyna Dam. Internal report, Pune
World Bank (1989) Staff appraisal report. Maharashtra Power Project, Washington, DC

Chapter 16
Resettlement and Rehabilitation: Lessons from India

Mukuteswara Gopalakrishnan

16.1 Introduction

India is one of the few developing countries in the world with a sizeable number of dams. The building of dams is still ongoing in as much as the storage per capita is far below requirements, especially when we consider the temporal and spatial variations in the pattern of rainfall, and the arid and semi-arid areas of large parts of India. The per capita energy availability is also far below the minimum need. Thus, notwithstanding the new emphasis on 'management' and 'efficiency improvement' of existing assets, the creation of additional and new storage through dams is an absolute necessity. The planning of new hydro projects for energy demands will continue to enhance the share of hydropower from 26% (at present) to about 40%. The Government of India has recently made its own ambitious plans, both short and long term, in the water sector. The Prime Minister's Bharat Nirman Plan, announced in 2006, proposed the harnessing of an additional 10 million ha of irrigated area within a period of 4 years after its announcement. In addition, the hydropower sector has set a target of 50,000 MW capacity addition and for a few years in the past launched a concerted action programme that could yield further development at the rate of 2,000 MW each year; an ambitious programme indeed.

Success in achieving some of these ambitious goals, however, hinges upon the achievements of a 'well-conceived and implemented rehabilitation and resettlement plan' for water resources development projects. This is especially true since the

---

The views and opinions expressed in this chapter do not necessarily reflect, nor are to be construed as, those of the organisations with which the author is associated, either at present or in the past.

M. Gopalakrishnan
Indian Water Resources Society, New Delhi, India

New Delhi Chapter World Water Council, New Delhi, India

International Commission on Irrigation and Drainage, New Delhi, India

bottlenecks in achieving desired acceleration rates are essentially caused by impediments in the implementation of resettlement and rehabilitation plans or even faulty conceptual frameworks. This is an important social issue.

While there is broad agreement on the significant role that dams play as a development tool, bearing in mind their overall benefits and costs, a key factor is inundation. Addressing the problems and issues of inundation can help the overall cause of dams and reservoirs for the welfare of society at large. Identification of related issues and proposing viable solutions to address them will help avoid the forced delays in the implementation of projects in the developing world. The pace of dam construction increased relatively late in these countries, and needs to be pursued till such time that enough storage is assured to enable food, energy and environmental security.

16.2 Inundation due to Water Projects in India and a Historical Review of Traditional Approaches

India has a rich history of dams and other diversion structures like barrages, which were called *anicuts*. The development of irrigation facilities for agriculture is certainly ancient. Large tracks of land, well-irrigated and drained, made for strong kingdoms. We see in the Grand Anicut (second century AD) one of the finest examples. This diversion structure across the Cauvery River in the peninsula near Tiruchirappalli was built by the great kings of the Chola dynasty. In the nineteenth century, during the British era, some marvellous structures were built: the Mettur Dam, Periyar Dam and Wilson Dam are a few examples. Apart from serving irrigation purposes, these projects helped in power generation.

Dams with significant storage capacities sprang up only after independence in 1947. Issues relating to inundation, resettlement and rehabilitation were a part of the overall perspective in the planning exercise and the project estimate, based on surveys. However, implementation was to be handled by the revenue authorities, which had the responsibility of matters relating to land acquisition.

If we go beyond the 15 m height criterion of the International Commission on Large Dams (ICOLD), and look for dams of significant height in India, we find that they are not large in number. So far, only two dams over 200 m in height have been constructed in the country, the Bhakra-Nangal Dam in Punjab and the recently completed dam at Tehri in Uttaranchal state. There are nine storage reservoirs with capacities larger than 5 billion cubic metres (BCM), including the just completed Indira Sagar in the Narmada Basin. Some relevant statistics of major projects are provided in Table 16.1.

Dams with a height between 100 and 200 m are 20 in number. The dams in India, despite an impressive total of 4,636 in number, are mostly in the category of 30–100 m height (about 400) and the balance are less than 30 m in height (but above 15 m). Saddle dams and similar structures further reduce the number of projects that create inundation areas in the country.

16 Resettlement and Rehabilitation: Lessons from India

Table 16.1 Statistics of some major dams in India

Name	Year of completion	River	Max. height (m)	Live capacity (MCM)	Benefits: Irrigation (1000 ha)	Benefits: Power (MW)	Rehabilitation: Village	Rehabilitation: Population/families	Area submerged (ha) Forest	Area submerged (ha) Total
Hirakud	1957	Mahanadi	59	5,378	251	270	283	18,000 (F)	N.A.	73,891
Gandhi Sagar	1960	Chambal	64	6,827	273.3	115	8	302 (F)	2,144	12,350
Rihand	1962	Rihand	91	8,967	–	300	1,989	–	–	–
Bhakra Nangal Dam (Gobind Sagar)	1963	Satluj	226	6,655	2,880	1,204	375	36,000 (P)	5,746	17,984
Ukai Dam	1972	Tapi	81	7,092	191.3	300	138	16,000 (P)	22,222	60,307
Nagarjuna Sagar	1974	Krishna	125	6,841	895	810	57	4,830 (F)	12,442	28,490
Pong Dam (Beas project)	1974	Beas	133	7,119	–	360	–	–	–	–
Indira Sagar	2006	Narmada	94.4	9,750	1,230	1,000	249	39,173 (F)	40,332	91,300
Tehri	2006	Bhagirathi	262	2,610	820	1,000	Old Tehri Town+35+74 (partial)	–	–	5,200
Srisailam	1984	Krishna	145	4,250	–	1,670	–	–	–	–
Almatti	1999	Krishna	52	6,000	622	297	–	–	–	–

The Indira Sagar Project in Madhya Pradesh State, which was recently completed and has a water spread area of 913 km^2, has taken the mantle of the largest reservoir from Hirakud which was the largest artificial lake for more than five decades.[1] The Sardar Sarovar Project, when impoundment reaches its full planned storage, will join the list of large reservoirs in India and rank tenth.[2]

16.2.1 Land Acquisition

With regard to the historic development of land for governmental or public purposes, the basic law that guided land acquisition in India was the Land Acquisition Act of 1894. The government was empowered to acquire any land for 'public purpose' and to pay cash compensation determined by it according to a prescribed procedure. The Land Acquisition Act of 1894 specified the following procedure for all matters relating to acquiring land for public necessities, which was followed till the last few decades of the twentieth century despite having become outdated:

- Land identified for the purpose of a project was placed under Section 4 of the Land Acquisition Act. This constituted notification. Objections were to be made within 50 days to the Collector (the highest administrative officer) of the concerned district. The Coal Bearing Act 1957 allowed 30 days for objections.
- The land was then placed under Section 6 of the Land Acquisition Act (or Section 7 of the Coal Bearing Act). This was a declaration that the government intended to acquire the land. The Collector was directed to take steps for the acquisition, and the land was placed under Section 9. Interested parties were then invited to state their interest in the land and their price. According to Section 11, the Collector was to make an award within 2 years of the date of publication of the declaration, or the acquisition proceedings were deemed to have lapsed.
- In case of disagreement on the price awarded, under Section 18, the parties could make a request within 6 weeks of the award to the Collector to refer the matter to the courts to make a final ruling on the amount of compensation.
- Once the land was placed under Section 4, no further sales or transfers were allowed. However, since the time lag between the land being placed under Section 4 and subsequent stages was several years in some cases, land transfers were often observed.
- Compensation for land and improvements (such as houses, wells, trees, etc.) was paid in cash by the project authorities to the state government, which in turn compensated landowners.
- The price to be paid for the acquisition of agricultural land was based on sale prices recorded in the District Registrar's office, averaged over the 3 years preceding notification under Section 4. The compensation was paid after acquiring

---

[1] Water Spread Area is also used synonymously for 'area inundated' by the reservoirs created by dams or barrages in the official documents of India.

[2] See the chapter on Sardar Sarovar Project in this book.

the area. An additional 30% was added to the award, along with an escalation of 12% per year from the date of notification to the final placement. In cases of delayed payments after placement, an additional 9% per annum was paid for the first year and 15% for subsequent years.

It was seen that the provisions of the Land Acquisition Act, coupled with the manner in which cash flow for projects was released, made the process of land acquisition, particularly in submergence areas, rather slow, which resulted in project cost escalation. Delays led to disparities in the real value of money received, and as a consequence projects faced resistance from affected parties who felt that the compensation was inadequate. Since it was paid in cash, the compensation was often spent on consumption, and people were under distress soon thereafter. With the passage of time it became clear that although the Land Acquisition Act made an attempt to compensate for the loss of assets, the loss of people's livelihoods was not being addressed, leading to little compensation for the poor. The Act had no links with resettlement and rehabilitation, which was an acknowledged component in all water resources development projects. The lessons learnt from this procedure in the past few decades, especially when applied to dam-related inundation, were therefore bitter.

16.2.2 Guidelines for Preparation of Detailed Project Reports from India's Central Water Commission

The guidelines issued by India's leading agency, the Central Water Commission, for the preparation of Detailed Project Reports for water resources development projects attempted to provide some way out. It brought in desirable provisions, by including activities and tasks for resettlement and rehabilitation in the requirements for Detailed Project Reports. This was, of course, a good preliminary step in the right direction. The need today is to further refine and improve some of the provisions so as to make them more appropriate for particular situations and the characteristics of affected populations, such as the rural poor and tribal groups. In other words, the policy has to be more humane and show a greater human face. Can it develop a procedure to ensure that the displaced are 'better off,' and not just not worse off?

16.2.3 National Water Policy and Issues Related to Inundation

Clause 10 of the National Water Policy of 2000, which deals with resettlement and rehabilitation of those affected by large dam projects, states:

> 10. Optimal use of water resources necessitates construction of storages and the consequent resettlement and rehabilitation of population. A national policy in this regard needs to be formulated so that the project-affected persons share the benefits through proper rehabilitation. States should accordingly evolve their own detailed resettlement and rehabilitation policies for the sector, taking into accounts the local conditions. Careful planning is necessary to ensure that the construction and rehabilitation activities proceed simultaneously and smoothly.

One of the earlier provisions for resettlement in a few water resources development projects articulated the principle of 'land for land'—in other words, land lost to submergence to be compensated by other land. This was also attempted in a few earlier projects. However, with consideration of availability of land for land, especially in hilly regions as the Himalayas, the National Water Policy left it to the individual states to evolve their own detailed resettlement and rehabilitation policies, taking local conditions into account. This has enabled states to deal directly with those affected by the project and to evolve suitable and acceptable packages.

16.2.4 National Policy on Resettlement and Rehabilitation

The National Water Policy and the provisions contained therein should be read in conjunction with the National Policy on Rehabilitation and Resettlement of 2003 (Ministry of Rural Development 2004). The core objectives of this policy are broad based. It aims to create harmony between the project and project-affected families through full community participation and transparency. The need to handle all issues of those affected by inundation with utmost care is reflected quite fully. In India, the people affected in rural areas are mostly small and marginal farmers and, in quite a few cases, tribal groups whose welfare is a significant concern of the government's programmes. Of course, inundation also impacts the welfare of rural women.

The national policy is promoted by the Ministry of Rural Development, as well as the Ministry of Social Justice and Empowerment and the Ministry of Tribal Welfare. In the preamble to the policy, one sees full recognition of many interesting aspects which were the key impediments to progress. The policy is also concerned with appropriate measures for the implementation of dams and associated project works. Thus, there is an overlap of the missions and objectives of several ministries and departments in dealing with inundation. These cover, inter alia, the following issues:

1. Offering cash compensation does not by itself enable affected families to obtain cultivable agricultural lands, homesteads and other resources surrendered to the state. On the other hand, this problem could be worse for landless agricultural workers, forest dwellers, tenants and artisans whose distress and destitution could be more severe, if there is no cash compensation.
2. There is a need to provide succour to the rural poor who have no assets, and support the rehabilitation efforts of the resource-poor sections, namely, small and marginal farmers, Scheduled Castes and Tribes,[3] and women who have been displaced.

---

[3] 'Scheduled Castes and Scheduled Tribes' are Indian population groupings that are explicitly recognised by the Constitution of India, previously called the 'depressed classes' by the British, and otherwise known as untouchables. Together, they comprise over 24% of India's population, with Scheduled Castes at over 16% and Scheduled Tribes over 8% as per the 2001 Census. Their proportion in the population of India has steadily risen since independence in 1947.

3. The aim is to provide a broad canvas for an effective dialogue between project-affected families and the administration for resettlement and rehabilitation. Such a dialogue should enable timely completion of projects with concretely defined costs and adequate attention being paid to the needs of the displaced persons, especially resource-poor communities.
4. There is an attempt to impart greater flexibility for interaction and negotiation so that the resultant 'package' gains all-round acceptability in the shape of a workable instrument providing satisfaction to all stakeholders.
5. The 2003 National Policy on Resettlement and Rehabilitation shall guide resettlement and rehabilitation of all projects with displacement of over 500 families.
6. The states that undertake the projects and others like national power utilities which carry out dams and reservoir works have been given freedom to adopt their own 'further liberal packages', keeping in view the national policy guidelines.

One can see in the 2003 National Policy on Resettlement and Rehabilitation a positive, socially sensitive, forward movement in addressing not only the issues related to inundation, but also the entire spectrum of development issues for displaced peoples. This is often neglected when arguments on Indian resettlement and rehabilitation policies are discussed without proper mention of the positive steps spelt out in these recent guidelines.

16.2.5 National Rehabilitation Policy

A later version of the National Rehabilitation Policy (Ministry of Rural Development 2007) has recently been developed. This has been floated by the Union Ministry of Rural Development, Department of Land Resources. A comparative statement from the ministry clearly demarcates the significant advances and liberal provisions in the new guidelines contrasting and comparing it with the earlier 2003 National Policy on Resettlement and Rehabilitation.

Never before was so much attention paid to revisit policy guidelines in such a short span of time (3 years). This in itself should be acknowledged as a most encouraging step of the central government. It reflects a significant and new dynamism, and it may not be long before the old stories are replaced by some recent examples of best practice that are noteworthy. The next section discusses a few of them.

16.3 Case Histories: National Hydro Power Corporation, Tehri and Indira Sagar Projects

Two major water resources development projects, viz. Bhakra-Nangal and Sardar Sarovar are discussed in other chapters in this volume. In this chapter, therefore, we examine a few lesser-known projects.

16.3.1 National Hydro Power Corporation

The National Hydro Power Corporation (NHPC) is a major player in the power sector which, in addition to the generation of hydropower, is involved in multipurpose projects. Such cases could be of interest in gauging the progress made in satisfactory handling of resettlement and rehabilitation issues.

Project proponents such as NHPC are supposed to systematically follow measures that commence right at the stage of project formulation and submission for investment clearance. The provisions include, inter alia:

a. Socio-economic and ethnographic survey: A detailed socio-economic survey for the formulation of a resettlement and rehabilitation plan for project-affected persons and their families.
b. Special considerations for tribal groups and other communities: In places where ethnic minorities dominate, such as Sikkim, a separate ethnographic survey is advised to understand the local culture and behaviour of the people so as to take on board pertinent value-based resettlement solutions.

The formulation of a resettlement and rehabilitation plan follows after the process outlined above and involves the agencies answerable to the government for administering all land matters. Thus, the resettlement and rehabilitation plan is to be formulated by project proponents, the state revenue department, the district administration and representatives of the local people. After it has been developed, the concerned state government has the right to approve or/and modify the plan, as required. The revised plan is processed in the Ministry of Environment and Forests for final approval. All project proponents aim to contribute to the socio-economic upliftment of affected people through opportunities presented by the planned water resources project and to enhance the quality of life of project-affected people.

The broad framework of resettlement and rehabilitation in recent water resources development projects is substantially wider than that in historic projects which followed the old Land Acquisition Act of 1894. The broad resettlement and rehabilitation package encompasses many of the following components that are incorporated in a holistic fashion, as required for funding support, and implemented *pari passu* with construction:

- Compensation for land, houses, shops and other properties
- Homestead land
- Transportation charges for household items, cattle, etc
- Construction of houses
- Solatium amounts
- Financial assistance for construction of cattle shed or poultry farm
- Agricultural land, depending on availability, or landless grant
- Subsidy for seeds/fertilizers/land management
- Development of public health centre, school, community centre, etc
- Basic amenities such as roads, drinking water, electricity, medical facilities, etc

- Vocational training
- Preference in allotment of shops in new and modern shopping complexes
- Renovation/relocation of archaeological and religious monuments and structures

The Government of India has constituted a Grievance Redress Cell at the project level and a Monitoring Cell at the national level to monitor the progress of various resettlement and rehabilitation projects. This was done in order to ensure desired compliance. Transparency is a key element in the process and the reports of the Grievance Redress Cell form the basis for allowing progress in project construction *pari passu* with the dam and associated works after the approval of the project for funding and construction. This aspect is also monitored by a Central Level Monitoring Committee with representatives from the Ministry of Environment and Forests, constituted to ensure overall environmental safeguards.

Several projects have been successfully handled by the NHPC with regard to resettlement and rehabilitation.

16.3.1.1 Key Features of Resettlement and Rehabilitation of the National Hydro Power Corporation Projects—Some Recent Case Histories

The National Hydro Power Corporation (NHPC) has projects in various parts of India. As such, project-specific Resettlement and Rehabilitation plans are drawn in consultation with the respective state governments. Important features of Resettlement and Rehabilitation packages in two of the projects are as below:

a. Teesta HE Project, Stage V (Sikkim)
   - Land for construction of house @ 0.02 ha per family for 46 families.
   - Land development charges for every plot @ Rs 5,000/family.
   - Land for common facilities like panchayat building, primary health centre, etc.
   - Land for drainage system.
   - Land for approach road to the rehabilitation colony.
   - Construction of a panchayat building.
   - Construction of primary health centre including furniture, equipments and medicines.
   - Construction of two nos. of shopping complex.
   - Strengthening of project hospital in the main colony to cater the need of project oustees.
   - Providing power connection/electric supply in the rehabilitation colony.
   - Providing water supply system in the rehabilitation colony.
   - Special grant for SC/ST/OBC at Rs 10,000/family.
   - Grant for construction of house at Rs 100,000/family.
   - Grant for fertilizer and seeds at Rs 5,000/family.
   - Disturbance allowance at Rs 7,000–10,000/family
   - Subsistence allowance will be paid to the family from the date they are ousted till one member of the family gets a regular service.
   - Transportation charges for animals and household items at Rs 10,000/family.

- Children of the affected families will be provided with education in the project school.
- Technical training will be provided to the affected people to improve their skills and necessary training would be provided for their capacity building.
- Scholarship would be provided to the wards of the full oustees as per the norms of the Corporation.
- Preference for jobs to eligible members. (Out of 46 numbers of oustees, 39 have been identified for jobs and their case is in process).

b. Subansiri Lower Project (Arunachal Pradesh)

- Villagers of both the villages have decided to shift from their villages to a new location to be decided by the District/State Administration due to submergence of their cultivable land in the reservoir of Subansiri Lower Project.
- Cultivable land. It has been decided that NHPC will provide at the most only 1 ha land to each PAF (total 38 PAFs) and they will be compensated for remaining cultivable land coming under submergence.
- House and Homestead land. Each PAF will be provided homestead land of 150 m^2 and another 50 m^2 would be given to construct animal shed and granary.
- A lump sum amount of Rs. 250,000 (Rupees, two lakhs fifty thousand only) will be paid by NHPC to each PAF as grant/assistance/subsidy for land development costs of 1 ha of land and seeds and fertilizers for it; house construction and its plot development; rehabilitation grant such as subsistence/maintenance allowance for up-keeping of cattle, poultry and piggery, etc.
- The above lump sum assistance shall be paid after 6 months of getting possession of the land where rehabilitation is to be done and fencing of entire land.
- The PAF will be allowed to retrieve materials from their original house. The NHPC would provide free transportation facilities to shift the household belongings and other retrievable materials to the new location site.
- The compensation of land, trees and other immovable property shall be paid by NHPC to the landlords after due assessment/verification and approval by the District Authority of State Government of Arunachal Pradesh.
- About 40 ha of land nearer to a road and water source will be required for rehabilitation of both villages (20 ha for each village), which have to be provided to NHPC Ltd. by the state government of Arunachal Pradesh totally free of cost. The PAFs from the two villages would be rehabilitated at two different sites as per their choice preferably.

Sanitary facilities. Each house would be provided with a low-cost sanitary latrine with a common septic tank or soak pit that can cater to five houses.
School building. One primary school including a playground would be provided by NHPC in the Rehabilitation colony.
Vocational training. Training would be provided to the village youth in animal husbandry, horticulture, weaving and other activities. Other facilities to be provided at the resettlement colony:

1. One small building for Community Centre
2. One small building for Primary Health Centre
3. Approach road to the new relocation site
4. Piped water supply system

Important Features of Resettlement and Rehabilitation Packages in Other Projects of NHPC (Earlier Completed)

In Tanakpur HE Project (Uttaranchal), five crossings (bridges) across the power channel have been constructed for proper communication between the resettlement colony and the local market.

In Uri HE Project (Jammu and Kashmir), compensation for the mosque affected at Buniyar was paid and has been fully renovated. A new mosque was also constructed at Kanchan village. Both are being used by the locals.

In Rangit HE Project Stage III (Sikkim), besides from paying compensation for the land and building of existing junior high school, an area of 1.41 ha was provided free of cost for construction of a new school building and play ground. The project also constructed a DAV Public School, presently up to 10th standard, in the main colony of the project and admission to this school is available to the wards of the displaced/local residents.

Also, after detailed consultation with state government officials and as well as religious teachers, a cluster of religious monuments (Mane) falling within the submergence area were shifted to another location.

Social Benefits Observed at Hydropower Projects

Hydropower projects rejuvenate the economy of the entire area bringing prosperity and raising the standard of living for the inhabitants of the area due to educational facilities, public health benefits, roads, electric power and other infrastructural development that takes place during the construction of the projects. It has a multiplier effect on the benefits accruing to the area. The Chamera HE Project Stage-I in Himachal Pradesh has boosted the development of the entire Chamba region. Dalhousie, which once used to be a sleepy town in winter, now hums with tourist activity around the year. The road from Pathankot to Banikhet/Dalhousie was so narrow that traffic used to be allowed from one side, causing long queues of vehicles at the other end. Today the approach is a smooth/metalled two-lane road, all due to construction of Chamera Stage-I.

A study conducted to see the benefits revealed the following trends in development.

Chamera-I (540 MW), Himachal Pradesh
- Increase in population of Chamba District from 311,147 in 1981 to 393,286 in 1991 (26%).

- Literacy in Chamba District increased from 64,495 in 1981 to 105,692 in 1991 (64%).
- Availability of potable water increased from 886 villages in 1985 to 1,095 villages in 1992 (24%).
- Increase in production of construction material from 162 MT in 1985–1986 to 1,919 MT in 1992–1993 (1,085%). Also an increase in slate production from 19,856 MT in 1985–1986 to 316,350 MT in 1992–1993 (1,493%).
- An increase in actual irrigation area from 4,079 ha in 1984–1985 to 5,669 ha in 1991–1992 (39%).
- An appreciable increase in educational institutes, e.g. 84 higher secondary schools in 1992–1993 compared to 56 in 1985–1986 (50%); 82 middle schools in 1992–1993 compared to 74 in 1985–1986(11%); and 773 primary schools in 1992–1993 compared to 673 in 1985–1986 (15%).
- Fish production increased from 64 MT in 1985–1986 to 268 MT in 1992–1993 in Chamba District (319%).
- Number of factories increased from 48 in 1984 to 130 in 1992 (171%).
- Number of hospitals has increased from 69 in 1981 to 87 in 1991 (26%).

Uri HE Project (480 MW), Jammu and Kashmir
- During the construction of the project, the stretch of the national highway from Sheeri to Rajarwani (about 34 km), including Bailey Bridges, was upgraded and widened to a two-lane highway. The national highway between Srinagar and Sheeri was upgraded to 70 tonnes capacity.
- As a welfare measure, a motorable RCC bridge was constructed across Mundri nallah about 3 km upstream of Sheeri village, which has improved the mobility of the surrounding villagers.
- Water supply scheme with a reservoir capacity of 10,000 gallons constructed at Buniyar village benefits about 1,000 persons, including two mosques and a school in the area.
- One mosque at village Buniyar was fully renovated and a new mosque was constructed at Kanchan village.
- A beautiful childrens' park and view point was constructed at Buniyar, which is now an attraction point for the locals as well as tourists.

Dhauliganga HE Project, Stage-I
- NHPC is upgrading the 258 km road from Tanakpur to Tawaghat to 5.5 m carriage way.
- NHPC is upgrading 6 km long Tawaghat-Chirkila road to the same standard as above.
- NHPC is constructing 7 km long road to surge shaft top, connecting about 5 villages.
- Social welfare measures in Dharchula and adjoining areas of NHPC colony including developing toilets in local areas are being taken up.
- Medical facilities are being extended to the locals.
- Self-employment opportunities have increased.

16.3.2 Tehri Dam, Uttaranchal: Tehri Hydro Power Corporation[4]

Due to construction of the Tehri Project, a total area of 5,200 ha will be inundated. Old Tehri town and 35 villages, including those affected due to infrastructure works, are being fully submerged, while 74 villages will come under partial submergence.[5] Under the category of urban rehabilitation, 5,291 families living in Old Tehri town are treated as fully affected, while in the category of rural rehabilitation 5,429 families are considered fully affected and 3,810 are partially affected, who are not to be relocated.

The basic principles followed in the resettlement and rehabilitation policy adopted is as summarised below:

- To minimise displacement and to identify non-displacing or least displacing alternatives.
- To plan the resettlement and rehabilitation of project-affected families including special needs of women, tribal groups, Scheduled Castes and Schedules Tribes, and other vulnerable sections.
- To provide a better standard of living for project-affected families.

The rehabilitation policy for the Tehri Dam Project was formulated with stakeholder consultation. The state government of Uttar Pradesh coordinated the interaction with representatives of the local population when the project commenced as a state sector project. The policy was improved from time to time based on the recommendations of a special committee of the Government of India headed by Professor C. Hanumantha Rao. Additional measures to address the demands of the local population were incorporated and implemented pari passu with the engineering works of the project without changing its basic features, to ensure maximum cooperation of the project-affected persons. The two-phase resettlement and rehabilitation is still in progress. Phase I covers families affected by the construction of the coffer-dam at Tehri, including those in Old Tehri town. The urban rehabilitation is complete and Old Tehri town has been vacated. New Tehri town (for resettling 5,291 urban families) is a fully developed urban town with modern facilities. Phase II focuses on all the remaining rural families affected by the filing up of the Tehri Dam Reservoir (Wadhwa and Khan, Rehabilitation and resettlement aspects of large storage Tehri Dam Project, unpublished document).

Phase I of rural rehabilitation is already complete for 2,064 families, while 2,863 of the 3,365 rural families covered by Phase II have been rehabilitated or

---

[4] See Tehri Hydro Development Corporation (2007).

[5] In case reservoirs coming under submergence due to dams, the population affected belongs to two categories: urban (like Old Tehri Town which had a sizeable population before coming under water) and sporadic settlements (also affected with the reservoir creation). Sporadic settlements fall in a different group and are covered under 'rural rehabilitation'. The differences in compensation inherent to the particular category could then be better addressed.

paid compensation. Thus about 91% of rural rehabilitation is complete, and land acquisition proposals for all the villages to be submerged have been initiated. Rehabilitation benefits have been disbursed up to El 790 m, and rehabilitation from El 790 m to El 835 m is in progress. The resettlement colonies provide all basic amenities such as schools, roads, electricity, irrigation, piped drinking water, *panchayat ghar*,[6] etc.

The various studies and reviews by different monitoring agencies have commended the sufficiency of the resettlement and rehabilitation package as well as other socio-economic aspects of the resettled population affected by the Tehri Project. The socio-economic study conducted by the Administrative Staff College of India concluded that the new resettlement colonies have been provided with better facilities and amenities than earlier resettlements (Water for Welfare Secretariat 2008). The new settlements have *pucca* buildings, electricity, drinking water facilities, good medical facilities, connections for cooking gas (liquefied petroleum gas), etc. In addition, the study found that the values of assets and income generation from agriculture have increased considerably for displaced people; the annual income of households after rehabilitation has risen by 34.67%.

The Estimates Committee of the Legislature of the state of Uttar Pradesh visited the resettlements and found that the facilities provided in these colonies are better than in any village in the country.

In relation to the 2003 National Policy on Resettlement and Rehabilitation, the Tehri Hydro Power Corporation has included the following provisions for compensation:

- Each family including landless agricultural labourers is to receive 2 acres of developed irrigated land or Rs 5 lakhs (hundred thousand) cash in lieu of land as per their preference.
- The allotment of agricultural land is limited to the extent of actual land loss, subject to a maximum of 1 ha of irrigated land or 2 ha of un-irrigated land, subject to the availability of government land in the district, as per the National Policy.
- Landless agricultural labourers are also to be allotted 2 acres of developed irrigated land in Tehri. The National Policy provides for financial assistance equivalent to 625 days of minimum agricultural wages.
- Each rural shopkeeper is to be paid cash compensation between Rs 80,000 and Rs 120,000 depending upon the location. There is a provision for financial assistance of Rs 10,000 for the construction of shops or working sheds, as per the National Policy.
- All additional family members of fully affected rural families and dependent parents will receive an ex gratia amount equivalent to 750 times the minimum

---

[6] Panchayat Ghar is the village level headquarters, a terminology in Hindi and used as such in English. The village level decision-making office (a mini scale parliament or white house that decides what is best for their village community) seats here.

agricultural wage. As per the National Policy, each project-affected family whose entire land has been acquired for the project will be given a one-time financial assistance equivalent to 750 days of minimum agricultural wages for the 'loss of livelihood', while project-affected families whose entire land has not been acquired will receive 375–500 days of minimum agricultural wages.

- A house plot of 200 m^2 is to be allotted to each family at nominal cost, along with house construction assistance of Rs 1 lakh, which is a greater amount than was envisaged in the National Policy. Project-affected families whose houses have been acquired are to be allotted house sites free of cost, and a grant of Rs 25,000 for house construction only to families below the poverty line.

Indira Sagar Project (Narmada Dam), Madhya Pradesh: Narmada Hydroelectric Development Corporation (NHDC)[7]

The Indira Sagar Project (ISP) has come up on the River Narmada in the state of Madhya Pradesh, and is an ambitious large project of the state. The Saint Singaji Reservoir is part of this mega project and is spread over an area of 913 km^2, the largest in the country and the second largest in Asia. The project involves the relocation of about 150,000 persons and rehabilitation and resettlement of 39,173 families. The town of Harsud, founded by King Harshwardhan, has come under submergence due to this project. The total area under submergence includes 44,741 ha of agricultural land, 40,332 ha of forest land, 779 ha of *Abadi* (habitation) land and 6,653 ha falling under other land categories. The process of rehabilitation and resettlement is nearly complete. About 38,809 project-affected families were rehabilitated and resettled by August 2005.

The project proponents (NHDC) have taken due care to adequately compensate project-affected families (Ministry of Rural Development 2004). In fact, the people of Harsud preferred submergence and relocation over the construction of a protective wall to save the town. Once compensation had been disbursed for Indira Sagar Project, Harsud became the second town in India where there was a sharp increase in motorcycles, which were purchased by project-affected families from the lucrative compensation they received; Harsud also had a record number of daily sales of motorcycles.

To calculate compensations for agricultural land, rates of the adjoining Tawa command area of Harda District were taken into consideration. A special rehabilitation grant was also introduced in order to ensure adequate compensation, to enable project-affected families to purchase land of their own in areas of their choice. Engineers from the Public Works Department wing of the NHDC evaluated all the pre-existing structures of project-affected families. Each of these families received the replacement value for their structures; at the same time, they were also allowed

---

[7] See Narmada Hydro-electric Development Corporation (2005) and Sharma (2005).

to take old building material away with them to their new settlements. NHDC provided a free transport facility to project-affected families for transporting household goods and old building materials, while families that chose to make their own transport arrangements were paid Rs 5,000 each. The rehabilitation policy of the Indira Sagar Project includes provision for a free residential plot at one of the 34 government rehabilitation sites developed by the NHDC, or Rs 20,000 in lieu of such a plot. Of these 34, one rehabilitation site Chhanera (New Harsud), was developed exclusively for the project-affected families of Harsud town. All the new rehabilitation sites have been well planned and developed, also to ensure that they are more aesthetically pleasing than the old submerged villages and Harsud city, for the people's satisfaction and comfort. The sites are provided with internal and approach roads, both of Water Bound Macadam category, and if there are more than 100 plots in a resettlement site, the internal roads are upgraded to black top.[8] New Harsud (Chhanera) is the only town in the state of Madhya Pradesh, which is well planned and has all the basic required facilities, including a modern water treatment plant and sewerage. It is also possible for the common man to judge the claims of the project proponents that the resettlement and rehabilitation has been handled well—while travelling by train on the Itarasi–Bombay route, New Harsud is easily identified by its beautifully aligned street lights.

People were also given the freedom to settle in privately purchased rehabilitation sites. If more than 50 project-affected families settled in such a private location, the project proponents developed that site by providing all the basic facilities as in other sites developed by the government. Eight such private rehabilitation sites were developed by NHDC. The quality of new houses constructed at these locations was ensured to be superior than that of the old homes of project-affected families in the submergence area. A rehabilitation grant (Rs 18,700 or 9,350), as per the entitlement of project-affected families, was paid to them to support the process of settlement. An employment grant (Rs 49,300 or 33,150), as per an eligibility norm, was paid to project-affected families to enable them to begin their professions at the new resettlement site. In order to assist in the purchase of property anywhere in Madhya Pradesh, a certificate was provided by NHDC, enabling project-affected families to get an exemption from stamp duty and registration fees in relation to the purchase of land (the duty and fees are paid by NHDC to the Deputy Registrar of the concerned district as needed).

It is pertinent to mention a few success stories of individuals resettled by the Indira Sagar Project (Sharma 2005). Shri Ramadhar Meena of Jalwamafi owned 20 acres of unirrigated land in the submergence area. After receiving compensation,

---

[8] Water-Bound Macadam road is a cheaper road construction methodology for village roads. The Water-Bound Macadam roads, in case of more frequent use, are protected with a spray of asphalt and then called 'black topped' road. Asphaltic road is superior and is not meant in the instant cases. Asphaltic concrete will require following a superior construction technology like coarse and fine aggregates of good-quality stones as per standards and mixing them with quality asphalt laid in layers and compacted with rollers. All these are not done in black topped roads: just a spraying of tar with spraying of fine sand on it so that the tar acts as binder and black topped smooth surface road is made available for users.

he purchased 15 acres of irrigated land in Deepgaon of Harda District. His income increased from Rs 15,000 to Rs 100,000 per annum. Shriram, son of Shri Fattu of Bandhania II, was a landless agricultural labour. Today he owns 12 acres of unirrigated land and a residential plot of 0.3 acre on which he has constructed a better house. Kamal Singh of Sindhkhera owned 15 acres of unirrigated land and now possesses 42.5 acres of irrigated land in Khamkhera of Damoh District. Seven landless families, belonging to the Scheduled Tribes, of Segwa village collectively purchased 20 acres of unirrigated land in Rajur village of Khandwa District and are now cultivating agricultural crops after setting up irrigation facilities; these former farmers have thus become labourers. There are many such stories from the project, where project-affected families have significantly improved their standard of living and enhanced their income by using their lucrative compensation wisely.

Care has also been taken with regard to the religious sentiments of project-affected families. All religious spaces in the submergence area, whether private or public, have been adequately compensated to facilitate the building of a better religious structure at the new settlement location. So far, an amount of Rs 1.840 million has been paid by NHDC for buildings for religious purposes. Special care has been taken for the Saint Singaji Samadhi, a holy site of great importance for the people of this region. The Samadhi (mausoleum) has been protected by erecting a reinforced cement concrete shaft around it up to El. 265 m, and by building an earthen embankment around this shaft and also an approach road. The Saint Singaji Project cost NHDC Rs 50 million. Similarly, monuments of archaeological importance have also been taken care of. There are 10 such monuments in the project area, of which six have been relocated/conserved and the rest are in the process of relocation/conservation by the project under the supervision of the Archaeological Department of the Government of India.

In cases where approaches to the fields of project-affected families have been cut off because of project construction, alternative approaches have been provided. If the approaches of the remaining villages are affected due to water from the dam, such villages are also provided approaches by NHDC.

16.4 Inundation and Water Resources Projects

16.4.1 Report of the World Commission on Dams and the India Case Study of World Commission on Dams

The India Case Study of World Commission on Dams (ICS-WCD) is exhaustive and touches upon several aspects, since India is one of the developing nations with a sizeable population of large dams. What is not widely known is the fact that the ICS-WCD is not a report agreed by any author or set of authors or the Government of India (Rangachari et al. 2000). It came about as a product of the compilation of separate chapters by different members of a team with a chapter titled '*Some* Agreed

Conclusions' (emphasis added). The study by the World Commission on Dams (WCD) (2000b) on the subject in respect of the India Country Study Report was split into a number of components. Each component was directly assigned by the WCD to different individuals and institutions. The WCD also commissioned an overall country report by a group of experts. In the preface of the report presented to WCD in June 2000,[9] this group pointed out that it could not be a joint report of the group as a whole but a collection of papers by different members. The responsibility for each chapter was reflected as that of the respective author conveying in essence that the others were not necessarily in agreement with.[10]

One of the members of the ICS-WCD team attempted to work out the total inundation area as well as people affected due to submergence based on many unrealistic assumptions. These inter alia apply a *questionable* average value that is multiplied by the reported number of dams as per the Dam Register of India. This yielded an estimate of the area inundated by all dams as over 37 million ha, and a figure of population affected by inundation 600% over the official values (Gopalakrishnan 2001). Not all the other authors agree with these figures. The agreed conclusion of the ICS team was only that dam 'have displaced a large number of people and submerged large areas of forest and other lands'.[11]

The response from the Government of India to the WCD Report (World Commission on Dams 2000a) was exhaustive, and facts and figures were furnished in a transparent manner so that the report of ICS-WCD could be suitably amended (Gopalakrishnan 2000b). However, these responses were relegated to an annex and the main report was not revised to take cognisance of the formal comments of the Government of India, despite their status as a significant stakeholder in the dams of the country (Gopalakrishnan 2002).[12]

Not infrequently, when it comes to inundation issues in water resources projects, people still seem to believe some of the sweeping observations of the Report of the World Commission on Dams. The WCD Report (2000) relied on the two Country Reports of India and China—in the case of the former, with direct quotations. The WCD upscaled some of the data to estimate the overall number of project-affected persons in India's dams and reservoirs in a most unrealistic manner. In Chap. 4, the WCD Report stated that 'Large dams in India displaced an estimated 16–38 million

---

[9] See the India Country Report in the World Commission on Dams Knowledge Base available at http://www.dams.org/docs/kbase/studies/csinmain.pdf.

[10] See the paragraph preceding the penultimate one on the acknowledgement in the Preface in the India Country Report.

[11] See the Final Summing Up in Chap. 7 'Some agreed conclusions', and paragraph 2 in the penultimate page of the Indian Country Report.

[12] As can be seen from the World Commission on Dams web site covering the Knowledge Base in which the India Country Study Report has been brought out, the Annex 7 details the comments from the Government of India whom the author represented as the Forum member and supplied the comments. That an important observation from the major stakeholder of a country of population of over a billion people (in 2000) stood relegated, is thus obvious.

people' (WCD 2002: 104) and ascribed it not to the India Country Report but another source (Fernandes and Paranjpye 1997[13]). Official data from governmental agencies of India places the figure around 6.5 million (of all cases prior to 2000), as mentioned in a rejoinder consisting of comments on the India Country Report (Gopalakrishnan 2000a, 2001).

16.4.2 Among the Effective Tools for Development, Do Dams Alone Cause Displacement?

The lack of development is indeed a worse scenario. The distress migration from Nepal to India that has continued for decades could have been averted if some of the large dams had come up in Nepal territory (besides Indo-Nepal territory for projects like Pancheshwar Dam which will lie in a common river that form the boundary between them) for a win-win situation favouring both India and Nepal. Dams and their associated impacts could have changed the economy of Nepal for the better. The lack of development forced a lot of Nepalese to engage in different forms of labour, working as watchmen, cooks and labour in far-flung areas in southern and north-eastern India. Distress migrations[14] are not uncommon and these are essentially caused by floods and other damage, and many times due to droughts. Dams could ameliorate all of these hardships and the projects could help to improve the welfare of the displaced as development directs investment into these areas in a more focused manner. We do have typical examples of poor labourers from Bihar in eastern India participating in the economic and cultural operations in the west, in Bhakra command[15]; special trains such as the Shramjivi Express (*shram* refers to the support of labour) have been provided to facilitate this. Dams can and should be considered as providing an equally golden opportunity for those affected by inundation, addressing their welfare issues while undertaking resettlement and rehabilitation as enshrined in the spirit of the new policies. We do see excellent examples in the Indira Sagar and Tehri projects but this approach needs better publicity and media support. Project proponents should factor in this dimension to ensure a well-articulated programme for disseminating accurate information about the way inundation issues are being addressed during implementation of projects.

---

[13] The reference is quoted in chapter 4, p. 104 of the main report of the WCD (2000).

[14] The distress migration is like the ones at the time of partition of India into Pakistan and India in 1947, where 7 million Muslim fled to Pakistan while 8.5 million Hindus fled to India. Many more instances are the ones like what was seen in Vietnam, labour migration from Africa to Europe and elsewhere all in twentieth century.

[15] Bhakra Dam command lies in Punjab (India), Haryana and Rajasthan. These lie to the west and northwest of the country. Bihar is an Indian state in the east of the country where a large number of labour moves to the Bhakra irrigated areas in Punjab seasonally.

One can go further. In his Chiman Bhai Patel lecture, B.G. Verghese (2006) advocated the following:

> Compensation for project-affected families in any project, whether dams, mines, steel plants or power stations, who lose their lands, livelihoods and other assets must be fair, even generous. They should also benefit from the processes of employment and income generation set in motion by the particular investment. A variety of stockholding, guaranteed employment, training and community development options can be considered to formulate a new paradigm of development. This is an important and urgent necessity.
>
> It may not be enough to compensate the project-affected families, howsoever handsomely this may be done. The population in the larger project influence area must also be given a stake and sense of ownership in the project. This is where resettlement and rehabilitation, catchment area treatment in the case of dams, and official poverty alleviation programmes under a large variety of headings can be brought together to ensure comprehensive area development in keeping with stated national goals and priorities.
>
> Take the SSP [Sardar Sarovar Project] for example. A generous compensation package has been handed out to PAFs above the dam. But what of the remaining Bhil [tribal] population in the Gujarat portion of the Narmada catchment? Should life for them continue to be what it has been for the past hundreds of years? Can there be a small betterment levy on the beneficiaries of irrigation, industry and power below the dam, to fund the social and economic development of this deprived community? In the Upper Volta Project in Ghana, a small percentage of the project's earnings from electricity and other heads, is annually returned to a community development fund for the project-affected area. Could a small part of the electricity generated at the SSP canal-head and river bed power stations be earmarked for the upper catchment? Should the SSP create new income disparities in Bhil society because PAFs benefit while life continues unchanged for the rest?
>
> Again, there are considerable numbers of tribals in the SSP command over and beyond the PAFs resettled therein. Are these simple tribals, some of whom have hitherto practiced subsistence agriculture, equipped to maximise their gains from the new opportunities for irrigated agriculture now open to them? Are special extension, credit and other on-farm and post-harvest support services in place to help speed them on their way? If not, can specific efforts be made in this direction? The same may be true of other backward and marginalised categories in north Gujarat, Saurashtra and Kutch. They too may need special assistance and support for a while.

Recent pronouncements from the highest official levels in India are very reassuring in moving forward and offering stakeholdership in projects, but since the stakes could either be positive or negative, the people concerned are yet to be enthused by such offers.

16.5 Conclusions

Once we are able to bring the inundated peoples' welfare into mainstream consideration, examining all other possible options including stakeholdership (if it is of interest to project-affected families) in the project itself, dam projects could be seen as an opportunity by project-affected families and others concerned, and could be welcomed given the possibilities in particular regional settings. The possibilities of better education and vocational training for gainful employment are already being advocated. Overall, these are positive indicators and perhaps slowly but steadily we will move in the right direction. The principle that justice should be ensured to the maximum extent for people affected by inundation due to water resources projects has already attracted adequate support from all quarters. But challenges remain, since some activists are opposed to storage projects and stall the significant progress that is otherwise possible in this sector in India. Even in the case of well-conceived projects, project proponents and concerned agencies, including the government, are often held hostage by anti-development activism as soon as projects are announced, despite the stipulated public hearing process embedded in the required Environmental Impact Assessment clearance. It would seem that such activism is aimed at gaining publicity and international encomium from quarters keen to see India's progress retarded, if not stopped altogether. To avoid the waning of media attention, such activists repeatedly undertake various actions, including 'hunger strikes unto death', on the charge of poor handling of commitments on issues like inundation. It is for these reasons that the implementation of policies announced for project-displaced people has to be handled with sensitivity and utmost care. We are waiting for the time when enabling institutions can join hands with project implementers, once the basic rules of the game are set and agreed upon, when there is change in the current attitude of activists of blaming others and disrupting progress.

Perhaps we have to move further in handling these issues, beyond water engineers and economists. A new brand of education that is holistic would be a necessity in the water sector. In addition, the society at large has to be involved. The right information and right to information through media will play a significant role. Are we doing enough?

References

Gopalakrishnan M (2000a) Dams and development – the final report of the World Commission of Dams – impressions and comments. Deutsches Talsperren Komitee (DTK), ICOLD Annual Meeting, Dresden

Gopalakrishnan M (2000b) Comments on India Country Report on behalf of Government of India. A Demi Official (personal) Communication to WCD (Dr. Achim Steiner) from Central Water Commission, Government of India, New Delhi

Gopalakrishnan M (2001) Impressions and comments on dams and development. Final report of the World Commission on Dams. Conference on Water Resources Development – Irrigation & Hydropower, jointly organised by ICID, IHA, ICOLD and CPU, New Delhi

Gopalakrishnan M (2002) Independent review of WCD — comments of Government of India in Central Water Commission. World Resources Institute, Washington DC

Ministry of Rural Development (2004) National Policy on Resettlement and Rehabilitation for Project Affected Families, Department of Land Resources, Government of India, New Delhi. Published in the Gazette of India, Extraordinary Part-I, Section 1, No. 46, 17 February 2004

Ministry of Rural Development (2007) Comparative Statement of National Policy for Resettlement and Rehabilitation of Project Affected Families (NPRR-2003) and National Rehabilitation Policy (NRP-2006). Department of Land Resources, Government of India, New Delhi

Narmada Hydroelectric Development Corporation (NHDC) (2005) A special issue on Indira Sagar Project (Narmada Dam-Madhya Pradesh). Water and Energy International 62(4). Central Board of Irrigation and Power, New Delhi

Rangachari R, Sengupta N, Iyer RR, Banerji P, Singh S (2000) Large dams: India's experience. A report for the World Commission on Dams. (Mimeograph for restricted circulation)

Sharma JP (2005) Rehabilitation and resettlement of project affected families in Indira Sagar Project of Madhya Pradesh on Narmada River. Water and Energy International 62(4):281–285. Central Board of Irrigation and Power, New Delhi

Tehri Hydro Development Corporation (THDC) (2007) Special issue on Tehri Hydro Project, Water and Energy International 64(1). Central Board of Irrigation and Power, New Delhi

Verghese BG (2006) Dams, development, displacement, tribals. Why water lies at root of everything? Chimanbhai Patel Memorial Lecture, Ahmedabad

Water for Welfare Secretariat (2008) Impact of Tehri Dam. Lessons learnt. Indian Institute of Technology, Roorkee

World Commission on Dams (2000a) India Country Report. World Commission on Dams Knowledge Base. http://www.dams.org/docs/kbase/studies/csinmain.pdf

World Commission on Dams (2000b) Dams and development: a new framework for decision-making, Report of the World Commission on Dams. Earthscan, London

Chapter 17
Impacts of the High Aswan Dam

Asit K. Biswas and Cecilia Tortajada

17.1 Introduction

Without any doubt, the High Aswan Dam in Egypt has been the most well-known dam in the world in the past five decades. From Argentina to Australia and from Tokyo to Timbuktu, professionals in water management and development are often familiar with the High Aswan Dam, even though they may not be aware of the names and locations of major dams in their own countries. The logical question that arises is why are numerous people aware of a dam in a distant part of the world, which they have never visited, when they often lack even elementary knowledge of important dams in their own regions? What is so special about the High Aswan Dam that has made it one of the world's most well-known, if not *the* most well-known, and most discussed water infrastructures in the world?

There is also a fundamental issue that needs to be discussed. The vast majority of Egyptians have been fully aware that without the High Aswan Dam, their country would have faced serious socio-economic and political crises during the last 50 years. They are not only proud of the dam, but also strongly believe that the structure has significantly contributed to the nation's prosperity as well as economic and social stability. In sharp contrast, the prevailing international view outside Egypt has consistently been that the dam is an unmitigated disaster for the country in

---

A.K. Biswas (✉)
Third World Centre for Water Management, Atizapan, Mexico

Lee Kuan Yew School of Public Policy, Singapore

C. Tortajada
International Centre for Water and Environment, Zaragoza, Spain

Lee Kuan Yew School of Public Policy, Singapore

Third World Centre for Water Management, Atizapan, Mexico

social, economic and environmental terms. Myths relating to the dam's adverse impacts—many of which are propagated by so-called 'experts' who have never visited the dam and its environment, let alone spent any time seriously studying its impacts, both positive and negative—circulate around the world. In fact, if the average environmentalist outside Egypt is asked to identify the worst dam in the world, chances are very high that the person will name the High Aswan Dam for this dubious distinction. Even the World Commission on Dams vilified the dam and failed miserably to consider its overall impacts in an objective fashion. All these perceptions abound, despite the fact that any unbiased and objective person would be hard pressed to identify a dam anywhere in the world whose overall social and economic benefits to the country concerned have been so overwhelming positive.

In order to answer the fundamental question as to why the perception of the world is totally different from the reality on the ground, the Third World Centre for Water Management undertook a study of the dam, with the support of the Arab Fund for Economic and Social Development in Kuwait. The present chapter is a small part of this overall study (Biswas and Tortajada 2012).

17.2 Methodological Considerations

Some facts and methodological issues of the study should be noted. Prior to the completion of this assessment, a detailed, comprehensive and objective impact analysis, covering both positive and negative aspects, of the High Aswan Dam was not available. It should, however, be pointed out that the absence of an objective impact analysis for this particular dam is not an exception. In fact, the number of large dams anywhere in the world for which economic, social and environmental impacts have been scientifically and objectively monitored on a regular basis after 10–20 years of operation and then evaluated, can be counted on the fingers of one hand, and still leave some fingers over! (Biswas and Tortajada 2001)

At present, thousands of studies on Environmental Impact Assessments (EIAs) of large dams are available, some of which are quite good; but many others are not even worth the paper they are printed on. It should also be noted that all EIAs are invariably predictions and until dams become operational, their impacts (types, magnitudes, spatial and temporal distributions, facts about who receives the benefits and who pays the costs, etc.) are not certain, and thus remain in the realm of hypotheses. Even the best pre-project impact assessments conducted anywhere in the world can reliably forecast only about 70–75% of the actual impacts in terms of time, space, magnitude and the nature of beneficiaries. The EIAs that were conducted in the past for an average dam did not properly identify or assess nearly 40–50% of impacts (Biswas 2004).

The High Aswan Dam was planned, designed and constructed during an era when the world was significantly less environmentally conscious than it is at present. During the late 1950s and throughout the 1960s, no country in the world had an environmental ministry, nor was an EIA required for any type of development project anywhere in the world. In fact, methodologies for conducting EIAs were not even available when the High Aswan Dam was formally opened in 1970. The National Environmental Policy Act of the United States (or NEPA) became operational only in 1971. This pioneering legislation made EIAs mandatory in the United States which, in turn, contributed to the accelerated development of EIA methodologies, and also to the collection of data on which EIA depends. Other countries came to require EIAs later.

However, for a dam that was conceived, planned and constructed before the era of EIAs, the High Aswan Dam received considerable environmental and social attention from its planners and designers, which was somewhat unusual for its time, especially in a developing country. This means that the experts who planned and designed this structure were aware of, and sensitive to, environmental and social issues. More than 45 years after its completion, it is extremely difficult to identify a large dam anywhere in the developing world whose resettlement was carried out (in contrast to the planning phase alone) as sensitively and properly as was the case with the High Aswan Dam. Similarly, issues such as erosion and sedimentation were carefully analysed and considered by the designers. Thus, in many ways, the planners and designers of this dam were well ahead of their time.

Two other issues should also be mentioned. First, the EIAs, as conducted in the past and at present, mostly consider only negative ones—positive impacts are often ignored. In both conceptual and real terms, the impact analysis of a large dam is significantly more than consideration of the negative environmental developments alone. It should also consider all types of impacts due to improvements in the availability of water supply, transportation, hydropower generation, agricultural production, industrial development, control of floods and droughts, as well as all their social, cultural, economic, political and environmental impacts. These assessments are seldom carried out.

Some have erroneously claimed that the World Commission on Dams prepared numerous comprehensive assessments of large dams in different parts of the world. Regrettably, the vast majority of these assessments are somewhat superficial and often skewed to prove the dogmatic and one-sided views of the authors who prepared them. It is most unfortunate that mainly anti-dam analysts were selected by the Commission to carry out such analyses of selected dams from different parts of the world. Many scholars, professionals and governments have seriously questioned the objectivity and reliability of these studies. No serious peer reviews of these studies were ever carried out and, as a result, if there was some 'wheat' among what is mostly 'chaff', it has remained very well hidden. Thus, not surprisingly, these studies have been largely ignored by the world.

One important methodological issue should also be considered whenever ex post impacts of large development projects are to be analysed, which relates to the period over which such assessments are to be conducted. There is now general agreement among experts that in the initial years after a dam's construction, its direct and indirect impacts can be identified, and their cause-and-effect relationships can also be established with a reasonable degree of confidence. However, as time passes, it is difficult to say with any degree of reliability what percentage of the impacts can be attributed to the dam and what to other development factors, some of which may have been unleashed by the dam. Generally speaking, as the years pass, the attributable impacts of dams become hazier and hazier, because of changes brought about in other development sectors. Furthermore, after a certain period, most of the impacts of a new dam stabilise and they come into some form of equilibrium. As a general rule, most of the impacts of any large dam stabilise from 2 to 15 years after it becomes operational. Thereafter, the changes are often not significant, and their attributions to the dam become less and less relevant.

Because of these reasons, this study of the High Aswan Dam considered developments for some 35 years after the construction of the dam, even though the later impacts that could be attributable to the dam, especially during the third decade, are not definitive. As time progressed, especially during the third decade and after, the impacts of the dam were progressively overshadowed by other external development factors.

Any large infrastructure like the High Aswan Dam invariably has many impacts, some of which are positive while others are adverse, and some are very significant while others are minor. It is simply not possible that for any major project there will be only benefits and no costs, and only beneficiaries with no one having to bear costs. There is always a mixture of costs and benefits, and there will always be beneficiaries and people who will have to pay the costs. Ideally, determined attempts should be made to transform those who are likely to pay the costs to become beneficiaries of the project.

The impacts of large infrastructure projects like the High Aswan Dam are not easy to determine on a very reliable basis for many reasons. First, whereas many impacts can be estimated with a considerable degree of confidence, several others cannot be reliably estimated since our knowledge base at present, let alone at the time the dam was planned, leaves much to be desired. Second, the world of development is very complex—often cause-and-effect relationships cannot be established with any degree of dependability. This is because there could be several interacting forces that contribute to a net impact, the dam being only one among many factors. It is often very difficult to estimate which specific forces contributed to what percentage of impacts. Third, while many impacts can be estimated in quantifiable terms, others are intangible and so cannot be measured. Estimates of intangible benefits and costs depend on the experience, knowledge and bias of the analysts. Thus, they could vary, often very significantly, from one analyst to another.

Fourth, the world is not static. Accordingly, the boundary conditions of a project invariably change considerably within the period of the impact estimation, which may have no relation to the project, or only indirect relation. For example, in the case of the High Aswan Dam, land-use changes in Egypt may have been at most an indirect result caused by the dam's construction. This is because the population of Egypt has steadily increased since the dam was built, and industrial, agricultural and urban development have also progressed. The Egyptian population would have grown with or without the dam, and the land-use patterns would have changed because of substantial increases in human population and their activities. Of course, there is no question that the dam contributed to the improvement of the economy very significantly, especially during the early years of its operation when its impacts on the country and its people were the highest.

Fifth, there is also a time dimension to the change. Some impacts may be visible immediately after construction of the dam, like changes in erosion and sedimentation. These changes often stabilise with time, since after a certain period, the new forces unleashed reach an equilibrium with their environment. In contrast, some impacts take time to develop, for example, increase in groundwater or salinity levels due to over-irrigation and lack of adequate drainage facilities. When such impacts become visible and are considered significant, appropriate remedial measures are generally introduced, which unleash new and different forces. These forces again come to an equilibrium with the passage of time.

Finally, in all countries, including Egypt, many developments take place concurrently, along with the dam, which affect the final impacts. For example, the increasing incidence of schistosomiasis immediately after the dam and canal systems were constructed can be directly linked to the absence of sewage treatment works, education level of the population and level of health services available at that time. Each of these and other associated factors could explain the increasing incidence of schistosomiasis in the country. It is difficult to estimate what may be the relative contributions of each of these factors, and what percentage of these impacts could be attributed to the dam directly and exclusively.

Making realistic and accurate estimates of all the impacts pose serious methodological challenges, including data availability and reliability. However, a serious attempt was made in this study to estimate them as accurately as possible with the current state of knowledge.

17.3 Impacts of the Dam

The overall impacts of the construction of the High Aswan Dam can be classified into four broad categories: physical, biological, economic and social. This classification is somewhat arbitrary since many of the impacts are often interrelated, and some of them may spawn effects of other types.

Among the main physical impacts of the dam were the following:

- Changes in the level, velocity and discharge of the flow in the Nile River both upstream and downstream of the dam
- Increase in groundwater levels due to the introduction of year-long irrigation
- Changes in soil salinity and water logging
- Erosion of the river banks, beds and delta
- Sedimentation in the river and Lake Nasser
- Possible earthquakes
- Reclamation of desert for human habitat and agriculture

There were several major biological impacts, which included:

- Incidence of schistosomiasis
- Implications for fish production
- Changes in flora and fauna

Among the economic impacts were:

- Generation of hydropower
- Increase in industrial activities and industrial diversification, because of the availability of electricity
- Increase in agricultural land, as well as crop intensification and diversification
- Impacts on brick-making industry

There were several social impacts, which included:

- Peace and stability of the country due to increased economic activities and higher standard of living
- Eliminating the ravages of floods and droughts downstream of the dam
- Resettlement issues

It is impossible to cover this range of impacts in any meaningful fashion because of space limitations. The detailed analyses of all these impacts can be found in Biswas and Tortajada (2012). In this chapter, only one key issue each from the physical, biological and economic sectors has been selected for discussion.

17.4 Erosion and Sedimentation

It has been widely acknowledged for centuries that whenever there is human intervention in river flows, because of the construction of hydraulic structures such as dams, barrages or weirs, the hydrological and morphological characteristics of the watercourses are invariably changed. The High Aswan Dam is no exception to this rule. The dam irreversibly changed many of the prevailing characteristics of the Nile River. Among the major physical changes was the transformation of the river upstream of the dam from having a riverine to lacustrine nature, and the flow was very strictly controlled and regulated downstream of the structure.

The dam created a major lake at Aswan (Lake Nasser), which is nearly 500 km long and 10 km wide. The entire flow of the Nile was stored in the 3,830 m long dam from 1967. The storage of this enormous quantity of water in the reservoir ensured that as the river flow approached the reservoir, its velocity began to decline. When the flow entered the reservoir, its velocity fell to zero. As the flow velocity declined, so did its carrying capacity of suspended sediments. Consequently, suspended sediments began to precipitate as the water approached the reservoir. For all practical purposes, the river lost nearly all the sediments it carried as the flow approached and entered the reservoir. This meant that when water was discharged downstream of the dam, it was clear and free from sediments. Even during flood seasons when the sediment loads in the river were at their maximum, virtually all the sediments were captured upstream of the dam. In other words, the sediment-trap efficiency of the dam has been almost 100%. Since the water discharged downstream of the dam was sediment free, in the initial years of the dam's operation it contributed to erosion of the river bed and banks. The flow thus steadily picked up sediment as it progressed downstream. These changes were observed all the way to the Nile delta.

Historically, some 90% of the annual sediment flow in the river occurred during the flood season. The average suspended sediment load passing Aswan before the dam was constructed was estimated at 134 million tons (Abu-Zeid and El-Shibini 1997). A very significant percentage of this load is now deposited in the approach to the High Dam Lake. The rest is then deposited in the lake itself. During the planning of the dam, it was estimated that the annual sedimentation in the lake would average at about 60 million m³. Consequently, the reservoir was designed for a dead storage capacity of 31.6 km³, and the levels between 85 m and 147 m were reserved for dead storage. The reservoir has a live storage capacity of 90.7 km³, up to level 175 m, and then a provision for flood storage of 39.7 km³ up to level 182 m. Monitoring of the actual sedimentation in the reservoir indicates an annual deposit of 60–70 million m³, which is well within its design considerations. Studies have also indicated that the sedimentation is mostly occurring within 250 km upstream of the dam.

Predictions of erosion and sedimentation downstream of the dam, especially in the mid-1950s when the available knowledge and analytical capability were not as advanced as at present, were very difficult and complex. The problem was further compounded by the lack of reliable data and absence of a consensus among the erosion and sedimentation experts of the world on which method for estimating rates would best suit the Aswan case. Since different methods, with different assumptions, were used by different experts for these predictions and knowledge of the geological aspects of the river bed was grossly inadequate, these predictions, not surprisingly, varied widely. The predictions of drop in water levels and overall degradations downstream of each barrage ranged from 2 to 10 m.

During the late 1980s and the early 1990s, the National Water Research Centre of Egypt undertook detailed studies on water level to determine erosion and sedimentation changes following the dam's construction. Monitoring was undertaken at El Gaafra, Isna Barrage (downstream), Naga Hammadi Barrage (downstream) and Assiut Barrage (upstream and downstream). These studies (Abdelbary et al. 1990; Abdelbary 1996) showed some important changes before and after the construction of the dam. First, the suspended sediment content of the Nile waters decreased remarkably after the dam was constructed. For example, the mean annual suspended load declined at El Gaafra from 129 m/year in pre-dam conditions to 26.3 m/year in 1964, 4.20 m/year during 1965–1967, and has stabilised at around 2.27 m/year during the post-1967 period. If the peak annual sedimentation concentrations are considered, the changes are even more stark. This declined from about 3,000 ppm during the pre-construction period to about only 50 ppm later, representing a reduction of nearly 98.5%.

Second, as the Nile flowed downstream of Aswan in the pre-dam era, its sediment load declined steadily as it reached the delta that drains into the Mediterranean Sea. This pattern was completely reversed after the dam was constructed. This is not surprising, and was in line with the planners' forecast. Prior to the operation of the dam, the sediment content of the river flow fell as it progressed downstream because some

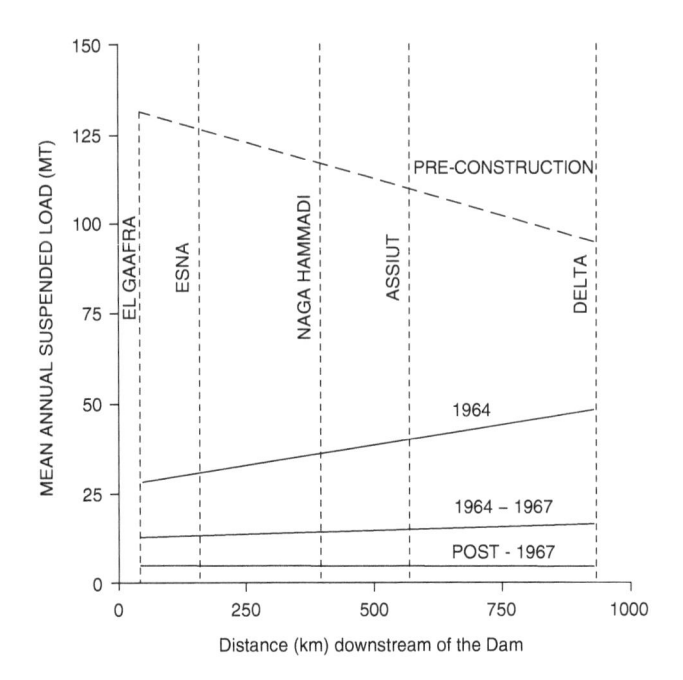

Fig. 17.1 Suspended sediment loads in the Nile downstream of Aswan (*Source*: Abdelbary 1996)

of the sediments precipitated upstream. Thus, the sediment content of the flow was the lowest when it reached the delta. Once the dam became operational, the relatively sediment-free water travelled downstream of the dam and steadily picked up sediments. At present, the sediment content of the water increases as it reaches the delta. The pre- and post-dam sediment contents of the flow are shown in Fig. 17.1.

It should be noted that the gradual increase during the post-dam period in the sediment load of the river, as it travelled downstream, might not be solely attributable to the construction of the structure. It is likely that the sediment load also increased because of other natural factors, such as sediment inflows from the wadis (in certain Arabic-speaking countries, a valley, ravine, or channel that is dry except in the rainy season), wind-blown sand and inflows to the river from the drainage system. Without a detailed analysis of the sediment balance from various sources, it is not possible to determine what percentage of the sediment can be attributed to erosion that occurred because of the dam. Detailed sediment balance of the river after construction of High Aswan Dam is not available.

Another impact caused by erosion and sedimentation is a reduction in the total bank length of the river after the High Aswan Dam, which is calculated by estimating the river length multiplied by two, and adding the bank lengths of the islands. The islands that are considered are those which have permanent vegetation. These do not include sand bars. Based on data available (Abdelbary 1996), the total bank length before the construction of the dam was estimated at 2,409 km. By 1978, the total bank length had fallen to 2,047 km, and by 1988, it marginally declined further

17 Impacts of the High Aswan Dam

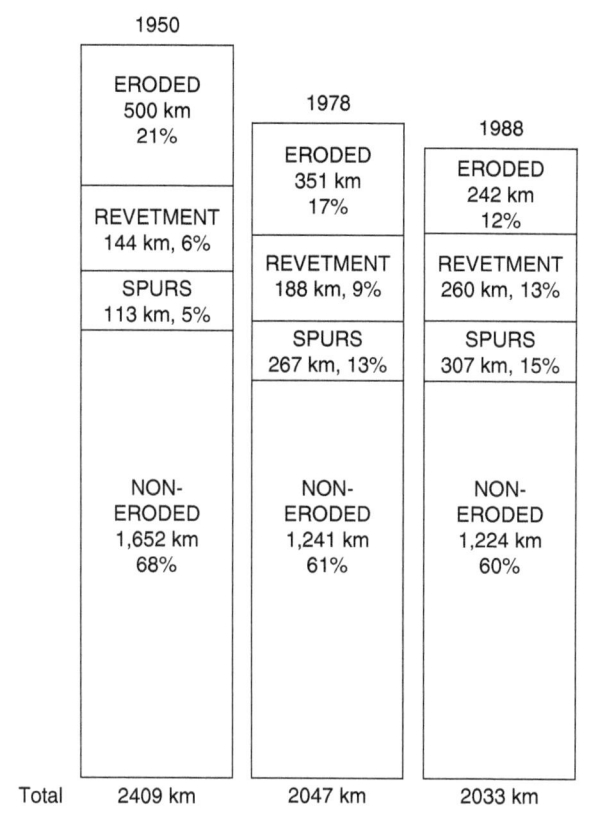

```
                    1950
              ┌─────────────┐
              │   ERODED    │
              │   500 km    │      1978
              │    21%      │  ┌─────────────┐
              │             │  │             │      1988
              ├─────────────┤  │   ERODED    │  ┌─────────────┐
              │  REVETMENT  │  │   351 km    │  │   ERODED    │
              │ 144 km, 6%  │  │    17%      │  │   242 km    │
              ├─────────────┤  │             │  │    12%      │
              │    SPURS    │  ├─────────────┤  ├─────────────┤
              │ 113 km, 5%  │  │  REVETMENT  │  │  REVETMENT  │
              ├─────────────┤  │ 188 km, 9%  │  │ 260 km, 13% │
              │             │  ├─────────────┤  ├─────────────┤
              │             │  │    SPURS    │  │    SPURS    │
              │             │  │ 267 km, 13% │  │ 307 km, 15% │
              │             │  ├─────────────┤  ├─────────────┤
              │    NON-     │  │    NON-     │  │    NON-     │
              │   ERODED    │  │   ERODED    │  │   ERODED    │
              │  1,652 km   │  │  1,241 km   │  │  1,224 km   │
              │    68%      │  │    61%      │  │    60%      │
              │             │  │             │  │             │
              └─────────────┘  └─────────────┘  └─────────────┘
    Total       2409 km           2047 km          2033 km
```

Fig. 17.2 Status of Nile River banks in 1950, 1978 and 1988. *Note*: Eroded bank length for 1950 is an estimate. 1978 and 1988 lengths are based on actual observations (*Source*: Abdelbary 1996)

to 2,033 km. This also indicates that the erosion and sedimentation activities have now basically reached an equilibrium. This is shown in Fig. 17.2.

The post-dam Nile has created a problem in terms of coastal erosion. Before the dam was constructed, the Nile carried a high sediment load during the flood period, which was deposited in the delta. During the winter months, the sediments were removed by wave action. Over the centuries, an equilibrium was established between the amount of sediment that was deposited during the flood season, and its removal by wave action during the winter season. This contributed to a reasonably steady shore line.

However, as soon as the first hydraulic structure was built on the Nile over a century ago, the sediment equilibrium was disturbed. Those built prior to the construction of the High Aswan Dam were not major, and thus the disturbance to the sedimentation–erosion equilibrium in the delta was not high. The dam had a serious impact on this equilibrium, as a result of which significant erosion was witnessed along the Mediterranean coast after it became operational. Considerable research and monitoring (Smith and Abdel-Kader 1988) were conducted to determine the extent of erosion of the coast over time, and to identify preventive steps. Appropriate remedial measures have now been taken.

17.5 Impacts on Fisheries

Impacts of large dams on fisheries, especially over the long term, are hard to predict. They are also difficult to estimate after the dams are constructed and become operational. This is because, as with any large infrastructure project, dams unleash a series of interacting forces of all types, over both space and time, which are difficult to forecast and to quantify.

17.5.1 Fish Catches Prior to the Dam

The data currently available on fish catch for the 1950s and 1960s in the Nile River and its delta can be used at best as an indicative estimate. For example, the Yearbook of Coast Guards and Fisheries of the Ministry of Defence (1960–1967), collated by the institution that was responsible for collecting fisheries data when the dam was being constructed, records a total fish catch from the Nile in 1955 as 850 metric tons. However, another available estimate of the fish catch for the same year was 6,700 metric tons. Even though the Ministry of Defence (specifically its Coast Guard) was the official governmental custodian of the fisheries data of Egypt during the 1950s, their estimates of the annual Nile fish catches appear to be consistently low. This underestimation can be appreciated by the fact that two sets of official data for the total Nile fish catch are available for the period between 1963 and 1967—one from the Ministry of Defence (MOD) (Saleh 2004), and the other from the then newly established civilian authority, General Enterprise for Aquatic Resources (GEAR). They are simply not comparable. For example, the GEAR estimate for 1963 is nearly 38 times that of the MOD. Similar divergences can be observed for other years.

Statistics for annual Nile fish catches are available from 1985 onwards from the statistical yearbooks of the General Authority for Fish Resources Development (GAFRD), another civilian institution. It is difficult to draw authoritative conclusions on the reliability of the GEAR or GAFRD data. For example, for the 1966–1975 period, GEAR consistently provided an annual fish catch at the same neat, round number of 20,000 metric tons, which is highly unlikely to be accurate.

Construction of a dam creates a barrier which reduces nutrient flows in the river downstream of the structure, and also to the estuary. This, in turn, reduces primary production, of organisms like phytoplankton and zooplankton, on which fish life depends. Accordingly, the construction of dams generally reduces fish production downstream of the structure, at least in the initial years. However, large hydraulic structures invariably trigger different forces that are often difficult to predict in terms of time, space and magnitude. For example, electricity generation and availability of water from the new dam often contribute to employment creation, which may further accentuate urbanisation. In the entire developing world, where wastewater treatment left much to be desired during the 1960s and 1970s, increasing urbanisation often contributed to higher nutrient loads to the rivers because of discharges of untreated, or partially treated, human waste. While the level of waste treatment in Egypt has improved somewhat since 2000, it is still inadequate, and the nutrient discharges to the Nile and its delta continue to remain high.

In addition, with reliable availability of irrigation water from the newly created reservoir, two to three crops can be grown each year instead of only one using floodwaters, as was the case under the pre-dam conditions. Increased cropping intensity has in turn significantly increased the total fertiliser use. Also, much more land has been converted to irrigation than before the dam, which again has contributed to additional input of fertilisers. Because of these developments, more and more phosphates are leached into the river downstream and on to the delta. This nutrient enrichment, over time, has generally increased primary production. The increase in primary production is followed by increased fish production, and thus fish catch.

Official data on fish catches are also somewhat unreliable because the price of fish is controlled, and also because the imposed fixed price is lower than the prevailing free market price. The price differential has resulted in extensive fish smuggling that does not get recorded in official statistics. As a general rule, the higher the differential between official and free market prices, the higher the level of smuggling. This factor, for example, may be the reason for the apparent decline in annual fish catches from Lake Nasser in the early 2000s, when the official fixed fish prices were a fair bit lower than the free market price.

The complexity of the fish catch statistics is further enhanced by other factors. For example, after the Egypt–Israel Peace Treaty was signed in 1979, security considerations in the Mediterranean changed significantly. Large areas of fishing ground that were off-limits to the fishing vessels for over a decade suddenly became available. These areas had considerably more fish since they had not been exploited during the previous decade. In addition, the eastern zone of the Mediterranean (east of Port Suez) was also opened up for fishing from 1980 onwards. Thus, the area over which fishing was possible expanded remarkably within a short period of time. The expansion of the fishing ground further changed the economics of fishing. This resulted in new investments in terms of adopting better fishing technology which had not been economically attractive earlier.

Reductions in fish catch from about mid-1967 and increases from 1978 onwards were partly caused by military considerations. Therefore, part of the decrease in fish catch during the earlier phase and the increase thereafter had nothing to do with the construction of the High Aswan Dam, but to reasons linked with national security. It is difficult to estimate what percentage of these changes in the fish catch from the Mediterranean during the 1967–1980 period can be attributed to military reasons. These factors should, however, be carefully noted in any comparison of pre- and post-dam fish catches. Many authors (e.g. Ibrahim and Ibrahim 2003) have claimed that fisheries in the area collapsed completely immediately after the construction of the High Aswan Dam (Thomas and Wadie 1989; Caddy 1993). While there is no question that fish catches declined following construction of the dam, the most important question is what was the extent of the decline that can be attributed directly to the dam?

The pre and post-dam fish catches from the Mediterranean are shown in Fig. 17.3. The available data show that fish catches increased rapidly from 1958 to 1960, with a

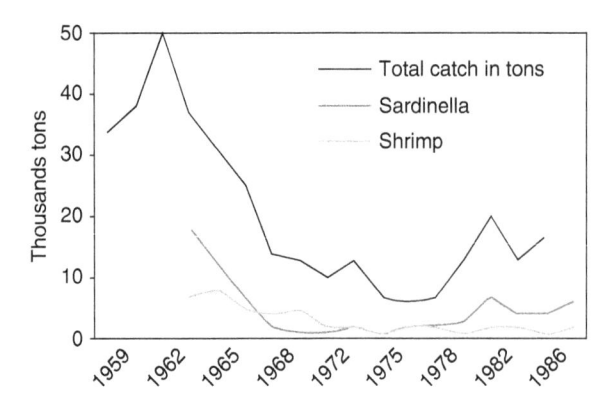

Fig. 17.3 Total catch, 1958–1986; sardinella and shrimp catch, 1962–1986 (*Source*: Fishery Statistics, Institute of Oceanography and Fisheries, Alexandria)

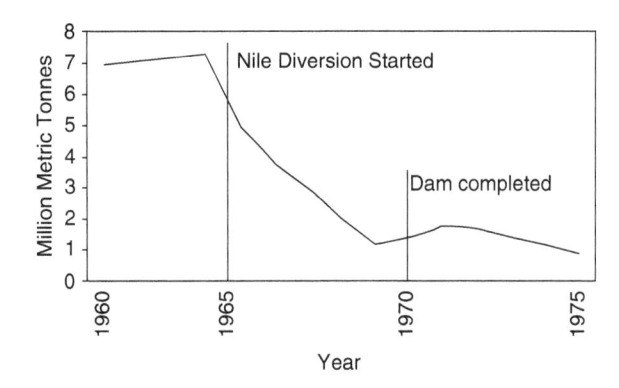

Fig. 17.4 Annual Mediterranean shrimp catch (*Source*: FAO 2005)

very pronounced peak in 1960. Unfortunately, data are not available for 1961, and thus it is not easy to say if a sharp decline began from 1960 or 1961, well before the Nile diversion was put into place in 1964. By 1965, the total catch had declined by nearly 70%. Much of this decline, going by current evidence, cannot be attributed to the dam.

This rise and then the sharp decline in fish catches can be directly linked, from 1958 to at least 1964, to government policies on fishing that were implemented in the late 1950s. The policies aimed at increasing fish catch, and included supporting fisheries cooperatives and subsidising the mechanisation of fishing vessels and gear. Because of these policy interventions, between 1958 and 1961 the number of motorised fishing boats increased by nearly 50%, from 428 to 622 (El-Zarka and Koura 1965). The policy ensured high fish catches during 1958–1961, which in turn contributed to overfishing and a resulting rapid decline in fish catch from 1962 onwards. Therefore, well before the dam was constructed, the fish catch from the Mediterranean had begun to decline very significantly. This is one aspect that nearly all the environmental impact analyses of post-dam developments have failed to consider.

Figure 17.4 shows detailed information on the changing nature of the annual shrimp catch from the Mediterranean for the period 1960–1975. It illustrates that

17 Impacts of the High Aswan Dam

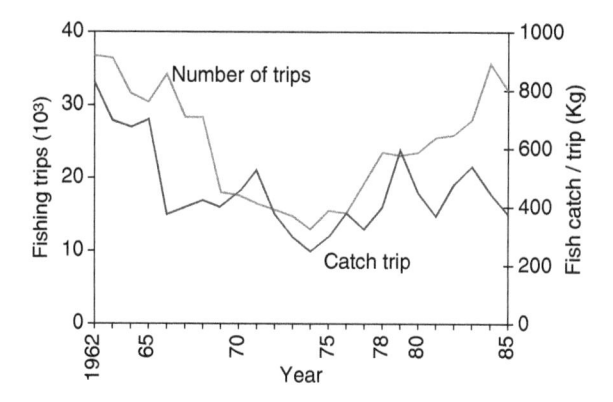

Fig. 17.5 Number of trips and average annual catch per trip (*Source*: Saleh 2004)

the rapid decline in the shrimp catch could be observed well before the Nile diversion was built in 1964, and then continued for another 4 years before reaching some sort of an equilibrium for a while.

17.5.2 Fish Catches After the Dam

This discussion will separately consider fish catches from the Mediterranean and those from Lake Nasser.

17.5.2.1 Mediterranean Catches

The reduction in fish catches after 1965 from the Mediterranean can thus be attributed to several reasons, among which were the decrease in primary productivity of the coastal areas, the reduction in fishing area after 1967 due to military considerations and earlier government policies which contributed to overfishing. The catch declined from 680 to 380 kg/trip, a 44% fall, within a 1-year period of 1965–1966 (Fig. 17.5). By 1973–1975, the catch had declined even further, to about 250 kg/trip, a level that was only 37% of the 1965 value and about 31% of the 1961 value (Saleh 2004).

These figures are not directly comparable since the number of fishing trips declined rapidly, from about 37,000 in 1962 to about 9,000 at the lowest points of 1973–1975, when the fish catches became some of the lowest (Fig. 17.5). As yield per fishing trip started to increase during the post-1977 period, so did the number of trips, because of the higher catch and thus the improved economics of fishing.

What is evident, however, is that the fish catch from the Mediterranean started to increase dramatically from about 1977. The most likely hypothesis could be the increasing nutrient loads that began to reach the Mediterranean (Nixon 2004; Oczkowski et al. 2009) because of increasing urbanisation (and the resulting discharge of untreated or partially treated domestic wastewater to the coast), and also due to higher levels of fertiliser leaching into the Mediterranean as more and more land was

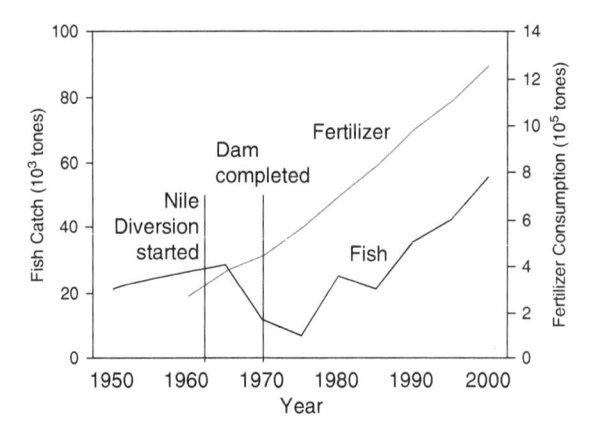

Fig. 17.6 Fish catch from the Mediterranean and fertiliser use in Egypt (*Source*: FAO 2005)

brought under irrigation as well as increasing fertiliser use per unit area to obtain higher yields (Fig. 17.6). While the discharges of this higher nutrient load have been beneficial thus far, especially in terms of increased fish catch, care needs to be taken to ensure that this load does not continue to increase progressively in the future. Otherwise, at some stage, such increased nutrient loads will over-fertilise the coastal waters, which in turn will adversely impact the total fish catch from the Mediterranean. It is worth noting that even though the Nile diversion was started in 1964, its dynamic implications are still being observed some 45 years later.

Thus, if total fish catches before and after the High Aswan Dam are compared, there is no question that after the dam was built, Egypt had more fish available from the Nile (including Lake Nasser) and the delta, compared to the pre-construction period.

17.5.2.2 Catches from Lake Nasser

The total annual fish catch from Lake Nasser in 1968, as expected during its initial years, was quite low at about 2,662 metric tons. The catch steadily increased from that time and reached a maximum of 34,206 metric tons by 1981. Records indicate that 1980–1983 were the most productive years, when the annual fish catch varied between 28,667 and 34,205 metric tons. Thereafter, the catch began to decline. The total annual catch fluctuated around 20,000 metric tons between 1993 and 1998 (Saleh 2004).

17.6 Hydropower Generation

A very major economic benefit of the High Aswan Dam is the electricity it generated when the country needed it the most. Thanks to this generated renewable energy, it was possible to provide electricity to much of the country's rural area. This simple factor of development changed the social and environmental conditions

17 Impacts of the High Aswan Dam

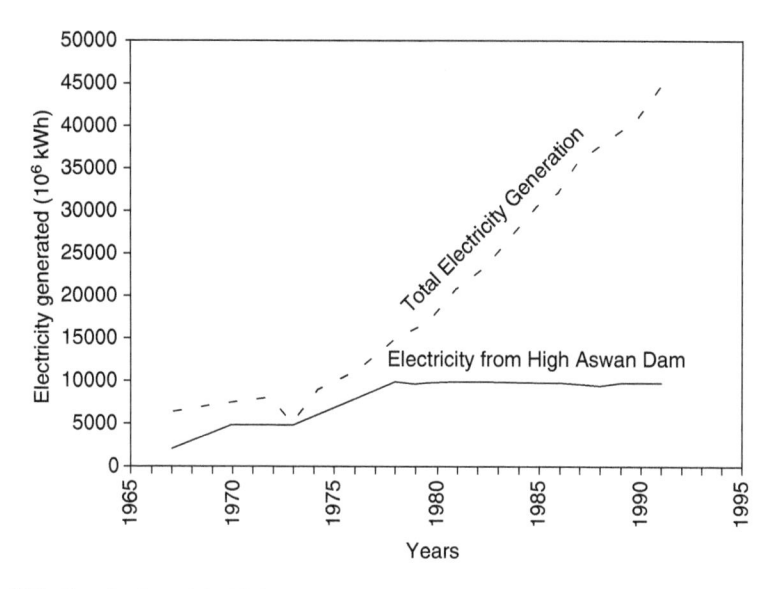

Fig. 17.7 Contribution of the High Aswan Dam to Egypt's electricity generation

of much of Egypt, and also substantially improved the quality of life of a very significant number of its people.

The 12 generating units of the Aswan are capable of producing 10 billion kWh of electricity per year. Before the dam became operational, Egypt had to import fossil fuel. The electricity generated by the dam reduced in a most significant way the potential energy import bill of the country. Figure 17.7 shows the contribution the dam has made to the generation of electricity in the country. Up until 1979, the dam contributed to nearly half of the electricity generated in the country. In some years, its share was significantly more than 50%. As time progressed, and especially after 1980, the dam's share began to decline because of increasing contributions from newly constructed thermal power plants which met the burgeoning electricity requirements of the country.

If one looks solely at the electricity generated by the dam since its beginning, the accrued benefits from this sector alone paid for the entire construction cost of the structure quite some time ago.

17.7 Overall Benefits and Costs of the Dam

According to the detailed benefit-cost analysis carried out by the Egyptian Ministry of Irrigation, at the time of the dam's construction its total cost, including subsidiary projects and the extension of electrical power lines, amounted to Egyptian £450 million (Abul-Atta1978). These costs seem to be realistic since they are somewhat similar to World Bank estimates. The then Irrigation Minister, Abdul Azim Abul-Atta, estimated that this cost was recovered within only 2 years, since the dam's annual return to national income was estimated at E£255 million, consisting of E£140 million from agricultural production, E£100 million from hydropower generation, E£10 million from flood protection and E£5 million from improved navigation. Detailed estimates of these costs were prepared by the Egyptian government, and since no one has challenged them, they can be assumed to be reliable. Thus, our focus has been to assess appropriate impacts, especially environmental and social, about which there has been considerable global controversy ever since the construction of the dam was initiated.

The importance of the dam to Egypt's economic survival was clearly demonstrated, at least qualitatively, during the 1980s. It is not too difficult to hypothesise what would have happened to the Egyptian economy and socio-political conditions if the dam had not been there to protect the country from, first, the potentially catastrophic impacts of a prolonged drought from 1979 to 1986, and immediately thereafter the abnormally high summer flood of 1988, which had devastating effects on Egypt's upstream neighbour on the Nile Basin, Sudan. Even with the Aswan High Dam in place, Egypt had come perilously close to experiencing the catastrophic impacts of the drought by early 1988, due to a dangerously low water level in Lake Nasser.

17.8 Conclusions

The analysis of selected impacts, both positive and negative, made in this chapter, should provide a clear indication of the complexities and difficulties of assessing impacts of any large hydraulic structure. There are several levels of complexities, the most important of which is the fact that it is difficult to assign accurately the benefits and the costs that can be attributed to any specific impact.

The problem is further compounded by the fact that the development of any country is never static. There are continuous changes, government policies are often dynamic, and the ranges, extents and magnitudes of human activities are constantly changing. It is thus very difficult to isolate and attribute benefits and costs to one specific activity—they are often impacted by many other factors. This challenge is made more complicated by the fact that the necessary data are often not available, or reliable, in developing countries, and many of the impacts are not quantifiable. This makes authoritative impact predictions for any dam or large structure a very difficult task under the best of circumstances.

The three important impacts of the High Aswan Dam analysed in this chapter indicate the complexities of the issues, as well as the danger of drawing simplistic conclusions because they often prove to be wrong. The case of the High Aswan Dam is an excellent example of how numerous people from all over the world have failed to understand and appreciate the difficulties associated with such simplistic analyses and linear thinking, which have led to totally erroneous conclusions. Based on the study carried out by the Third World Centre for Water Management, it is evident that the High Aswan Dam is one of the most successful hydraulic structures of the world, and not one of the worst as is often presently contended at the international level.

References

Abdelbary MR (1996) Effect of the High Aswan Dam on water and bed levels of the Nile. In: Shady AM, El-Moattassem M, Abdel-Hafiz E, Biswas AK (eds) Management and development of major rivers. Oxford University Press, Delhi, pp 444–463

Abdelbary MR, Attia K, Galay V (1990) River Nile Bank Erosion Development below the High Aswan Dam. National Seminar on Physical Response of the River Nile Interventions, 12–13 November 1990, Cairo, Egypt

Abul-Ata AA (1978) Egypt and the Nile after the construction of the High Aswan Dam. Ministry of Irrigation and Land Reclamation, Cairo, p 222

Abu-Zeid MA, El-Shibini FZ (1997) Egypt's High Aswan Dam. Int J Water Resour Dev 13(2):219–217

Biswas AK (2004) Dams: Cornucopia or disaster? Int J Water Resour Dev 20(1):3–14

Biswas AK, Tortajada C (2001) Development and large dams: a global perspective. Int J Water Resour Dev 17(1):9–21

Biswas AK, Tortajada C (2012) Hydropolitics and Impacts of High Aswan Dam, Springer, Berlin, forthcoming

Caddy JF (1993) Contrast between recent fishery trends and evidence of nutrient enrichment in two large marine ecosystems: the Mediterranean and the Black Seas. In: Sherman K, Alexander LM, Gold BD (eds) Large marine ecosystems: stress, mitigation and sustainability. AAAS Press, Washington DC, pp 137–147

El-Zarka S, Koura R (1965) Seasonal fluctuations in the production of important food fishes of the UAR waters of the Mediterranean Sea. Alexandria Institute of Hydrobiology, Fisheries Notes and Memoirs No 74, 69 p

Food and Agricultural Organisation (FAO) Egypt's country profile, 2005 version. Available at www.Fao.org/ag/w/aquasata/countries/Egypt/indix.stm

Ibrahim FN, Ibrahim B (2003) Egypt: an economic geography. I.B. Tauris, London

Nixon WS (2004) The artificial Nile: the Aswan High Dam blocked and diverted nutrients and destroyed a Mediterranean fishery, but human activities may have revived. Am Scient 92(2):158

Oczkowski AJ, Nixon WS, Granger SL, Al-Sayed AFM (2009) Anthropogenic enhancement of Egypt's Mediterranean Fishery. Proc Natl Acad Sci 106(5):1364–1367

Saleh M (2004) Fisheries of the River Nile. Report to Advisory Panel Project on Water Management. Ministry of Water and Irrigation, Cairo, Egypt

Smith SE, Abdel-Kader A (1988) Coastal erosion along the Egyptian Delta. J Coast Res 4(2):245–255

Thomas GH, Wadie WF (1989) Trends of fish catch and factors responsible for its fluctuations in the Egyptian Mediterranean Sea Waters. Bulletin, High Institute of Public Health, Alexandria (19)4, pp 885–908

Index

A
Abdelbary, M.R., 385–387, 395
Abdel–Kader, A., 387, 395
Åberg, J., 88, 92
Abril, G., 88, 92
Abu–Zeid, M.A., 385, 395
Adams, A., 50, 65
Adelman, I., 24, 35
Africa, 1, 3, 9, 17, 38, 47, 63, 66, 67, 75–79, 81, 83–85, 326, 375
Agência Nacional de Águas (ANA), 161, 163, 165, 168, 169, 170
Agência Nacional de Energia Elétrica (ANEEL), 158–160, 170
Agricultural commodities, 19, 20, 25
Agriculture, 12, 20, 27, 28, 30–32, 55, 57, 63, 67, 73, 95, 104, 126, 138, 161, 163, 174, 179, 184, 186, 187, 193–195, 197, 224–226, 233, 235, 247, 248, 251, 252, 254, 260–262, 265, 278, 287, 288, 300, 307, 315, 316, 321, 327, 333, 350–352, 358, 370, 376, 383
Akbulut, N., 180, 197
Akca, E., 188, 198
Akcay, A., 174, 197
Aksit, B., 174, 197
Akyürek, G., 182, 183, 185, 187, 197
Algesten, G., 88, 92
Almaca, A., 188, 189, 198
Almeida, D., 88, 93
Altinbilek, D., 171–199
Amazon Region, 160
ANA. *See* Agência Nacional de Águas (ANA)
Anderson, M., 201–218
ANEEL. *See* Agência Nacional de Energia Elétrica (ANEEL)
Antarctica, 75, 76, 78, 79, 83

Aquaculture, 55, 64, 163, 166, 167, 213, 214
Arab Fund for Economic and Social Development, 380
Aravena, R., 88, 93
Argentina, 1, 44, 48, 61, 67, 123–152, 379
Arizona, 45, 60
Aronsson, I.L., 53, 65
Artaxo, P., 70, 93
Asia, 3, 17, 38, 47, 63, 67, 75–79, 81, 83–85, 256, 316, 351, 371
Asian Development Bank, 38, 230
Asunción, 136, 137, 144
Australia, 1–3, 75–79, 81, 83, 84, 201–218, 379
Averyt, K.B., 70, 93
Aydemir, S., 188, 189, 198
Aydin, M., 188, 198
Aylward, B., 21, 35

B
Balazote, A., 149, 152
Ballester, M., 89, 94
Bambace, L., 70, 74, 81, 94
Bangladesh, 330–332
Baranger, D., 149, 150
Barrage
  Assiut, 385
  Isna, 385
  Naga Hammadi, 385
  Nangal, 302–303, 315
Barrios, L., 149
Bartolome, L.J., 50, 65, 123–152
Bastviken, D., 81, 92
Bayar, S., 180, 197
Bayram, M., 187, 192, 197, 198
Beijing, 122, 241, 243–257

Beleli, O., 174, 196, 198
Bell, C., 21, 35
Benefits, 2, 4, 6–9, 13, 21, 28, 30–33, 37, 41, 46, 55, 58, 61–64, 86, 95, 98–110, 112–122, 130, 131, 136, 139, 144, 151, 159, 167, 169, 175, 179, 187, 191, 192, 197, 202, 208, 214, 234, 264, 265, 271, 277–280, 282–284, 287, 293, 297, 299–301, 309, 314–324, 326, 329, 348, 350, 355, 358, 359, 361, 367–368, 370, 380, 382, 393–394
Berger, T.R., 264, 269, 276
Bergström, A., 88, 92
Berntsen, T., 70, 93
Betts, R., 70, 93
Bhalla, G.S., 26, 35
Bhatia, R., 19, 35, 62, 65, 324, 326
Bhutan, 330–332
Bihar, 26, 321, 324, 375
Bilharz, J.A., 45, 65
Bischof, R., 95, 122
Biswas, A.K., 1–18, 37, 67, 87, 179, 188, 198, 199, 253, 256, 331, 355, 379–395
Bodaly, R.A., 88, 93
Bonvin, J.M., 99, 110, 112, 118, 122
Braga, B., 153–170
Brazil, 1, 3, 5, 9, 19, 21, 22, 25, 31–35, 38, 48, 65, 82, 94, 125, 143, 153–170, 326
Briones, C., 151
Buckley, C., 255, 256
Buenos Aires, 124, 125, 136, 137, 149, 152

C
Caddy, J.F., 389, 395
Cakmak, B., 188, 198
Callander, B.A., 80, 198
Canada, 1–3, 48, 61, 72, 82
Canuel, R., 70, 88, 89, 93, 94
Capacity building, 172, 174, 193, 366
Caraco, N.F., 73–75, 81, 86, 92
Caribbean, 75, 76, 78, 79, 83
Catullo, M.R., 150
Centre for Policy Research (CPR), 301, 314, 319, 327
Cernea, M.M., 38, 39, 42–43, 45–47, 49, 57, 58, 60, 66, 67, 150
Cestti, R., 19–35, 65, 326
CGE. See Computable General Equilibrium (CGE) models
Chadha, G.K., 35, 271, 276
Chamberland, A., 88, 93
Chen, S., 229, 241

Chen, X., 232, 241
Chen, Z., 70, 93
CHESF. See São Francisco River Hydropower Company (CHESF)
China, 3, 5, 8, 9, 42, 48, 58, 67, 86, 122, 214, 219–241,
  land acquisition, 219–227, 229–233, 240, 241
  land legislation, 220
Choucri, N., 150
Cimbleris, A., 88, 94
Clarke, C., 47, 50, 66
Cluigt, J., 150, 151
CODEVASF. See Companhia de Desenvolvimento dos Vales do São Francisco e do Parnaíbam (CODEVASF)
Cole, J., 81, 92
Colson, E.F., 39, 45, 66, 67
Companhia de Desenvolvimento dos Vales do São Francisco e do Parnaíbam (CODEVASF), 166, 170
Compensation, 51, 52, 60, 95, 98, 109, 110, 124, 127, 128, 130, 133–136, 141–145, 147–148, 169, 192, 220–229, 232–236, 240, 241, 250, 253, 257, 261, 264, 265, 267, 268, 273, 274, 294, 310–312, 346, 347, 360–362, 364, 366, 367, 369–373, 376
Computable General Equilibrium (CGE) models, 21, 24, 25, 28, 29, 34, 35
Cook, C., 38, 66
Costa Rica, 46, 48, 64
Costs, 4, 6, 7, 23, 41, 95, 132, 192, 203, 224, 244, 266, 284, 300, 329, 358, 380
CPR. See Centre for Policy Research (CPR)
Crutzen, P.J., 86, 94
Çullu, M.A., 188, 189, 198

D
da Costa, S.V., 153–170
Dam
  Aleman, 48
  Alta, 48
  Arenal, 46, 48, 55
  Atatürk, 6, 13, 97, 171–199
  Batman, 181, 183
  Bayano, 44, 48
  Bhakra, 3, 20–22, 25–28, 33, 34, 299, 302, 303, 305, 307–309, 311, 315, 317, 319, 321, 323–325, 358, 359, 375

Birecik, 181, 183, 184, 195
Cahora Bass, 48
Cetian, 11, 247, 254
Ceyhan, 48
Chixoy, 48
Cirata, 49
Crotty, 202, 205, 206, 209
Danjiangkou, 11, 253
Darwin, 205, 206
Dicle, 181, 183
Emosson, 97
Fort Randall, 48
Garrison, 48
Göscheneralp, 97
Grand Coulee, 47, 48
Grande Dixence, 97
Guanting, 244–249, 251, 254, 255
High Aswan, 3, 13, 14, 18, 21, 22, 25, 28–31, 34, 40, 46, 49, 51, 55, 62, 97, 379–395
Hirakud, 3, 48, 264, 359
Ilisu, 181, 182
Itaparica, 32, 48, 156, 157, 162, 164
Jatiluhur, 64
Kainji, 46, 48, 51, 64
Kangsabati, 282, 294, 295
Karakaya, 181, 183
Kariba, 3, 40, 44, 45, 48, 49, 56, 57, 63
Karkamis, 181, 183, 195
Katse, 48, 65
Keban, 180, 181, 183
Kedungombo, 48
Khao Laem, 48
Kiambere, 49
Kinzua, 45, 48
Kolkewadi, 339, 342, 343, 346
Kossou, 46, 48, 53
Koyna, 329–356
Kpong, 48, 52
Kralkizi, 181, 183
La Grande, 48
Lac Gruyère, 97
Laforge, 88, 91
Lokka, 86, 93
Maguga, 39
Manantali, 48
Mangala, 305
Mattmark, 97
Mauvoisin, 97
Mettur, 358
Miyun, 245–246, 251, 252
Mohale, 48
Morazan, 48

Nan Ngum, 48, 64
Nangbeto, 48
Narayanpur, 48
Narmada (see Sardar Sarovar)
Oahe, 48
Orme, 45
Pancheshwar, 375
Pantabangan, 48
Paulo Afonso IV, 22, 32, 157
Periyar, 358
Petit Saut, 88
Piedra del Águila, 131–133, 149
Pimburetewa, 46, 48
Pong, 22, 48, 359
Porttipahta, 86, 88
Rajjaprobha, 63
Ramial, 48
Saguling, 49, 64
Saint Singaji, 371
Sardar Sarovar, 48, 259–276
Shanxi, 247
Shuikou, 46, 48, 55
Sihlsee, 97
Skinnmuddselet, 88
Sobradinho, 21, 22, 25, 31–34, 153–170
Tarbela, 48, 97, 305
Tehri, 358, 359, 369–373
Three Gorges, 65, 86, 97, 227, 232–239
Tucurui, 48, 88, 89
Ukai, 48, 359
Valle di Lei, 114–120
Victoria, 46, 48
Wilson, 358
Wonogiri, 64
Xiaolangdi, 227
Yantan, 46, 48
Youyi, 247
Zimapán, 48, 53, 65
Danklmaier, C.M., 123–152
Dean, W.E., 74, 92
De Fillipo, R., 88, 94
de Janvry, A., 32, 35
Delmas, R., 88, 92
Demir, H., 187, 199
Dentener, F.J., 86, 94
Dervis, K., 23, 35
Dettli, R., 99, 112, 122
Development, 1, 22, 38, 69, 95, 123, 154, 171, 201, 219, 243, 259, 277, 300, 329, 357, 379
Dezincourt, J., 88, 93
Dibble, S., 151
Dieringer, A., 149

Dincer, B., 193–196, 198
Displacement, 38, 43, 45, 49, 50, 65, 67, 123–152, 233, 246, 259–262, 266, 270, 275, 278, 294, 297, 326, 348, 349, 363, 369, 375–376, 378
Döll, P., 75–78, 85, 93
dos Santos, E.O., 70, 94
dos Santos, M., 88, 92
dos Santos, M.A., 70, 94
dos Santos, M.S., 70, 94
Downing, J.A., 73–75, 81, 86, 92
Downing, T.E., 44, 45, 66
Duarte, C.M., 73–75, 81, 86, 92
Duchemin, E., 70, 74, 75, 81, 88, 89, 93
Duhl, L., 40, 66
Dumestre, J.F., 88, 93
Durocher, C., 45, 62, 93
Dyck, B., 88, 93

E

Economic growth, 143, 161, 169, 171–174, 178, 191, 193, 197, 219, 262, 333, 334
Edwards, G., 88, 93
Egré, D., 45, 62, 66
Egypt, 3, 13, 19, 21, 22, 25, 28–31, 34, 35, 39, 48, 65, 326, 379, 380, 382, 383, 385, 388, 389, 392–395
Egypt–Israel Peace Treaty, 389
Electricity generation, 179, 191, 193, 195, 388, 393
El–Shibini, F.Z., 385, 395
El–Zarka, S., 390, 395
Employment, 12, 20, 23–25, 29, 32, 41, 46, 55, 62, 64–65, 104, 109, 112, 113, 172–178, 190, 191, 193, 196, 203, 214, 227, 228, 231, 250, 260, 264, 266, 267, 277, 287–288, 297, 312, 320, 323, 324, 326, 345, 347, 354, 368, 369, 372, 376, 388
Enders, S., 14, 250
Environmental protection, 16, 172, 173, 193, 235, 269
Ercin, A.E., 184, 186–188, 193, 195, 198
Erosion, 64, 167, 168, 210, 244, 246, 247, 249, 254, 294, 295, 297, 308, 344, 381, 383–387, 395
Escay, J., 151
Eswaran, H., 188, 198
Europe, 2–5, 32, 72, 75–79, 83–85, 92, 197, 375
Evaluation, 21, 34, 35, 38, 45, 50, 60, 65–67, 73, 92, 95, 110–112, 118, 119, 121, 122, 130–131, 135–136, 143, 144, 146–148, 151, 154, 167, 174, 178, 188, 198, 199, 230–232, 240, 241, 278–280, 298, 301

F

Fahey, D.W., 70, 93
Fearnside, P.M., 70, 85, 93
Federación, 125–131, 148
Federal Power Board of the Central African Federation, 44
Ferguson, A., 45, 66
Ferradas, C.A., 151
Filho, J.G.C.G., 153–170
Finland, 74, 75, 79, 87
Fisher, E., 44, 67
Fisheries, 44, 55, 63, 64, 163, 167, 210–212, 214, 215, 235, 249, 277, 286, 353, 388–392
Floods, 1, 2, 6, 11, 12, 19, 21, 28–30, 73, 75, 77, 84, 86, 95, 104, 105, 107–113, 118, 120, 121, 125, 126, 130, 136, 137, 139, 140, 154–156, 159, 163–164, 166–168, 170, 179, 182, 202, 203, 206, 211, 236, 243–245, 247, 251, 255, 260, 268, 279, 282, 283, 298, 300, 301, 308, 314, 318–320, 323, 326, 333, 350, 355, 375, 381, 384, 385, 387, 389, 394
  control, 2, 21, 29, 77, 84, 155, 156, 163–164, 166, 170, 243, 244, 251, 255, 301, 314, 318–320, 323
Forster, P., 70, 93
Francioni, M., 150, 151
Fried, M., 40, 66
Frischknecht, R., 110, 122
Fujii, T., 187, 198

G

Gagnon, L., 91, 94
Gallin, R., 45, 66
Galy-Lacaux, C., 88, 92, 93
Gat, J.R., 74, 94
Gay, J., 37, 45, 46, 50, 52, 55, 57
Ghana, 3, 48, 49, 52, 63, 64, 82, 376
Ghanekar, S.K., 341, 350, 356
Giles, J., 70, 81, 93
Gill, R., 201–218
Global Ecological Zones, 75
Global Lakes and Wetlands Database (GLWD), 70, 71, 75, 82–84
Goodland, R., 37, 60
Gopalakrishnan, M., 357–378
Gorham, E., 74, 92
Gosse, P., 88, 93
Gowariker, V., 269, 276
Greenhouse Gas Emissions, 69–92
Grubb, M., 151
Guatemala, 48
Guerin, F., 88, 92

Index   401

Guggenheim, S., 38, 66
Gulhati, N.D., 304, 327
Gulluoglu, S.M., 188, 199
Gupta, R., 196, 199

H

Habitats, 104–106, 110–113, 119, 168, 212, 260, 261, 346
Haggblade, S., 21, 23, 32, 35
Hagin, B., 122
Hamilton, S., 151
Hamlin, B., 23, 94
Härkönen, S., 69–94
Harran Plains, 181, 183, 184, 186–188
Harris, L.M., 187, 198
Harris, N., 80, 93
Haryana, 26, 306, 307, 316, 317, 320–324, 375
Hasankeyf, 181, 182
Hashimoto, T., 172
Hauenstein, W., 95–122
Haywood, J., 70, 93
Hazar, T., 192, 197
Hazell, P., 21, 35
Heikkinen, M., 69, 87, 88, 93
Hellsten, S., 69, 87, 88, 93
Herrán, C., 123, 150
Hewings, G.J.D., 23, 35
Heyes, A., 88, 93
Hidronor, S.A., 131, 133, 135, 136, 151
Hoffman, S., 21, 35
Hofstetter, P., 110, 121
Honduras, 48
Houel, S., 70, 94
Houghton, J.T., 80, 93
Huddar, S.N., 337, 356
Human development, 172, 196, 301, 321, 350
Huttunen, J.T., 69–94
Hydropower development, 32, 166, 167, 201, 215, 314
Hydro-Quebec, 45, 61

I

Ibrahim, F.N., 389, 395
ICOLD. *See* International Commission on Large Dams (ICOLD)
Imboden, D.M., 74, 94
Impacts
    economic, 2, 17, 19–34, 60, 131, 159, 164, 186, 188–192, 197, 280, 284–287, 298, 384
    environmental, 13, 17, 87, 95, 97, 166–169, 186, 188, 278, 280, 298, 301, 353–354

    social, 17, 37, 60, 124, 126–127, 140, 159, 188–186, 188–192, 197, 280, 284–287, 345–348, 353, 384
Impoverishment, 39, 41–44, 46, 49, 51, 52, 57–58, 60, 191, 261
India, 2, 3, 5, 9–11, 13, 19–22, 25–28, 33–35, 38, 48, 61, 65, 66, 259–327, 329–355, 357–378
Indonesia, 3, 48, 49, 63, 64
Industrial commodities, 20, 323
Indus Waters Treaty, 303–305, 315
Infrastructure, 3, 4, 6, 12, 28, 39, 41, 43, 61, 108, 109, 119, 121, 126–128, 133–136, 140–142, 158–160, 166, 169, 172–175, 177, 179, 181, 190, 192, 193, 217, 225, 227, 231, 236, 259, 260, 270, 274, 287, 293, 296, 297, 300, 319, 365, 379, 382, 388
Input–output (I/O) models, 21–24, 26
Institutions, 2, 8, 16, 23, 38, 39, 41, 42, 47, 56, 65, 124, 141, 144, 148, 175, 188, 190, 195, 196, 225, 229, 230, 232, 233, 236, 237, 239, 260, 261, 266, 270, 293, 316, 370, 374
Inter-American Development Bank, 124, 125, 132, 138, 141, 142, 149
Inter-basin Water Transfer, 166
Intergovernmental Panel on Climate Change (IPCC), 70, 81
International Commission on Large Dams (ICOLD), 37, 50, 51, 58, 74, 77, 358
International Network on Displacement and Resettlement, 38
International River Networks, 6, 50
Investment, 23, 29, 30, 32, 41, 43, 60, 100, 102–104, 110, 117, 138, 144, 145, 173, 177, 178, 182, 184, 188, 192–196, 235, 236, 254, 301, 323, 324, 349, 352, 364, 375, 376, 389
IPCC. *See* Intergovernmental Panel on Climate Change (IPCC)
Iran, 179–181
Iraq, 82, 180, 181
Irrigation, 6, 10, 11, 16, 20, 21, 28, 32–34, 42, 44, 46, 47, 55, 57, 62–64, 77, 79, 81, 84, 86, 95, 107, 126, 130, 134, 135, 137, 153, 155–159, 164–167, 170, 171, 173–175, 177–179, 184, 186–188, 190–194, 197, 231, 247, 251, 252, 263, 264, 278–284, 287, 290, 293–298, 300–303, 306, 307, 309, 312, 314–317, 319–321, 323, 330, 333, 334, 344–347, 349–353, 355, 358, 359, 368, 370, 373, 376, 383, 389, 392, 393

Itaipu hydropower plant, 154
Italy, 114
Ivory Coast, 46, 48, 53
Iyer, R.R., 265, 276, 378

J
Jacques, R., 89, 94
Jagdale, R., 353, 355
Jain, L.C., 269, 276
Jambert, C., 88, 93
James Bay Cree Nation, 45
Jansson, M., 88, 92
Jiao, Z., 253, 256
Jia, Y., 220, 241
Ji, Z., 254, 257
Jungner, H., 69–93

K
Kalff, J., 74, 93
Kapur, S., 188, 198
Kattenberg, A., 80, 93
Kavaso lu, T., 193, 194, 196, 198
Kayasü, S., 177, 196, 198
Keeley, J., 70, 94
Keller, M., 88, 93
Kelly, C.A., 74, 88, 93, 94
Kendirli, B., 188, 198
Kenya, 49
Khan, A.K., 305, 366
Khera, S., 45, 66
Knoepfel, I., 110, 122
Koenig, D., 45, 66
Kortelainen, P., 72, 87, 92, 94
Koura, R., 390, 395
Krishna Water Dispute Tribunal, 329, 344–345, 349, 350, 354
Kulandaiswami, V.C., 269, 276
Kulkarni, S.Y., 20, 355, 356
Kummu, M., 69–94
Kundat, A., 187, 198

L
Labroue, L., 88, 93
Lafitte, R., 95–122
Lake
  Burbury, 201, 204, 205, 207–209, 211–212, 214, 215
  Gordon, 204
  Kainji, 64
  Nasser, 383, 384, 389, 391, 392, 394
  Pedder, 202, 204
  Sobradinho, 31, 161
  Volta, 64
Lambert, M., 91, 94
Land acquisition, 219–229, 233–235, 240, 265, 309–314, 346, 348, 358, 360–361, 364, 370
Landscape, 31, 75, 103–106, 110–113, 118–121, 209
Laos, 48, 54, 64
Larose, C., 70, 89, 94
Latin America, 17, 38, 47, 75, 77, 79, 81, 92
Lean, J., 70, 93
LeDrew, L., 91, 94
Lee, Y.F., 243, 246, 251, 255, 256
Lehner, B., 75–78, 85, 93
Lehr, J.H., 70, 94
Lelieveld, J.O.S., 86, 94
Leontief, W., 21, 23, 24, 35
Lerman, A., 74, 94
Lesotho, 48, 60, 65
Li, G., 248, 251–253, 255, 256
Li, H., 244, 256
Lima, I., 70, 74, 81, 94
Lindqvist, O., 69, 87, 88, 93
Liu, C.M., 253, 256
Liu, H., 254, 257
Liu, W, 248, 254, 257
Living standards, 39–43, 46, 49–51, 54, 55, 57–60, 62–64, 172, 224, 227, 228, 232, 239, 266, 287
Li, Y., 254, 256
Lowe, D.C., 70, 93
Lucotte, M., 70, 88, 93, 94
Lu, L., 252, 256
Lu, Z., 252, 256

M
MacIsaac, E., 88, 94
Mahabal, A.S., 329, 356
Mali, 48
Malik, R.P.S., 19–35, 65, 326
Mandela, N., 326
Mane, P.M., 356
Manning, M., 70, 93
Mapuche Indians, 131–136
Mariella, P.S., 45, 66
Marquis, M., 70, 93
Martikainen, P., 69, 87, 93
Maskell, K., 80, 93
Masuhr, K., 110, 122
Mathur, H.M., 38, 66
Matvienko, B., 88, 92, 94
McDowell, C., 38, 39, 66

Index

McDowell, W.H., 73–75, 81, 86, 92
Mehta, L., 45, 66
Meira Filho, L.G., 80, 93
Melack, J., 81, 92
Mexico, 4, 48, 49, 53
Meybeck, M., 74, 94
Middelburg, J.J., 73–75, 81, 86
Middle East, 38, 47, 190
Migration, 3, 20, 25, 64, 65, 106, 109, 113, 119, 121, 126, 139, 154, 159, 166, 167, 175–177, 186, 190, 191, 217, 260, 264, 291–292, 310, 322, 326, 375
Miller, H.L., 70, 93
Mills, R., 151
Mitigation measures, 124, 166–167, 169
Miyata, S.1, 187, 198
Modak, D.N., 341, 350, 356
Monitoring, 49, 73, 85, 87, 143, 144, 146, 167, 174, 175, 178, 206, 211–213, 217, 230, 232, 240, 269, 349, 365, 370, 385, 387
Monte Carlo simulations, 29
Montevideo, 125
Moore, E., 151
Moore, V., 88, 93
Morse, B., 264, 269, 276
Morton, A.J., 64, 66
Mouvet, L., 122, 167, 170
Mozambique, 48
Murti, N.G.K., 356
Myhre, G., 70, 93

N

Nam Theun 2 Electricity Consortium (NTEC), 54
Nanjing, 252
Narayan, J., 271, 276
Narmada Bachao Andolan (NBA), 268–273
Narmada Water Dispute Tribunal, 61, 263–266, 271, 272, 274, 275
Navigation, 6, 19, 28, 30, 126, 130, 137, 155–159, 161–163, 168, 170, 243, 394
NBA. *See* Narmada Bachao Andolan (NBA)
Nehru, J.L., 3, 263, 271, 299, 305, 315
Nenonen, O., 69, 88, 93
Nepal, 9, 330, 332, 371, 375
Nganga, J., 70, 93
NGOs. *See* Non-governmental organisations (NGOs)
Nickum, J.E., 243–257
Nigeria, 10, 46, 48, 64
Nippon Foundation, 301
Nishizawa, T., 32, 35
Niskanen, A., 69, 88, 93

Nixon, W.S., 391, 395
Non-governmental organisations (NGOs), 6–9, 16, 42, 47, 133, 143, 145, 146, 177, 190, 230, 260, 267, 273
North America, 75, 77, 79, 84, 92
Norway, 15, 48, 87
Novo, E., 89, 94
NTEC. *See* Nam Theun 2 Electricity Consortium (NTEC)
Nykänen, H., 69, 88, 93

O

Obot, E.A., 64, 66
Oceania, 75
Oczkowski, A.J., 391, 395
Olivera, M., 151
Ometto, I., 89, 395
Organisation for Economic Co-operation and Development (OECD), 42, 58
Ozaslan, M., 193–196, 198
Ozdogan, M., 187, 199

P

Pace, M., 81, 91
Paddock, P., 315
Paddock, W., 316, 327
Pakistan, 48, 302, 304–306, 315, 322, 330, 332, 375
Panama, 44, 48
Pandey, B., 44, 67
Paraguay, 44, 48, 61, 124, 136–143, 145, 154
Parasuraman, S., 45, 66
Patil, J., 269, 276
Paulo Afonso Complex, 155, 156
Pearce, F., 70, 74, 94
Peisert, C., 245–250, 254, 257
Pendse, M.D., 337, 356
Peralta, C., 151
Pereira, H., 88, 93
Philippines, 48
Picciotto, R., 38, 60, 66
Poggiese, H., 150, 151
Poverty alleviation, 16, 34, 288–289, 320, 321, 350, 376
Prairie, Y.T., 81, 92
Price, C., 152
Prinn, R., 70, 93
Project
    Bhakra-Nangal, 13, 18, 299–327
    Chambal Valley, 300
    Chamera, 367
    Damodar Valley, 300

Project (*contd.*)
   Dhauliganga HE, 368
   Great Whale, 45
   Hinterrhein development, 95, 114–122
   Hirakud, 300
   Indira Sagar, 360, 363, 373, 375, 378
   Kangsabati, 277–298
   King River Power Development, 201–218
   Lesotho Highlands Water, 60, 65
   Piedra del Aguila, 124
   Rangit HE, 367
   Salto Grande, 123, 125–131
   Southeastern Anatolia (GAP), 171–199
   South to North Water Diversion, 230
   Subansiri Lower, 366
   Tanakpur HE, 367, 368
   Teesta HE, 365
   Upper Volta, 376
   Uri HE, 367, 368
   Yacyreta, 124, 136–147, 148, 149
Punjab, 20, 26, 302, 305–312, 314–317, 319–324, 358, 375

Q

Qin, D., 70, 93
Quality of life, 4–6, 12, 173, 178, 187, 190–192, 197, 237, 240, 266, 308, 350, 364, 393
Queiroz, A., 88, 93
Qu, Z., 254, 257

R

Radovich, J., 149, 152
Raga, G., 70, 93
Rajasthan, 38, 263, 264, 306, 307, 314, 320, 322, 323, 375
Ramasamy, C., 21, 35
Ramaswamy, V., 70, 93
Ramos, F., 70, 74, 81, 94
Rangachari, R., 14, 18, 299–327, 355, 373, 378
Rantakari, M., 72, 94
Reconstruction, 38, 39, 42–44, 219, 231, 236, 305
Recreation, 19, 110, 112, 201, 298, 324
Regional development, 6, 44, 130, 171–173, 175, 177–179, 193, 196, 323
Rehabilitation, 54, 105, 124, 127, 133, 138, 142–149, 167, 191, 201, 210, 211, 228–230, 240, 259–262, 264–272, 274, 275, 310–314, 318, 322, 324, 325, 346–349, 357–377

Resettlement, 37–68, 123–131, 133–135, 137, 138, 140–148, 159, 181, 190–192, 194, 219–240, 246, 259–275, 310–314, 324, 325, 346–349, 357–377, 381, 384
Resettlers, 37–41, 43–52, 54, 55, 57, 58, 60–65, 134, 136, 142, 192, 219, 221, 224–229, 231–233, 235–237, 239, 266
Rew, A.W., 44, 67
Ribeiro, G.L., 44, 67, 152
Rice, E., 38, 60, 66
Richard, S., 88, 92, 93
River
   Adhaim, 181
   Andrew, 206
   Beas, 303–305, 319
   Botan, 180
   Chang Jiang (Yangtze), 253
   Chaobai, 243, 253
   Chenab, 304, 305
   Collon Cure, 131
   Diyala, 181
   Euphrates, 179, 181, 183
   Franklin, 203
   Ganga, 306, 330, 331
   Garzan, 180
   Greater Zab, 181
   Han, 253
   Hezil, 180
   Indus, 302, 304
   Islam, 305
   Jhelum, 304, 305
   Kangsabati, 279, 282
   Karasu, 179
   Karun, 179
   King, 201–218
   Koyna, 329, 336, 337, 339, 340, 349, 351
   Krishna, 329, 334–336, 345, 351
   Kumari, 279
   Lesser Zab, 181
   Limay, 131–133
   Murat, 179
   Narmada, 260–265, 371
   Negro, 131, 132
   Neuquen, 131, 132
   Nile, 383, 384, 387, 388
   Ocker, 336
   Panchganga, 336
   Parana, 130, 136, 137, 140, 154
   Queen, 208, 210
   Ravi, 303–305
   Sanggan, 248
   São Francisco, 31, 154–161, 163–170
   Suleimanki, 305
   Sutlej, 302, 303, 305, 308, 309

Tigris, 171, 173, 174, 179–181, 183, 196
Uruguay, 61, 125, 126, 130
Vaitarani, 330, 339
Vashisti, 339
Vazarde, 340
Yamuna, 306, 314
Yang, 248
Yerala, 336, 352
Yongding, 243, 244, 254, 255
Zambezi, 44
River Basin
  Amazon, 154, 160
  Brahmani-Vaitarni, 330, 331
  Cauvery, 330, 358
  Ganga-Brahmaputra-Meghna, 330, 331
  Godavari, 330, 331, 333
  Indus, 302–306, 315, 330, 331
  Krishna, 329, 334–336
  Mahanadi, 330, 331, 359
  Mahi, 330, 331
  Narmada, 260–265
  Parana, 130, 136, 137, 140, 154
  São Francisco, 154–156, 158, 161, 163–166, 169
  Tapi, 330, 331, 333, 359
  Tocantins, 154
  Warna, 235
  Yangtze, 65, 253
  Yerala, 336, 352
Robinson, S., 21, 23, 24, 35
Rodda, J.C., 74, 94
Roder, W., 64, 67
Rodrigues, V., 153–170
Roquet, V., 45, 62, 66
Rosa, E.A., 43, 67
Rosa, L.P., 70, 88, 94
Rosa, R., 70, 74, 81, 94
Rosenberg, D.M., 70, 74, 75, 81, 93, 94
Roulet, N.P., 88, 93
Roy, A., 300, 301327
Rudd, J.W.M., 70, 74, 75, 81, 88, 93, 94
Ruttner, F., 71, 94

S
Sadoulet, E., 32, 35
Sahin, Y., 189, 198
Saleh, M., 388, 391, 392, 395
Salinity, 167, 168, 186–189, 296, 317, 383
Salvucci, G.D., 187, 199
SAM-based multiplier analysis, 23–26, 28, 29, 34
Sanliurfa, 172, 174, 176, 177, 183, 184, 186–188, 190–192, 194, 197

São Francisco River Hydropower Company (CHESF), 31, 155, 157–159, 164, 167
Sardar Sarovar Grievance Redressal Authority, 269, 271–275
Saxena, R.P., 277–298
Say, N.P., 175, 199
Scatasta, M., 35, 62, 65, 326
Scenarios, 26, 27, 29, 30, 32–34, 148, 172, 277, 289, 323, 354, 375
Schellhase, H., 88, 94
Schiff, S., 88, 93
Schulz, M., 70, 93
Scott, K.J., 88, 93
Scudder, T., 37–67
Sedimentation, 41, 162, 247, 295, 297, 324–326, 341–344, 381, 383–387
Seismicity, 336–337, 354
Shapiro, J., 247, 257
Sharma, J.P., 367, 369, 378
Shi, G., 219–241
Shiklomanov, I.A., 74, 94
Shuilibu, S.Z.S., 252, 257
Shunglu, V.K., 271, 276
Sikar, E., 70, 88, 94
Simulations, 29, 34
Sinniger, R., 96, 122
Smith, S.E. 387, 395
Social Accounting Matrices (SAM) based models, 21
Social development, 1, 167, 173–175, 178, 179, 188, 300, 380
Söderback, K., 88, 92
Solomon, S., 70, 93
Soumis, N., 70, 89, 94
South Africa, 9, 81, 326
South Asia, 63, 316
Southeast Asia, 81
Southeastern Anatolia (GAP) Region, 171–197
Sri Lanka, 46, 48, 56, 62, 63
Srinivasan, B., 45, 66
Stakeholders participation, 43, 61, 128–131, 135, 145–146, 148, 214, 217, 234, 363
Stallard, R., 88, 93
Steen, N., 152
Sternfeld, E., 244–252, 254, 257
St. Louis, V.L., 70, 74, 75, 81, 88, 93, 94
Striegl, R.G., 73–75, 81, 86, 92
Subramaniam, C., 315, 316, 327
Sub-Saharan Africa, 9, 75, 77, 79, 81
Sudan, 39, 394
Sugai, M., 153–170
Sun, J., 253, 256
Suwanmontri, M., 63, 67

Swaziland, 39
Sweden, 79, 87
Switzerland, 14, 95–122
Syria, 180, 181

T
Tang, C., 240, 241
Tang, J., 220, 241
Tasmania, 201–204, 206, 208, 209, 211–217
Tassara, J., 151
Tautschnig, W., 110, 122
Tavares, de Lima, I., 89, 94
Taylor, L., 24, 35
Thailand, 3, 48, 63, 64
Thatte, C.D., 259–276, 329–356
Therrien, J., 89, 91, 94
Third World Centre for Water Management, 13, 301, 380, 394
Thomas, G.H., 389, 395
Tignor, M., 70, 93
Tigret, S., 196, 199
Togo, 48
Toke, A., 151
Tortajada, C., 3, 4, 14, 18, 171–199, 355, 379–395
Tranvik, J.L., 81, 92
Tranvik, L., 81, 92
Tremblay, A., 88, 89, 91, 92, 94
Tribal groups, 261, 262, 348, 361, 362, 364, 369
Tundisi, J.G., 70, 94
Turkey, 3, 13, 48, 171–196

U
Ucar, Y., 188, 198
Uitto, J.I., 32, 35
UN Declaration of Human Rights, 44
United States, 1–3, 5, 8, 9, 45, 47, 381
Unver, O., 174, 179, 184, 196, 199
Uruguay, 61, 125, 126, 130
Uttar Pradesh, 26, 324, 369, 370
Uyanik, S., 187, 199

V
Väisänen, T., 89, 93
Vakkilainen, P., 87
Van Dorland, R., 70, 93
Van Wicklin, W., 38, 60, 66
Varfalvy, L., 88, 92
Varis, O., 69–94
Veiga, D., 152

Verghese, B.G., 319, 372, 378
Vergolino, T.B., 32, 35
Virgolini, M., 152
Virtanen, M., 88, 93
Vischer, D., 96, 122

W
Wadhwa, H., 366
Wadie, W.F., 389, 395
Wali, A., 44, 67
Warner, B., 88, 93
Water logging, 188, 295, 297, 324, 352, 383
Water quality, 64, 72, 87, 108–110, 167, 201, 208, 210–213, 217, 249, 296
Water supply, 1, 16, 20, 21, 28, 32, 86, 153, 156, 166, 170, 171, 219, 237, 243, 249, 251–253, 255, 267, 277, 313, 314, 316, 332, 334, 339, 347, 351, 365, 367, 368, 381
WCD. *See* World Commission on Dams (WCD)
Weidig, I., 110
Weissenberger, S., 70, 94
West Bengal, 270, 279, 281
Wichmann, E., 91, 94
Winarto, Y.T., 64, 67
Wind energy development, 214
Woodcock, C.E., 187, 199
World Bank, 8, 14, 15, 19, 21, 37, 38, 42, 45–47, 49–51, 53, 54, 58, 60, 61, 124, 132, 138, 141–149, 230, 232–234, 262, 264, 267, 269, 305, 324, 334, 350, 393
World Commission on Dams (WCD), 13–16, 20, 21, 44, 46, 47, 50, 51, 60, 268, 301, 323, 326, 373–375, 380, 381
Wu, K., 152

X
Xiang, H., 220, 229, 241
Xie, J., 248, 254, 257
Xun, H., 220, 241

Y
Yang, D., 248, 254, 257
Yavapai Nation, 45
Yesilnacar, M.I., 184, 187, 188, 199
Yi, M., 248, 257
Yuan, B., 254, 256
Yucel, M., 175, 199
Yu, Q., 219–241

Z

Zambia, 3, 48, 49, 57, 63
Zhang, Y., 248, 254, 257
Zhou, J., 219–241
Zhu, R., 253, 257
Zou, Y., 234, 241
Zuo, D., 253, 256